INSIGHT GUIDES

SOUTHEAST ASIA WILDLIFE

DISCOVER NATURE

Produced by Hans-Ulrich Bernard
Edited by Hans-Ulrich Bernard with Marcus Brooke
Photography by Alain Compost, Morten Strange and others
Editorial Director: Geoffrey Eu

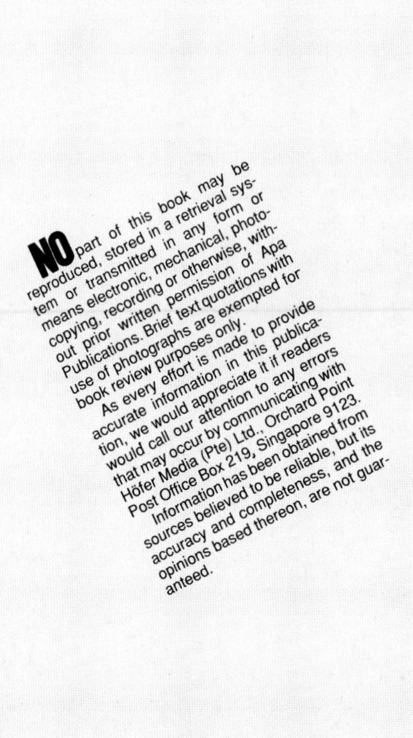

S.E. asia WILDLIFE

First Edition

Printed in Singapore by Höfer Press Pte. Ltd

ABOUT THIS BOOK

Welcome to *Insight Guide: Southeast Asia Wildlife*, a milestone in APA Publications Discover Nature series. This book is a culmination of work by an international team of the region's leading naturalists, and a product of their knowledge and love for nature.

Conceived by **Hans-Ulrich Bernard** and his wife Vera and supported by APA Publisher **Hans Höfer** and Editorial Director **Geoffrey Eu**, it is the first comprehensive guidebook on the nature and national parks of Southeast Asia. This book aims to inform and educate nature buffs in this part of the world, and to present the best that nature in Southeast Asia has to offer.

The Right Staff

Bernard, a scientist at the Institute of Molecular and Cell Biology of the National University of Singapore, is responsible for gathering a team of the region's outstanding naturalists to contribute to this book.

A keen "birder" and conversationist, he wrote the sections on natural habitats and man and nature, Bukit Timah and Pulau Tioman, and also contributed to the sypnosis on birds.

Phil Round, a leading ornithologist of the region at Bangkok's Mahidol University affiliated to the Worldwide Fund for Nature (WWF), contributed articles on several parks throughout Thailand and the devastating effect of the wildlife trade.

Pilai Poonsward, also at Mahidol University, has spent many years studying Khao Yai National Park and its hornbills as well as the storks of central Thailand. In this book, she documented her experience with photographs by her research partner, **Atsuo Tsuji**.

Richard Lair described the status of Thailand's elephants, beloved as domestic animals but endangered in the wild.

Allen Jeyarajasingam, teacher, nature writer and wildlife photographer from Kuala Lumpur, contributed the sections on Taman Negara, crown jewel of the region's national parks, Templer Park, and the mountain forests of West Malaysia. Places for wildlife observation on the tourist islands of Penang and Langkawi were identified and described by **Kanda Kumar**.

Yong Hoi Sen, from the University of Malaya in Kuala Lumpur, wrote the section on insects. Yong's photographs give a fascinating insight into the insect world, which constitutes an overwhelming portion of the region's wildlife.

Geoffrey Davison explored the latest addition to protected lands in Malaysia, the state park at Endau-Rompin.

Jeanne Mortimer, from the WWF in Kuala Lumpur and a leading turtle specialist with previous assignments in Costa Rica and Brazil, described the plight of marine turtles, whose breeding grounds continue to be poached.

Ho Hua Chew, from the National University of Singapore (NUS), described little-publicised pockets of nature in southern Malaysia and Singapore. **Chou Loke Ming**,

Bernard

Eu

Round

Poonsward

professor of biology at NUS, wrote the chapter on coral reefs.

Anthea Philipps wrote about the region's highest mountain, Sabah's Mount Kinabalu. Other members of the team from Kota Kinabalu include **Mahedi Andau**, **Anthony Lamb**, **Clive Marsh**, **Jamili Nais** and **Junaidi Payne** who assembled information on virtually every nature reserve in northernmost Borneo, and aspects of plant life there.

Payne, who works with the WWF on the life-history of orang-utans, is also author of a field guide to the mammals of Borneo.

Morten Strange, a Danish nature writer/photographer based in Singapore, explored Selangor and western Borneo.

Elizabeth Bennett works for the New York Zoological Society and contributed the piece on the proboscis monkey, one of the most unique animals of the region.

Wim Verheught, who wrote on the protected areas of Sumatra, is a Dutch ecologist who studied the swamp forests around Palembang for four years. He also undertook the task of writing the sections on mammals, birds and reptiles of the region.

Jan Wind and **Bastian van Helvoort**, Verheugt's colleagues and compatriots, and British journalist **Janet Cochrane** have studied central Indonesia and documented viable nature sanctuaries found in some of the world's most densely populated islands.

Colin Rees, director of an Environment Unit of the World Bank, Washington D.C. , and author of a field guide to the birds of the Philippines, described nature reserves in the Philippines.

Bernard first checked all the manuscripts for scientific accuracy before turning the text over to veteran APA contributor **Marcus Brooke** for further editing.

Picture Perfect

The fabulous collection of nature photographs were garnered from various sources. While many illustrations were provided by the authors themselves, two photographers deserve special mention.

Alain Compost, a French photographer currently living in Java, is a specialist on large mammals and birds of Indonesia, including one of the rare Javan rhinoceros. Morten Strange has one of the region's most impressive collection of bird photographs, and his fine work is well represented in this volume.

Finally, we wish to thank the Worldwide Fund for Nature in Kuala Lumpur for making available some particularly rare animal photos.

Jeyarajasingam

Strange

Ho

Verheught

Compost

CONTENTS

TRAVEL TIPS

A First Encounter

This book serves two purposes. It introduces the reader to the wildlife and the ecosystems of Southeast Asia, and it gives directions on where to find unspoilt nature in the region's national parks and protected reserves. Toward this goal, 30 leading naturalists and nature photographers of the region have brought together hitherto inaccessible information and rare pictures and created a unique documentation of the wealth of Southeast Asia's tropical nature.

The tropics of Southeast Asia, from the monsoon forests of northern Thailand to the rain forests of Malaysia and the mountain forests on the volcanoes of Indonesia, are home to the richest plant and wildlife on earth, equalled in diversity only by South America's Amazonia. The region is home to spectacular endemic animals, ranging from orang-utans and the Sumatran Rhinoceros to bizarre looking hornbills and colourful green pigeons, from Komodo dragons to the atlas moth. Here is a haven for the birdwatcher, the entomologist, the botanist and for all naturalists who have a love of ecological detail.

Some of Southeast Asia's natural treasures are protected within safe boundaries. However, most are threatened even more than elsewhere, since Southeast Asia is not thinly populated like Amazonia or central Africa, but is home to a fast-growing population of about 400 million people. It is hoped that *Southeast Asia Wildlife* contributes to an increased awareness and to an understanding of the value of nature in man's effort to be the steward, rather than destroyer, of his environment.

Preceding pages: Elephants and their *mahouts* set off for the day; bat activity is most frenzied at sunset; Giant Flying Squirrel "taking off"; termites at work; White-handed Gibbon enjoying the fruits of success. Left, nature took a cue from science fiction in creating the Sulawesi Tarsier.

The geographic setting: Southeast Asia is that region east of the Himalaya and south of China and stretches through Indo-China and the peninsula of Malaysia through the Philippine and Indonesian archipelagos to the island of New Guinea. This vast area which extends for 4,000 km in a north-south direction and for 6,000 km in an east-west direction is mostly covered by oceans. Only one-fifth, approximately five million square km, is land surface, roughly 10 times the size of France, or half the size of the USA. Southeast Asia is positioned in the tropics such that two-thirds of a rectangle drawn around its edges are north and one-third south of the equator. Politically, Southeast Asia is divided into 11 countries: Burma, Laos, Cambodia, Vietnam, Thailand, Malaysia, Singapore, Brunei, Indonesia, the Philippines and Papua New Guinea. Indonesia comprises nearly half of the total land surface. This book deals with the central part of the region from north Thailand to central Indonesia.

Climate: A constantly warm and moist climate is the most important determinant for the growth of tropical forests with their extreme diversity of plant and animal life. Southeast Asia's position at the equator leads to intense sunshine throughout the year and to temperatures which range between 23°C and 32°C. Only in northern Thailand and adjacent countries do temperatures drop below 15°C in winter and rise above 40°C in May and June.

The climate is dominated by two monsoon seasons. The northern monsoon blows from November to February and is activated by the high pressure systems that build up in winter over Siberia and China. This wind brings cool and dry air to northern Southeast Asia but heavy precipitation on eastern exposures from peninsular Thailand southwards. The southwest monsoon, which blows from April to September, is activated by the heating up of the Asian land mass and by cool air over Australia. It brings moderate precipitation throughout the region and is particularly important for the northern countries where it is the major source of rain. These major monsoon weather systems are interspersed with local convectional thunderstorms that occur throughout of the year but which are rare in northern latitudes from November to May.

The central part of Southeast Asia, namely Peninsular Malaysia, Sumatra, Borneo, western Java and parts of the Philippines, has a permanently moist climate with an annual rainfall of 2,000 to 4,000 mm spread fairly evenly over the year. In contrast, most areas north and east of this region have a seasonally dry monsoon climate. They receive less precipitation and it is unevenly distributed over the year. The plant and animal communities that live in ever moist or in seasonally dry climates differ dramatically from each other.

Two thousand millimetres precipitation, the typical rainfall of Southeast Asian countries, is two- to five-fold the amount of rain that falls in most parts of Europe or North America. It is usually deposited in heavy downpours and rarely gives rise to several days of continuously "bad weather". By western European standards, weather in this part of the world is fairly good for travelling purposes even during the monsoon seasons.

Natural and man-made landscapes: Today, Southeast Asia is a patchwork of landscapes which are either natural and unchanged or which have been modified by man for agriculture and homesteading. Other parts have been exploited and then abandoned. Before the arrival of man, Southeast Asia, with the exception of beaches, tidal flats and the tops of some high mountains, was completely covered by forests. This forest cover varies with climate, soil and human modification. Wildlife differs between the diverse tropical forest types and it is useful to become familiar with the different ecosystems recognised by plant ecologists.

Tropical Forests: An important classification attempts to separate forests that grow in permanently humid conditions from those that grow in a seasonally dry climate. The

Left, young couple are dwarfed by this giant member of the Southeast Asian forest.

forest in those regions that receive rainfall during every month of the year is called rain forest. In contrast, vegetation that has to cope with extensive dry spells between monsoon rains is termed monsoon forest. Many monsoon forest trees lose most of their leaves during the dry season in order to cope with the short water supply. Such forests are called tropical deciduous forests. Animals often prefer only one type of forest, and the transition zone between rain and monsoon forest frequently coincides with the border of the distribution of many species. For example, peninsular Thailand, where northern monsoon forest and southern rain forest mix, is the southernmost distribution of many Thai birds and the northernmost of many birds of Malaysia. A line across peninsular Thailand at the Isthmus of Kra is the border of two floral and faunal realms.

Climate not only changes with the latitude, the distance from the equator, but also with altitude above sea level. Temperature is lower at higher elevations and mountains receive more rain than lowlands since they favour the build-up of thunderstorms and stand in the way of weather front systems. Trees on mountains are smaller and sturdier than those on the lowlands. Transition stages of forest growing at different altitudes are given names such as lowland rain forest, and lower and upper mountain (montane) forest. Montane forests are home to a wildlife community completely different from that of lowland forests. Some species are even restricted to a narrow altitudinal ring around a given mountain.

Complicating the classification of forests is the fact that plant life is very dependent on the type of soil. Forests on sandy beaches, on limestone outcrops, on soils which are temporarily or permanently waterlogged (freshwater swamp forest and peat swamp forest), or on poor soils (heath forest) support different plant communities. However, few higher vertebrates are restricted to these special habitats. Another ecosystem, the mangrove, a tropical forest that grows in the salt water of tidal flats or river estuaries, has its own rich invertebrate fauna and also holds many particular vertebrates.

All these forests are either called primary (or virgin), when they have never been changed by logging, or secondary once this has happened. In the tropics, this distinction is more important than in colder climates. Tropical secondary forests have lost the major part of their plant and animal species, but in colder climates the biodiversity of secondary forests is comparable to that of the primary forest. Sometimes the dense growth of tropical secondary forests is called "jungle". It is best to avoid this expression since it does not refer to a particular habitat but rather indicates that the observer is bewildered by much dense greenery.

Lowland rain forest: It took millions of years of continuously favourable climatic conditions for the ecosystems with the greatest diversity of life forms on earth — the rain forests of Southeast Asia, Central Africa and the Amazon to evolve. These Pandora's boxes of plant and animal diversity are home to more than 50 percent of all species on earth. There is no easy explanation for this phenomenon. Could this be because these lush and rich habitats provide a supermarket of ecological niches that guarantee the survival of most life forms that evolve? Whatever the answer, nowhere does such a high density of species exist as in these forests, where often one square km houses several hundred different species of trees. In temperate zones a forest is normally composed of less then a dozen species.

It is impossible for the non-expert to bring order to his impressions by trying to identify the plants he encounters. More than 100 plant families with several thousand tree species grow in Southeast Asia's forests. Trees that belong to completely unrelated families have similar characteristics, in particular large leathery, oval leaves with an elongated tip, the so called drip-tip. It is the rule, rather than the exception, that any tree is surrounded only by unrelated species. The next member of its own species may stand 100 metres away. The visitor from northern climates will be surprised to learn that many huge trees belong to plant families that grow in his home country as annual herbs or small bushes, such as trees of the bean family (*Leguminosae*), the violet family (*Violaceae*), or the hypericum family (*Guttiferae*). Other tree families, such as the dipterocarps (*Dipterocarpaceae*), the most abundant trees in these forests, are restricted to the tropics. The unchallenged winners of a diversity contest are the palms (*Palmae*), of which more than 1,300 species grow in this region. They play a role as small trees of the understorey, as lianes (rattan), or as medium sized trees in special habitats but they are rarely part of the forest canopy.

Left, Bird's Nest Fern is characteristic of ever-moist forests. Above, Strangling Fig provides bizarre structural form.

Overwhelmed by plant taxonomy, the visitor may concentrate on special morphological features of his environment. The first impression of a tropical lowland rain forest is that of a cathedral of huge trees and not of impenetrable scrub. Trees with two-metre girths are the rule although those with a six-metre girth are not uncommon. The tallest trees, the emergents, attain a height of 50 to 80 metres. They tower above the canopy which closes at 30 to 40 metres above the ground. Below the canopy, the forest is often quite open, and this area can be called the "hall of the forest". The open space on the ground allows visitors to wander from the trail if their path is not obstructed by the dense growth of saplings. These young trees make up most of the undergrowth and replace the herbal plants and bushes that dominate in non-tropical forests. They often have buttressed or stilt roots that become so large in older trees that people can hide within them. Many tree species flower and fruit from buds on the trunk rather than from branches. *Cauliflory* is the scientific term for this phenomenon. Lianes, sometimes as thick as the trees, enrich the somewhat uniform pattern of strictly vertical trunks. Their branches intertwine with those of their support trees, and together with many epiphytes, plants that grow on trees, produce rich structures for habitats high above the forest floor.

Frequently encountered lianes are strangling figs (*Ficus spec.*). Their seeds germinate on the branches of large trees and the saplings send feeder roots to the ground, which eventually grow into nets which strangle the support trees to death. While this is happening the fig's roots gain strength and form a trunk at the time their support fails. A liane has turned into one of the emergents of the forest. Like all members of their family, strangling figs are particularly rich fruiting trees and a major source of nutrients for many forest animals. One of the best ways to observe birds and small mammals is to identify fruiting figs and then find a good vantage point from which to await their guests.

Most animals are restricted to single structural elements or to a layer of the lowland rain forest. Large ungulates and elephants live on the forest floor and gibbons in the upper canopy. A heavy tree dweller like the orang-utan lives on the thicker branches of the lower canopy but forages in the upper canopy in search of particular fruits. Most bird groups are adapted to life in one layer only. The upper canopy is the home of sunbirds, flowerpeckers, barbets and fruit doves; the lower canopy harbours trogons, woodpeckers, flycatchers and bulbuls, while in the undergrowth can be found families of pheasants, pittas and babblers.

The largest and most beautiful lowland rain forest in the region is Malaysia's national park Taman Negara, but most parks in Malaysia and Indonesia have beautiful representations of this ecosystem at elevations below 750 metres above sea level. Transitions of rain to monsoon forest are found in peninsular and southern Thailand, eastern Java, Bali, Sulawesi and on many islands of the Philippines.

Monsoon forests: This term covers all tropical forests of Southeast Asia that grow in a climate which has significant seasonal changes in precipitation. Such forests are much more heterogeneous than rain forests. This is predictable when one considers that the annual dry spell may last from two months in areas such as peninsular Thailand to six months in northern Thailand.

Monsoon forests are less massive than rain forests. The girth of the average tree is much smaller although gigantic emergent trees still occur. Trees with buttressed and stilt roots and epiphytes are less frequent than in more moist climates. Lianes are common. The forests are decidious to different degrees and the increased amount of light that reaches the ground can permit the growth of grass.

Tropical tree families such as dipterocarps are still abundant but species occur that are absent from the rain forest such as pines (*Pinus spec.*) and casuarines (*Casuarina spec.*). Teak (*Tectona grandis*) can be naturally abundant and is further spread through artificial planting. Different bamboos, a group of huge tropical plants belonging to the grass family (*Gramineae*), are present in primary monsoon forest but absent from primary rain forest. Bamboos are favoured by human disturbance.

Man has changed this type of forest for ages through regular burning, and unaltered monsoon forests are now even rarer than virgin rain forests. Savannas, such as that of Baluran national park in eastern Java, have been derived from this forest through intense human influence, mostly by fire. The sparsely populated hill country between Thailand and Burma has the best examples of this type of forest in Southeast Asia. Readily accessible monsoon hill forest is on Doi Suthep close to Chiangmai in Thailand. It has fairly large natural tracts on the hilltop and heavily altered terrain on the flanks of the mountain.

Mountain forests: As one climbs a tropical mountain, characteristics of the lowland rain forest such as great height, presence of huge emergent trees, buttressed and stilt roots, large leaves, cauliflory and lianes gradually disappear. Epiphytes remain abundant and lichens become a predominant feature at the

Left, Nipa Palms fringe this forest river. Above, fog, lichens and crooked trees feature in the mountain forest.

highest elevations. Conceptually, the border between lowland forest and lower montane forest is set around 750 metres above sea level while the transition to upper montane forest is around 1,500 metres. At this height, the temperature has decreased by 10 °C, and rain can fall uninterruptedly for days on end in contrast to the occasional showers characteristic of the lowland climate.

Typical upper montane forest attains a height of only 10 metres. Tree trunks and larger branches grow crooked rather than upright and straight and leaves are only a few centimetres in size. The forest ground is very open and often moss covered and it is possible to walk without following a path. Trees often belong to families typical of northern latitudes such as oaks (*Fagaceae*) or heather (*Ericaceae*) rather than to tropical families such as dipterocarps.

Most animal species of mountain forests are different from those in the lowlands. Some species have preferences for a fairly restricted altitudinal range. Surveys show a decrease of the density of different animal species with increasing height. This deficit is compensated to the advantage of the observer by the greater visibility of some animals, especially birds.

Mangrove forest: This forest formation of the tropics occurs in salt water on muddy tidal flats and river estuaries. The bizarre structures of the trees with their extensive stilt roots and breathing shoots (*pneumatophores*) and *viviparous* seeds (seeds that germinate on the tree and, falling down, sink the elongated heavy root into the muddy ground) have great aesthetic attraction. The muddy ground is normally solid enough to permit walking through this type of forest at low tide. With approximately 30 tree species, the Southeast Asian mangrove is the richest plant community of its type in the world, but botanically poor in comparison to other forest formations. Mangrove trees can reach heights of more than 20 metres and a girth of more than 3 metres. They can be very rich in mammal and bird life — for example, the extensive stands of mangrove on the east coast of Sumatra. However, this forest formation has been intensively exploited for charcoal production, prawn farming and land reclamation. The visitor to these disturbed areas can still observe wintering shore birds and explore one of the richest non-vertebrate faunas on earth.

Natural Changes: During geological eras continents and their segments, tectonic plates, drift to different positions relative to the poles and to the equator. On their path, they enter and leave different climatic zones. It is fortunate for Southeast Asia's nature that its western part, the Sunda shelf, remained geologically stable on the equator for tens of millions of years. This permitted an evolution of tropical plant and animal life unaffected by dramatic climatic changes. Consequently, tropical rain forest existed in the region throughout the tertiary period, the last geological era. Moreover, the tertiary period's 65 million years was also the period necessary for today's plants and animals to develop from simpler precursor forms. At different times in the past, tropical rain forests supported plant and animal communities completely different from today's.

Natural habitats change not only through evolution, but also through the migration of new plants and animals that spread after tectonic or climatic changes. One of the most consequential events for Southeast Asia's nature was the collision of the Indian subcontinent with continental Asia. It arrived as an island-fragment of the ancient continent Gondwanaland. It is likely that the dipterocarp trees arrived on this tectonic plate about 30 million years ago. They infiltrated the region so successfully that they became the dominant group of trees in humid forests.

The eastern half of Southeast Asia, namely eastern Sulawesi, the islands east of Bali and New Guinea, were added in form of another tectonic fragment of Gondwana-Land, the Sahul shelf, to the western region, the Sunda shelf. The mixing of the two region's floral and faunal elements over a period of 15 million years further enriched Southeast Asia's biological wealth. This mixing was never complete. A borderline between the two floral and faunal realms can still be observed and has been termed the Wallace line.

In recent geological history, not millions but thousands of years ago, the Ice Age changed much of northern Eurasia's vegetation because of a dramatic fall of temperatures at higher latitudes. Southeast Asia kept a basically tropical climate during these periods and so probably did not experience many changes in vegetation and wildlife. At times, however, the sea level fell by as much as 180 metres because of the large amount of water bound as ice at the earth's poles. This opened large corridors of dry land between what are now islands — for example, Sumatra, Java and Borneo — which became forested and served as a migratory route for animals. Consequently, the vegetation and the wildlife of these areas today show a marked similarity.

Left, Pandanus near the water's edge. Above, looking for food in a mangrove swamp.

Man And Nature

The emergence of Man: Ecosystems can be viewed as arenas of competition. Plants compete with each other for space and evolve defense mechanisms to protect themselves against herbivorous animals. Yet, these animals enjoy some success in their attempt to obtain food. For their part, they must defend their pastures against other herbivores and protect themselves against predators. This competitive framework is fierce for the individual but gains and losses of species as a whole are so small that little change of the ecosystem is visible over decades, centuries, even millenia. However, this argument no longer holds true after the emergence of a particularly successful and ruthless species — man.

Palaeontalogists have gathered evidence for the evolution of man in Africa three to four million years ago. From Africa, *Homo sapiens* and his precursor *hominids* invaded Eurasia in several successive waves. Some areas of the world, in particular America and Australia, were colonized as recently as 30,000–40,000 years ago and in many parts of Southeast Asia man probably also arrived fairly late. However, this view is difficult to verify in a climate that leads to the quick degradation of human fossils and cultural artifacts.

Hominids arrived much earlier in some localities of Southeast Asia. The most important findings come from the valley of the Solo River in central Java. Bones of Pithecanthropus found here may be 1.3 million years old. Interestingly, Pithecanthropus and his younger geographic neighbour, the Solo Man (*Homo soloensis*, possibly related to *Homo erectus*) shared the Java of their time with animals quite different from today's wildlife. These included a hippopotamus, a stegodont elephant, antelopes and sabretoothed cats. It is quite possible, though somewhat speculative, that some of these animals became extinct through the hunting pressure of small populations of primitive men. Extinctions of this extent certainly occurred on such islands as Madagascar, New Zealand and Hawaii. Man may even have caused the disappearance of mammoth and sabretoothed tiger in North America.

Left, agricultural land use in harmony with nature? Right, coastal splendour.

Man's first cultural stage: *Homo sapiens*, people of our species first appeared in Southeast Asia much later than the early hominids of Java, probably 50,000 years ago. *Homo sapiens* arrived in the form of ethnic groups that are still present as small minorities. Pigmy-like negritos, people of

very short stature with very dark skin and woolly black hair, are thought to be the oldest group. They lived for thousands of years on the primitive cultural level of nomadic hunters and gatherers of the rain forest.

In Peninsular Malaysia three ethnically different groups of aborigines are referred to as *orang asli*. One of these groups, of about 2,000 people, consists of small negritos who are regularly seen around Kuala Tahan, the headquarters of the national park. While living in the forest, these people had for long periods of time contact with the modern life of the coastal region, engaging in trade of minor forest products such as rattans. Another group of negritos, the Agta, live in the Philippines. One tiny group, the Tasadays, made international news in the 1970s, even becoming the subject of a cover story in an issue of *National Geographic*. This group was presumed to have survived in the forests of Cotabato, Mindanao without any contact with the modern world, living naked in a cave in the forest, feeding on the fruits of the forest. Recent research makes it likely that this whole story was a hoax. With the unglamorous disappearance of the Tasadays, it is unlikely that any other human group lacking contact with the modern world survives in the increasingly fragmented forests of tropical Asia.

The effect of hunters and gatherers on forest vegetation and on the general appearance of the landscape is negligible. However, their hunting pressure on large, slow breeding animals with low population densities may have been considerable over long periods, as is shown by well documented extinctions from various parts of the world. Today, Southeast Asia's wildlife suffers from man's hunting that has continued since those archaic ages. For example, in many forested areas of northern Thailand, large, "edible" birds like pheasants and hornbills have disappeared. Rare storks in the mangrove swamps of Sumatra and Borneo that are in danger of extinction have no higher value for local fishermen than the price of their meat. Ornamental birds like parrots are driven to the brink of extinction by the bird trade. Possibly, the nature tourism of today's world will provide an economic argument and demonstrate to these poor people the higher value of these animals.

The agricultural revolution: After the hunters and gatherers, human races which were ethnically different from the negritos, appeared in the region. Among these were groups with australoid physiognomy, possibly ancestors of the inhabitants of New Guinea and of the Australian aborigines. It is scientifically unresolved whether the ancestors of these australoid groups, of the proto-malays (another ethnic group of the region, for example, the Orang Hulu of southern Malaysia) or even of today's majority population in Indonesia and Malaysia migrated to Southeast Asia from different regions. Alternatively, they may have developed their racial differences in their present home countries. Nevertheless, it is certain that these groups participated in one of mankind's greatest cultural and technical achievements, the development of agriculture.

Agriculture is more than 10,000 years old,

and originated from three geographic centres, one of which was in Southeast Asia. The stimulus was the natural occurrence of many edible plants. Exact timescales for the domestication of many plant species are not clear but findings of fossil rice in northern Thailand have been dated 13,000 years back and pollen analysis from New Guinea is indicative of extensive agriculture 9,000 years ago.

Though unrecorded, it is easy to imagine how man first started systematic planting and later completely replaced natural vegetation with the selective growth of useful endemic plants. One of the oldest practices was the extraction of starch from the sago palm (*Metroxylon sagu*) and from taro (*Colocasia*) which grew in the swamp forests around fishermen's villages. Some will have found it burdensome to search for wild rice (*Oryza sativum*) and rather sowed it into a patch of soil next to the house. Bananas (*Musa*) appear naturally in forest clearings and were easy to maintain at the periphery of human habitations. The coconut (*Cocos nucifera*), a likely endemic of the beach vegetation of Southeast Asia before it was spread pantropically by man, germinates without any culture techniques from nuts dumped on the ground. Many trees of rain and monsoon forests produce highly prized fruits such as mango (*Mangifera*), mangosteen (*Garcinia mangostana*), nangka (*Artocarpus heterophyllus*), durian (*Durio zibethinus*), or rambutan (*Nephelium lappaceum*). It is difficult to understand why it took mankind thousands of years to "invent" growing fruits around their homes rather than searching for them in the forest. Possibly the life of the hunter and gatherer was fairly convenient and provided ample leisure time and, in the absence of hardship and hunger there was no urgent need to practise agriculture.

Left, Negrito natives at home in the forest. Above, evidence of the links between Man and nature.

Ultimately, man started to terrace the hillsides to create artificial swamps for the growth of rice and transformed the forest on the ridges and in the gullies between rice fields so they could be used for horticulture. These man-made fruit-tree forests are frequent elements of landscapes which have a long history of dense human habitation such as Java, Bali and parts of the Philippines, where they surround villages and fringe the rice fields. The first-time visitor to the tropics tends to believe that these orchards are remnants of the rain forest. Despite this mistaken belief, it is aesthetically satisfying to walk in this type of landscape which is home to such wildlife as flying squirrels, parrots and monitor lizards.

Although eliminating forest animals, clearance of the forest for agriculture opened new habitats for numerous species. Some animals that were naturally restricted to marginal habitats such as forest fringes could now spread, while others with a need for open land could migrate from remote geographic locations. Among the birds, favoured groups include larks, pipits, munias and weavers. Today, observation of birds can be quite satisfactory in man-made landscapes such as the rice fields of northern Thailand where, during winter, local birdlife is enriched by a host of migrants from as far away as Siberia.

The industrialised world: Some cultivated plants of Southeast Asia, such as the papaya (*Carica papaya*) have been introduced to the region. Two other introduced plants which have changed the region's landscape and become economically beneficial, but ecologically disastrous are the oil palm (*Elais guineensis*) of West Africa and the rubber tree (*Hevea brasiliensis*) of the Amazon. Both plants were known for their useful products in their countries of origin but growth in large plantations only began after their arrival in Asia. The rubber tree was first brought to Singapore in 1877 and was kept for years in experimental nurseries. H. N. Ridley, the director of the Singapore Botanic Gardens, was a principal proponent around the turn of the century for the potential for the growth of this plant in Malaya. Rubber plantations quickly replaced the country's dying coffee industry. The oil palm was introduced into the region in 1911, initially in Sumatra. The area open for the growth of this plant started to expand dramatically in the 1950s. Today, rubber and palm-oil plantations cover nearly 20,000 square km. The major centre is Peninsular Malaysia and significant plantations are found in Sumatra, East Malaysia and peninsular Thailand.

The establishment of these plantations has been an important economic factor and rubber and palm-oil exports constitute a major source of income, especially for Malaysia. However, aesthetically, the landscape has to put up with kilometre after kilometre of dreary, monotonous plantation. Ecologically, there is a complete loss of all biological diversity in these extreme monocultures. Unfortunately, in spite of today's environmental value systems, the arguments for agricultural profits still weigh heavier than the value of conserved nature. An example is the unfortunate destruction of the rain forest at Ulu Lapar in Malaysia where, during the

1970s, one of the largest herds of gaur was sacrificed in favour of the fairly moderate income from plantations.

Timber harvest and forest degradation: Virgin forest is locally opened under natural circumstances. The fall of a dying tree leaves a small gap while a thunderstorm or a typhoon may lead to extensive clearings. A small group of plant species has adjusted to colonize these clearings with properties such as wind-borne seeds, light-demanding saplings, and rapid growth. Wherever man clears the primary forest, leaving open land behind, these plants establish a secondary forest community. The widely spreading crowns of albizias (*Albizia sp.*), the fresh-green maple- like leaves of Mahangs (*Macaranga sp.*), the characteristic shape of the Fishtail palms (*Caryota mitis*), the beautiful yellow flowers and red seeds of the Shrubby Dillenia (*Dillenia suffruticosa*) and the impenetrable scrub created by resam ferns (*Dicranopteris sp. and Gleichenia sp.*) are familiar sights in the region. On the positive side, these forests protect the bare soil and are even somewhat pleasing to the eye, but the negative aspect is that most of the wildlife and the vegetation of the virgin forest have disappeared.

Selective logging and burn farming: What are the possible fates of a forest once the logger has set eyes on it? One system, selective logging, aims at the extraction of the most valuable timber. A dirt road is cut into the forest with the help of bulldozers; desired trees are felled and removed; the logger leaves. Short-term effects are disastrous: Most of the undesired trees have fallen together with the extracted timber, the scene left behind is reminiscent of the war of man against nature rather than of sophisticated forestry. The long-term effect, however, is not so bad. Many tree species of the primary forest had survived as saplings, and over the years much of the original plant community can regrow in the shadow of the secondary forest. The wildlife that had fled into the possibly unharmed surroundings can return. Some species, like gibbons or smaller birds, return quickly while others, like hornbills, only come back if old trees with breeding cavities remain, and some large mammals only if large tracts return to quietness. The nature of a large part of Southeast Asia is in this state, in patches on hill tops, in vast stretches along the southeastern coast of peninsular Malaysia, and on Sumatra and Borneo. Unfortunately, the significant eco-

Left, agriculture has its aesthetic value too. Above, even the most rugged landscapes can be quite conducive to human settlement.

logical and probably even touristic value of these areas receives insufficient attention. For example, the regrowing forest that surrounds the Malaysian resort of Desaru is home to such mammals as elephants, deer, siamangs and gibbons, and birds like hornbills, trogons and pittas. It is one of the few places where luxurious tourist facilities are immediately surrounded by rich wildlife habitats. Unfortunately, several proposed large scale developments will almost certainly destroy this fauna and eliminate Desaru's potential as an attractive resort for nature tourism.

The positive scenario of returning primary forest plants and animals only becomes a

reality in the absence of human interference. Most often, logging roads are the access for slash and burn farmers, often with governmental support as, for example, in the Indonesian *transmigrasi* project. The problem is two-fold. The peasant, often inexperienced in his new work, enters a way of life that is far away from villages, towns and civilisation in general. After a few years, the loss of the soil's fertility eventually leaves the peasant even poorer than at the time of his arrival. Exhausted land, useless for man or nature, remains.

In recent times, the demand for valuable tropical timber for furniture and construction work has been overtaken by the insatiable hunger of papermills and chopstick-producers for softwoods. Unfortunately, this is not accompanied by the planting of fast growing trees such as albizias which can be harvested after a few years. Rather, it further increased the aggressiveness of the logging of primary forest. Instead of selective logging, the whole forest now goes in one cut, leaving behind open ground that, at best, may be overgrown by secondary scrub, if the soil is not flushed away by the first monsoon storm. This type of land use is increasingly considered to be both an ecological disaster and economically unreasonable. The whole potential of a region is destroyed in one gigantic clearance and long-term use such as sophisticated forestry, harvesting of minor forest products and tourism are not even considered. Occasionally, examples of this type of destruction can be studied alongside optimal protection: the visitor to Sabah will gather impressions of this sort while travelling between the Mount Kinabalu, Sepilok, and Danum reserves.

Conservation of Southeast Asia's nature: The number of human beings on earth has grown rapidly: from one billion 140 years ago to five billion in 1987. It is safe to predict that there is an upper limit to the number of people that can inhabit this planet, but it is not possible to predict that figure. Nobody knows whether mankind will only stop growing once the density has become unbearable, or whether growth will slow before that in order to permit a more pleasant equilibrium between people and nature, between crowded places and open landscapes. Worldwide, an increasing part of the educated public advocates ethical, scientific and economic arguments in favour of the second option.

The proponents of the ethical argument state that it is incomprehensible that a civilised mankind may destruct in a few decades a major part of the diversity of animals, plants and landscapes that took millions of years to evolve. The opponents of this argument either simply deny that destruction is taking place, or emphasise that there is no higher value than human life, and pretend that nature and man are mutually exclusive. It is hoped that the growing appreciation of nature will eventually dominate.

The extraction of timber has become the

predominant form of forest usage because it requires technically unsophisticated infrastructures and because it earns easy profits. At least three other forms of forest usage are less destructive, long-term sustainable and economically superior. These are (a) the traditional extraction of minor forest products such as rattans, fruits and ornamental plants; (b) research to find novel pharmaceutical and agricultural products that exist in wild plants; and (c) the rapidly growing area of nature tourism. Proponents of these suggestions document that optimal use is possible in a virtually unaltered natural state.

While Southeast Asia has seen some of the world's most extensive nature destruction over the last four decades, all nations of the region have adopted conservation measures which are significant, at least when taken at face value. All countries have national parks or wildlife conservation areas where major ecological changes should be excluded. In practice, this only succeeds in some regions, while elsewhere the law is not enforced or logging concessions are even granted within protected areas. Change can be expected through today's growing ecological knowledge, which includes the awareness of potential ecological catastrophies. An example is Thailand's proclamation of a nationwide ban on logging after fatal floods in November 1988.

Beyond the enforcement of existing legislation, environmental issues will concentrate in the immediate future on three targets. One is the identification of entire ecosystems which are under-represented in the present park system, such as mangrove and swamp forests. A second is the lack of knowledge of the critical minimal size of an ecosystem that allows the survival of its various plant and animal members. Island-like protected areas will be prone to extinctions, as the study of the flora and fauna of oceanic islands suggests. A remedy against this may be the creation of corridors of habitats managed in a somewhat natural way, as in for example the creation of harvested but ecologically managed forest plantations. A third issue will be the monitoring of endangered animals and plants that occur only outside protected areas. A guiding help in these areas of concern are the activities of non-governmental organisations such as the identification of threatened birds by the International Council for Bird Preservation, or conservation initiatives by the Worldwide Fund for Nature and the Malayan Nature Society.

Left, sap from a rubber tree. Above, a large oil palm plantation.

LOGGING AND FOREST MANAGEMENT

Few topics today are as controversial among environmentalists as the selective logging of tropical forests. Throughout the world, and especially in Asia, large tracts of land are opened up each year by logging and are subsequently degraded or deforested by immigrant farmers. Timber-exporting countries point to the importance of timber revenues and employment to their economies and to their record of establishing protected areas and trying to manage other forests for sustainable production. They also point out that all developed countries were severely deforested in earlier times and that history is merely repeating itself.

Commerce in insular Southeast Asian timber manly involves trees in the family *Dipterocarpaceae* which dominate the forests. Although more than 300 species of dipterocarp grow in the region, most of the common ones can be conveniently grouped into a few commercial classes.

While administrative and industrial practices vary significantly between countries, the following account based on Sabah is broadly descriptive of the situation. Logging takes place within blocks of land allocated by the government under various forms of long-term concession or short-term licence. The operator builds a main road into the area with spur roads off it. Selected trees are cut subject to a minimum girth limit of 60 cm diameter at breast height. The logs are hauled out by bulldozer along skid-trials to a collection point, where they are measured and identified with a hammermark before being loaded onto trucks for transport to the coast or a navigable river. Once in the water, logs are rafted out to ocean-going ships or to a nearby port. Throughout this process the Forest Department keeps close control of the areas cut and, above all, of the volume, species and grade of logs produced.

In Sabah and Sarawak, most timber is still sold in log form to buyers from Japan, Korea and Taiwan, although the proportion processed locally is fast increasing. Most other major producers — Peninsular Malaysia, Indonesia and the Philippines — now process almost all their logs. Technical advances are gradually reducing the number of end-uses that require hardwood rather than softwood and with supplies of tropical hardwood dwindling rapidly throughout Asia, the timber boom — at least in its present form — is coming to an end.

Interestingly, birds and mammals almost always survive the direct effects of logging, albeit sometimes with major changes in relative abundance. For example, bearded pigs probably decline in abundance in logged forest while rusa deer thrive on the grasses and shrubs that characterise early secondary growth. However, improved access for outside hunters often proves more fatal, particularly for species such as the Sumatran rhino, which are already very rare and hunted for profit rather than subsistence.

Rural communities undergo massive transformations when logging companies work nearby. The industry brings in roads — at least temporarily — and some locals may obtain employment. However, hunting, fishing and rattan resources all decline, thus undermining the subsistence economy. These are serious costs of logging which are not generally acknowledged by the companies that cause them. Undoubtedly, some compensating benefits are derived from general development activities funded, in part, by government revenues from timber.

Left, transporting logs downriver. Above, harvesting palm fans.

But in many countries development initiatives in rural regions are mainly concerned with resettlement schemes or the promotion of commercial plantations.

When the timber industry began, all these problems looked less serious or simpler to manage. Forestry in Sabah was based originally on a version of the "Malayan Uniform System", in which all commercial trees were to be harvested at a single felling and the non-commercial species subsequently poisoned, so as to produce from existing seedlings an even-aged second rotation stand which would be richer in timber species than the original forest. To maintain a supply of timber in perpetuity, only one to two percent of a commercial estate should be logged each year thus giving a rotation period of 50 to 100 years.

Unfortunately, practice bears little relation to this classical prescription, or various local versions of it. First, independent governments heavily dependent on timber revenue have permitted much higher cutting rates than can be sustained. Second, as markets have come to accept more "lesser-known" species, extraction intensities have increased, with concomitant damage to young trees and seedlings. Third, since timber companies rarely have any long-term interest in the land, they have little incentive to not damage the forest or to replant it. The onus of regenerative work falls on Forest Departments which are invariably short of budget and manpower.

In recent years, as governments have come to recognise the real costs of the timber industry, attitudes and polices have begun to change. First, the importance of preserving some tracts of undisturbed habitat is now almost universably recognised, if not always acted on. Then there is the increasing recognition of the value of non-timber forest products. Many of these, such as traditional medicines, wild fruit, vegetables, building materials, game meat and fish, enter commerce only marginally, yet are vital to people in rural areas.

A third change is the development of plantation forestry as an alternative source of utility timbers. While a poor substitute for natural forest as a source of most non-timber goods services plantations have an important role to play in helping to reducing the rate of deforestation.

WILDLIFE

Southeast Asia's wildlife forms part of the Indo-Malayan realm, a zoo-geographical zone comprising most of tropical Asia. By virtue of its huge dimensions and its wide array of habitats, Southeast Asia is endowed with unsurpassed biological riches. It is home to such large mammals as the elephant, tiger and three species of rhinoceros. Bird-life, reptiles and amphibians abound with more than 1,500 species of birds and well over 1,000 species of amphibians and reptiles. Many of these are endemic with a very localised distribution.

In the west, the Indo-Malayan realm is bordered by Pakistan's Baluchistan mountains, the Indian sub-continent and all land south of the Himalaya; its eastern limits follow the Yunnan hills in China, the entire Philippine archipelago and most of Indonesia's islands.

This zoo-geographical realm can be divided into four distinct sub-regions. These are the Indian, Indo-Chinese, Wallacean and the Sundaic. The last of these includes the islands of the Asian Sunda shelf, Sumatra, Borneo, Java and Bali. Due to periods of partial isolation, each of these sub-regions has developed its own faunal and floral elements. A distinct boundary between the Indo-Chinese and Sundaic sub-regions is located around the isthmus of Kra in south Thailand. A transition zone covers most of Thailand, and some parts of Burma and Vietnam.

The Indonesian archipelago is inhabited by two distinct types of fauna. That in the west belongs mainly to the Indo-Malayan while towards the east a gradual transition to Pacific-Australian faunal elements can be noticed. Sir Alfred Russel Wallace, in 1858, identified the dividing line between these two zoo-geographical regions as lying between the islands of Bali and Lombok.

To the west of the Wallace line lie the Sunda continental shelf islands of Sumatra, Java and Borneo, all of which were connected to the Asian mainland during past periods of glaciation. Under these conditions, land-bridges were formed and immigrants such as mammals and birds could, by "island hopping", gradually colonize the islands as far as Bali. Lombok, part of the Sahul continental shelf, was disconnected by a deep sea which never dried up. Yet the Wallace boundary was later superseded by the Lydekker line as it became clear that the Asian influence reached much further eastwards. The region between those two lines,

with a mixture of species from both regions, is now referred to as Wallaceana. It includes the island of Sulawesi, which has a most distinctive fauna, particularly of mammals. Of its 127 indigenous mammal species 79 are endemic. New species continue to be found.

Several animal species are unique to Southeast Asia, and in some cases, entire sub-families and even orders are confined to this realm. Distinct species include leaf monkeys, gibbons and flying lemurs. Only one family of birds is restricted to this realm, the leaf birds and ioras. Otherwise, the Indo-Malayan fauna shares most of its taxonomy

Preceding pages: Slow Loris about to make a meal of a Stick Insect. Left, Sun Bear on a stroll. Right, a foliage-framed Barking Deer.

with either the Palaearctic or the Afro tropical regions.

The region has a rich reptilian fauna, with many species of lizards such as gekkos and the Komodo dragon, the world's largest lizard, colourful and numerous poisonous snakes, crocodiles and terrestrial and marine turtles. Marine life is abundant and includes rich coral reefs, especially in the sheltered Flores and Banda seas of Indonesia. Birds include colourful kingfishers, the maleo, an exotic megapode which buries its eggs in communal mounds, magnificent argus pheasants with tail feathers attaining one-and-a-half metres, and emerald ground-dwelling pittas.

Indonesia is a centre of mega-diversity, with not less than 1,500 species of birds, 530 species of mammals, and 3,000 species of fish. These totals probably constitute the largest list of animals for any single country in the world and amount to 17, 14 and 16 percent respectively of the world's total bird, mammal and fish species. However, it partly includes Pacific-Australian elements. The Sundaic sub-region boasts about 1,200 species of birds and 250 species of mammals. The latter figure includes aquatic whales and dolphins.

Evolution of species: The rising of the Himalayas, a rather recent geological event occurring only 10 million years ago, virtually cut off the free migration of species, including birds, between the Indo-Malayan realm and the Palaearctic. This geographic barrier, together with climatic changes with long periods of glaciation, led to periods of isolation, which have proven to be ideal for the evolution of new species. Species could take advantage of new niches, colonizing habitats which were not yet fully utilised by other animals.

The flora and fauna found in the island of the Sunda plate region are similar because of the past existence of land-bridges that occurred during these periods of global glaciation. During the Pleistocene Epoch — roughly the period from one million years B.C. to 12,000 B.C. — there were times where the sea-level in Southeast Asia was up to 180 metres lower than today, thus allowing a free interchange of species. However, for unknown reasons, not all animal species made it to the eastern part of the Indonesian archipelago. For instance, Sumatra, the largest island of Indonesia shares more species with Malaysia than with Borneo, and even less with Java, although the latter island is geographically closer to Sumatra than to Malaysia.

Tigers are known in Malaysia and Sumatra as well as in Java, but not in Borneo, although this huge island provides ample opportunities for tigers to thrive. In contrast, the orang-utan is known only in Sumatra and Borneo. Even on Sumatra, orang-utans are only found in the northern part and seem to have been unable to colonize the area south of Lake Toba while species such as the tapir are restricted to the southern part of Sumatra. Apparently, the Lake Toba area, after a colossal volcanic explosion that occurred some 75,000 years ago, must have represented an impenetrable barrier. Many animals in Southeast Asia are restricted to specific habitats, and are tolerant of only a certain range of conditions. The present distribution of many species might have been determined by events of extinction whose nature is not yet understood.

Primates: There are at least 31 species of primates in the Sundaic and Wallacean sub-regions, of which at least 23 are endemic. Indonesia, which has the richest primate fauna of any Asian country, is home to all but two of these primates — the Dusky Leaf Monkey and the Philippine Tarsier. In addition, 15 species are found in the Indo-Chinese sub-region, and seven are shared with the Sundaic area.

Of the various monkey families found in Southeast Asia, the macaques are the largest. No less than 10 species are found in the Sundaic sub-region. Only two are widely distributed: the Long-tailed and the Pig-tailed Macaque. Sulawesi is home to some of the rarest macaque species in the world, all seven of which are endemic. The Indo-Chinese sub-region includes three macaque species that can be observed in Thailand: Assamese, Stump-tailed and Rhesus Macaque. The last species was formerly heavily persecuted because of its use in bio-medical research. It can have very high population densities and live naturally in a wide variety of habitats, ranging from swamps to cities. The Pig-tailed Macaque has a short curly tail and is the most terrestial of all macaques. Male Pig-tailed Macaques or *beroks* are caught by villagers and taught to climb coconut trees and harvest coconuts.

Left, intense stare from a Black Macaque. Above, a Red Leaf Monkey is startled by an intruder.

The Long-tailed Macaque is a common primate of the lowlands and has adapted well to cultivation. At several places, especially on Bali, they are regarded as sacred animals and and fed with rice and fruits at special places near temples. In the wild they are most commonly found on the water's edge of rivers, coasts and lakes, and feed on a wide array of food items which include fruits, but mainly crabs and other animals. Its other common vernacular name is Crab-eating Macaque. They are excellent swimmers and divers. They have large cheek pouches where they can store food to be chewed more thoroughly at a safe place. Macaques are highly social animals and can be seen in large troops of up to 30 individuals consisting of two to four adult males, six to 11 adult females and their young. Large groups can be very noisy and are easily detected by their far-ranging "Kra" calls. Their daily range is fairly large and covers up to 15 hectares.

Eight species of leaf monkeys or langurs are found in the Sundaic sub-region. The Silvered Leaf Monkey, and the Banded Leaf Monkey have the widest distributions. A ninth species occurs in the Indo-Chinese sub-region and is commonly seen in Thailand. The Phayre's Leaf Monkey is found in evergreen forest, spreading into secondary forest where it even can reach higher densities. On Peninsular Malaysia the Dusky Leaf Monkey is found in lowland rain forest. It has conspicuous white skin round its eyes and mouth. Leaf monkeys are slender with hindlegs longer than front legs and a long thin tail. They are truly arboreal and seldom descend to the ground. They feed on leaves, shoots and fruits which they digest by means of a very large stomach which has several compartments resembling those of ruminants. The Batak people of northern Sumatra still worship leaf monkeys just as the Hindus worship the related langurs in India. Both feature prominently in the *Ramayana* epic.

The Silvered Leaf Monkey, the most common leaf monkey, is widely distributed throughout Indo-Chinese and Sundaic sub-regions. It has a black skin with grey tipped hairs, giving it a somewhat silvery appearance. Infants are bright yellow or orange. The group size is smaller than that of the Long-tailed Macaque but can reach up to 14. Each group consists of one or more adult males and a number of females and young ones.

Gibbons are tail-less slender monkeys with a small round head and long arms. They are strictly arboreal and are highly social animals. They are all extremely agile and able to make long jumps from branch to branch using their long, strong arms. Gibbons, of which nine species exist, are restricted to the Indo-Malayan realm.

The Siamang, whose fur is black, is the most heavily built of the gibbons, being twice as heavy as any of the other eight species. It is restricted to the rain forest of Peninsular Malaysia and Sumatra. Its diet consists mainly of fruits and young leaves which they forage for most actively during the early morning hours. Both sexes have a distinctive naked air pouch beneath the throat which they blow up when making their very powerful booming double-noted call. With these calls they mark out their territories that are fiercely defended against other groups and other species of gibbons. Siamangs have the first two digits of their feet webbed, a peculiarity found in primates only in this species.

Two species of gibbons are commonly heard and seen in the rain forests of Malaysia and Indonesia: the White-handed or *Wak-wak* by its local Malay name and the Dark-handed which is also known from the Malaysia/Thai border and Borneo. Both species vary in the colour of their fur from almost white to almost black, the main distinction being the colour of their hands and their different calls. Other well-known gibbon species include the Bornean, Pileated, and Javan.

Most unusual of all primates is perhaps the long-nose or Proboscis Monkey, a monotypic genus confined to Borneo. The orang-utan belongs to the family of *Pongidae* or apes which includes the gorilla and the two species of chimpanzees. All are closely related to man. The orang-utan is restricted to tropical rain forests, normally below 1,000 metres in Borneo and Sumatra, north of Lake Toba.

The Slow Loris, a nocturnal, well-furred, quaint looking and slow moving animal with large eyes and a very short snout is related to the monkeys. Its tiny tail is usually hidden by the thick fur. Its thumb and great toe are apposable and its broad fingertips permit it to obtain a better grip on branches. The skin colour varies from pale grey-brown to reddish brown. It has a striking dark-coloured ring around each eye, and a dark stripe running down the middle of the back. It feeds on insects, fruits, flowers, eggs and small vertebrates. When sighted, the prey is grabbed quickly with both hands while the Slow Loris holds onto the branch with its hind feet. It is widely distributed throughout Southeast Asia, and can be found from lowland to hill localities, living alone or in small family groups.

Another group related to the monkeys, are the tarsiers, a family truly unique to Southeast Asia. Three species are found, all of which are strictly arboreal and nocturnal: the Western Tarsier on Sumatra and Borneo, the Sulawesi Tarsier, and the Philippine Tarsier on the islands of Mindanao, Bohol, Leyte

Left, not a reptile but a mammal: the Pangolin or Scaly Anteater. Above, a large Treeshrew.

and Samar. Their most striking feature are very large eyes which occupy most of the face. They can turn their heads 180 degrees in any direction without changing the position of the body. Tarsiers are well adapted to life in the higher strata of the forest. With their long legs they are able to jump easily up to two metres, using their enlarged terminal sucking pads to attain a firm grip on the tree's branches. Their mostly naked tail is twice as long as their body and is used as an additional tool for holding onto the tree trunk.

Pangolins: Strangest of all mammals found in Southeast Asia is the Scaly Anteater or Pangolin which looks more like a reptile than a mammal, as it is covered in scales. Toothless nocturnal animals, they hunt exclusively for ants and termites in trees, on the ground and below the surface. They use their large (up to 40 cm long) sticky tongue to devour these creatures. They have extremely strong claws on the front feet which are capable of digging out deep termites' nests and climbing trees. When attacked they roll up into a scaly ball in order to protect the non-scaly underparts.

Rodents: Rodents, of which there are well over 3,000 species worldwide, form the largest order of mammals. Rodents include squirrels, rats, mice, and porcupines. Some 160 species have been described for Indonesia. Their main characteristics are large chisel-shaped incissors used for gnawing. Whereas rats and porcupines are mainly nocturnal and ground-dwelling, squirrels are diurnal and arboreal.

At least 40 species of squirrels are found in the Indo-Malayan region. The giant squirrel, with a combined body and tail length of 80 cm is the largest. It lives high in the trees and is capable of making huge leaps from tree to tree. It feeds on seeds, leaves and bark. The somewhat drab-coloured Slender Squirrel is less interesting than the giant squirrel but the mammal most often seen in nearly all wooded habitats. Flying squirrels inhabit tree tops, are mainly nocturnal, and are difficult to detect. They include the Spotted Giant Flying Squirrel, which like other flying squirrels, has developed a gliding membrane between the front and hind limbs. Flying squirrels are unable to truly fly like the bats but can glide for distances of up to 300 metres.

Six species of porcupines, which are rodents, are found in the Sundaic sub-region. The Common Porcupine, the size of a dog, is the largest. Its back and hindparts are covered by long hollow two-coloured quills. The smaller Brush-tailed Porcupine, with considerably shorter quills, is a common mammal of rain forests.

Insectivores: These are small mammals which resemble rodents. This order includes the moonrat, which is not a species of rat, as the name suggests, but is related to hedgehogs, shrews and Flying Lemurs. The latter, like flying squirrels, possesses a gliding membrane. When not gliding it can often be noticed clinging to the side of a tree trunk and looking like a dead leaf. It is not related to true lemurs, which are primates found only on Madagascar and which feed entirely on leaves.

Treeshrews were formerly included in the insectivore order but are now a group on their own. At a glance, they are easily confused with slender squirrels from whom they can be distinguished by their pointed snouts. Although agile climbers, they tend to spend most of their time on the ground in search of insects, seeds or bugs.

Carnivores: These are distinct in that they feed on other animals, which may include animals of a larger size. To capture their prey they have developed powerful claws and are able to attain great speeds. They are distinguished from all other mammals by their large sharp teeth. There are two groups of carnivores: the *canoid* which includes the dogs, martens and bears and the *feloid* which comprises cats and civets. Wild dogs are widespread in Asia although not in Borneo. They hunt in packs: their quarry are pigs and small deer species. The sunbear, also known as honey bear as it is fond of eating bees' nests, is one of the smallest bears. This species, which is confined to the tropics, does not require a wintersleep. It roams the rain forests but can sometimes be found in coconut groves near forested areas. By far the most common carnivore, occurring widely throughout the region, is the Yellow-throated Marten, which belongs to the weasel family. They are mainly diurnal, living both on the ground and in trees. Other representatives from the weasel family are the Ferret Badger and the Malay Weasel. There are four species of otters, Hairy-nosed, Small-clawed, Common and Smooth, all of which are widespread throughout the region, but in small numbers. They are well adapted, with their waterproof fur and streamlined body, to living in rivers. The tails are long and used for both swimming and steering. Their main prey are fish and crabs. Civets, which are well represented, are slender, small carnivores. In Indonesia, 15 species have been recorded, including Palm Civet,

Left, the Clouded Leopard is endangered and elusive. Above, a Fishing Cat lives up to its name.

linsang, mongoose and binturong or Bear Civet which, unlike the other carnivores, feeds off fruits. There are seven cat species. The Leopard Cat is the smallest, the size of a domestic cat, and is also the most common, even being seen in villages. The leopard or panther is still commonly seen, as it has adapted well to man-induced environments. Its fur ranges from entirely black to yellowish-grey. It hunts mainly small mammals and, being a good climber, often preys on monkeys. The Clouded Leopard is slightly smaller and entirely arboreal, hunting squirrels, birds and monkeys, including young orang-utans. Confined to forests, which includes the mangroves, its survival is severely threatened. The tiger is the largest of the Asian cat with adults measuring over two and a half metres and weighing over 200 kg. It is among the most magnificent of the large mammals of the Indo-Malayan region.

Odd-toed ungulates: This includes tapirs, rhinoceroses and the hippomorpha. The last named do not occur in the wild in Southeast Asia. The tapir is a small-hoofed animal, the size of a pony. Its nose is prolonged into a trunk-like proboscis. It has a remarkably deceptive colour pattern. The startling defined black of the front part and limbs and white of the hind part of the body break up its outline and confuse predators. However, the camouflage of a newborn tapir is quite different: its brown body with tawny spots and streaks simulates the sunrays on the jungle floor.

The Sumatran Rhino which was once widespread throughout mainland Southeast Asia, is today confined to Sumatra, Thailand, Peninsular Malaysia and Sabah. Surviving animals, not more than 1,000, are scattered over a number of isolated areas. The Sumatran Rhino is the largest herbivore that lives exclusively in the tropical rain forest. It can live in densities of one per 10 square km which indicates the large areas required for a healthy population. The Sumatran Rhino is solitary for most of its life. Although the home range of a rhino overlaps with other rhinos and they occasional meet, they do not stay together for any length of time. Even rhino calfs are nursed for only a short period. The Sumatran Rhino is a typical browser eating leaves and stems of broad-leaved herbs, twigs of saplings, shrubs and trees. Occasionally it eats fallen fruits. It can walk for kilometres without feeding intensively and systematically from one source, merely taking a mouthful along the way. It is particularly fond of ginger stems, and the hearts of the giant Colocea palm, commonly found in the rich and dense undergrowth of mountainous forests. In areas of natural landslides and treefalls, regeneration can be rigorous, offering ample fodder and feeding opportunities for the rhino. Despite their varied diet, rhinos are regular visitors to mineral-rich springs and salt lick areas to offset an inbalance or a deficiency in certain minerals such as sodium and phosphorous. Wallows are a characteristic feature of any rhino area. The pits are often used for a long time, each being used by several rhinos at least once per day in order to prevent skin diseases.

Javan Rhino: Tucked away in Java's most westerly corner is Ujung Kulon National Park. This 761 square km area is home to one of the most globally endangered mammal species, the Javan Rhinoceros. This, the bigger of the two Sundaic rhinos, weighs up to 2,000 kg compared to the 1,000 kg attained by the Sumatran Rhino. The distribution of this magnificent beast used to cover the greater part of the Indian and Sundaic subregions, including countries such as Bangladesh, eastern parts of India, Burma, Thailand, Laos, Kampuchea, Vietnam and Peninsular Malaysia. Indiscriminate hunting and loss of large undisturbed forests led to a rapid decline. With only one or two specimens left in the Nam Cat Tien National Park in Vietnam and another 10 or so along the Vietnam/Laos border, its main stronghold now is the Ujung Kulon population. However, their numbers in the park have never been large. In 1955, during a first census, 35 were counted. This had been reduced to a mere 25 in a 1969 census, when poaching was still rife. In 1982, a serious gastric epidemic killed several of this small population and plans were designed to split the already small population into two groups, with one group being placed on Sumatra. Illegal hunting still continues to be a major threat, but the government has stepped up its efforts to combat poaching: those caught face stiff penalties. Rhinos have been relentlessly hunted for their horns, which are wrongly thought to have medicinal qualities. Rhino horns are wholly different in structure from those of other mammals for they are made up of consolidated hair.

Left, Sumatran Rhinos take mud baths to keep cool. Above, the Banteng is one of two species of wild cattle.

Population growth is restricted because of the relatively limited habitats still available for this species, and must have reached its point of saturation. Last censuses show a population with a healthy age composition. Unlike the African Rhino, this is a true forest dweller feeding on leaves and shoots. It seems that the rhino population has benefited from the habitat changes as a result of the Krakatua eruption in 1883. This volcano forms part of the Ujung Kulon Park. Huge tidal waves swept away most of the lowland forest and extensive alang-alang savannas appeared. This homogenous vegetation has gradually been replaced by a forest which appears to be very favourable for the rhinos. Medium-sized trees and many tree saplings are attractive sources of food.

Even-toed ungulates: There are nine families of even-toed ungulates four of which are present in Southeast Asia. They include the family of swine of which there are a few species including wild boar and babirusa. The latter is exclusively found on Sulawesi and its satellite islands. Its legs are very long and slender and it has very long and backward curved tusks, the upper ones growing through the snout. The mousedeer family comprises two species, the Large and Small Mousedeer which, unlike the deer, lack antlers. Although seldom seen, they are common small mammals which thrive in the dense undergrowth of the rain forests. The Small Mouse Deer, standing less than 20 cm at the shoulder, is Southeast Asia's smallest hoofed animal. The deer family is representated by five species including the muntjac and sambar. The latter, by far the largest Southeast Asian deer, is found in dense forests and swamps. An animal of the dense forest undergrowth, the muntjac is mainly diurnal. When alarmed, or during the mating season, it produces a very loud barking sound, which can be heard at great distances. Spectacular large herds of rusa once existed. In 1884, herds of up to 50,000 were estimated on the Yang Plateau in East Java, but in recent times, both range and numbers have seriously declined because of excessive hunting and habitat conversion. Other deer species include the Bawean Deer on the island of Bawean, Java.

The bovine family includes gaur, anoa, banteng and serow. The banteng is a beautiful bovid species, dark brown to almost black with white stockings, and whitish forehead used to be widely distributed throughout mainland Southeast Asia. Due to extensive hunting it is now confined to Burma, Thailand, Borneo and Java while it has been domesticated on Bali. Nowadays, it is mainly confined to a number of nature reserves, with the largest populations occurring in the Ujung Kulon and Baluran national parks on Java. Its main predator is the leopard, which is known to be able to prey on young banteng. On Java, the now-believed extinct tiger must undoubtedly have been an important predator. The banteng has been important as domestic cattle since 1,000

B.C., as a good source of food and leather. Banteng are mainly nocturnal and feed on grasses and young woody vegetation. There are two species of anoa, also referred to as dwarf buffaloes. They are found on the island of Sulawesi, comprising a lowland and mountain species. They resemble a crossbreeding of an antelope and a cow. The smallest bovid is the serow or mountain goat which mainly inhabits limestone hills and other mountainous areas. It has a wide range, although in Indonesia it is only found on Sumatra.

Elephants: There are two species of elephants: the African and the Asian. Both are considered to be the largest land mammals. The Asian Elephant is primarily a forest dwelling animal. On Sumatra, the Sumatran sub-species is found. Elephants usually wander in medium-sized herds, comprising several females and their young. They frequent a wide variety of habitats and are found from sea level up to altitudes of over 1,700 metres. Their main strongholds however, are the lowland forests. About 3,000 Sumatran Elephants still roam in the wild. With no large tracts of continuous forests left outside protected areas, current forested areas are too small to sustain any viable population of this magnificent beast in the long term. In addition, the elephant is known to regularly migrate. Wandering herds are often seen raiding crops near forested land. They frequent sugarcane and oil palm plantations as these crops are extremely palatable. Growing man-elephant conflicts, causing much economic damage, have forced the Indonesian government to translocate entire herds to conservation areas. Trained elephants are used to capture wild elephants.

Bats: There are well over 200 species of bats in the Indo-Malayan region of which many are seldom seen as all are nocturnal. Identification is difficult. Bats are the only mammals capable of flying, their wings consisting of a thin double layer of skin. Best known are the Flying Foxes, largest of all bats, weighing more than one kg and with a wingspread of more than 60 cm. They roost in tall trees, hanging downwards during the day. As many as 7,000 Flying Foxes can be seen at one roost.

Marsupials: Some peculiar faunal elements found in Australia have expanded westwards as far as the island of Sulawesi. There are about 250 species of pouch-bearing mammals, including cuscuses or phalangers, arboreal animals the size of a cat, of

Left, the Anoa, or Dwarf Buffalo, and above, the Babirusa, are endemic to Sulawesi.

which seven species are found in Indonesia; possums and kangaroos, of which six species found on Irian Jaya; wallabies, pouched rats and mice, bandicotts and the gumbem or marsupial cat. On the eastern islands of Indonesia four representatives of the monotremata are found. These egg-laying animals include the spiny anteaters.

Cetaceans: Mammalian life does not end at the edge of water. There are three families of dolphins, relatives of the whales: oceanic dolphins, true porpoises, and river dolphins. Most oceanic dolphins are beaked— except for the riverine Irrawaddy Dolphin which is round-headed— like all species of true porpoises. Not less than 21 species of oceanic

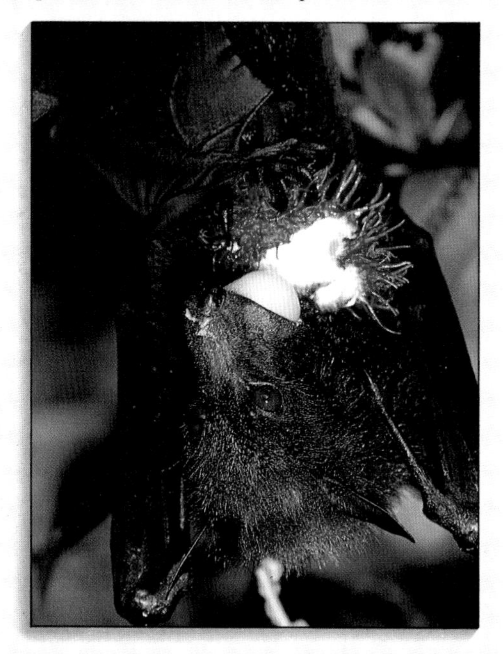

dolphins frequent the offshore waters of the tropical Pacific or Indian oceans. Only four species — Irrawaddy, Indus Susu, Ganges Susu and Beji Dolphins — can be found in estuaries of the Indo-Malayan region and are adapted to freshwater habitats, able to reach rivers far upstream. Dolphins are mainly fish eaters, although some are known to devour squid and crustaceans. Some species are only seen individually; others are gregarious. Oceanic dolphins like the Spotted Dolphin are known to make up enormous groups with herds of 1,000 to 2,000 individuals being quite common. Dolphins are among the most mobile creatures in the world, surpassed among the vertebrates only by long distance migrant birds.

Over the centuries, dolphins have managed to exploit virtually all types of marine, estuarine and riverine habitats, ranging from oceans to muddy coastal shores and crystal clear rivers. Although mammalian, they are remarkably adapted to a complete marine existence, so hostile to other mammals. They can only give birth in the water. Calves, which are born tail first, are generally large relative to the size of the mother. The hind limbs have been transformed into a tail like a fish but are set horizontally. Most of the mammalian senses cannot be used in water and thus taste and smell organs are either reduced or absent. Vision is developed to different degrees, depending on the environment which the species inhabits. The riverine species, which frequent murky estuarine waters, either have limited vision (such as the Beji), or are entirely blind (such as the two Ganges species). By contrast, the dolphin's accoustic sense is acutely developed. It is known to be able to dive to a depth of 300 metres and can stay underwater for more than eight minutes. During these dives its heart rate and peripheral blood flow are markedly reduced.

At least 10 species of whales pass regularly through the waters of the Indo-Malayan region. The opportunity to see any of them is slight unless travellers visit the whales' favourite migration areas such as Lembata island in the province of East Nusa Tenggara, Indonesia. Many tourists visit the villages of Lamalera and Lamakera to see sperm whales. Deep waters, rich in squid, close to this island's coastline attract the whales to the area. Sperm whales can also be spotted in the coastal waters of Meru Betiri National Park in East Java.

Sirenians: The Dugong is a marine species which ressembles, but is not related to, the seal. Its forelimbs have been transformed into flippers while its hind limbs have disappeared. Dugongs are entirely herbivorous, feeding on sea weeds and other vegetation in shallow waters at night. They are usually found in small groups, and are believed to be capable of travelling long distances to reach suitable feeding areas.

Hanging around, are a fruit Bat (left), and a Tube-nosed bat (right).

About 1,650 species of birds are found in the Indo-Malayan region. Of these, some 1,200 occur in the Indo-Chinese and Sunda sub-regions in Southeast Asia. The Wallacean sub-region has its own bird fauna which is of Australian affinity, and includes species of megapode and cockatoo, while a number of families commonly found on the Asian mainland are either entirely absent or have only a few representatives such as woodpeckers. With some 1,500 species, Indonesia is one of the most bird-rich countries in the world, and no less than 400 species are confined exclusively to this archipelago.

This chapter provides a short introduction to the various groups of birds found in Southeast Asia. The selection of species is somewhat arbitrary, based on those that are easily encountered by the casual observer, or those that are of special interest to the dedicated birdwatcher.

Species account: Cormorants are black, long-necked waterbirds that are extremely good swimmers and divers and which prey on fish. There are five species: that most commonly seen along waterways is the Oriental Darter. Herons, egrets and bitterns are wading bird species that are characterised by their long necks and stilts. Very common are the Cattle Egret which, as the name indicates, feeds near cattle, the Little Egret, common in rice fields and the Grey Heron. Three species of pond herons appear brown when sitting but are mostly white when flushed.

Southeast Asia hosts the richest stork diversity in the world. Eleven species are found. All Asian storks are highly endangered and all but two are difficult to observe. The exceptions are the Asian Openbill Stork and the Painted Stork. Both are resident in Thailand where they breed in large colonies such as at Wat Phai Lom temple close to Bangkok. The Painted Stork also occurs in Thailand where it shares colonies with the Openbill Storks. Indonesia provides the core population of the extremely rare Storm's Stork, Lesser Adjutant and Milky Stork, all of which can be seen at the Sembilang reserve in Sumatra.

Many duck species found in the Indo-Malayan region, such as the Common Teal and Northern Pintail are migratory and only stay during the winter. Strangely, only a few resident species occur in the region, such as the Wandering and the Lesser Tree Duck.

The largest bird of prey is the White-bellied Sea-eagle, a resident of coast and river estuaries. It preys on fish and sea snakes. In spite of its size and visibility, it is not endangered and frequently breeds close to man. Perhaps the most common of all raptors is the Brahminy Kite, a scavenger found near rivers and villages. The Sparrow-sized Collared and Black-thighed Falconet can be found in most areas without too much difficulty as can the Changeable Hawk-eagle with its beautiful banded wings.

Quails, partridges and pheasants are mainly ground dwelling birds. This group includes the magnificent Argus Pheasant, a bird more often heard by its "kuang" call

Preceding pages: a large flock of Wreathed Hornbills in northern Thailand. Left, the Crested Wood-partridge. Right, Little Spiderhunter inspects a *Heliconia*.

than seen, and the Red Jungle Fowl, the ancestor of the domestic fowl. Unfortunately, many members of this group, including about a dozen pheasant species, have become extremely rare as a result of hunting pressure.

No less than 46 species of shorebirds and 15 species of terns and gulls can be observed near waterbodies and on coastal mudflats. Most species are migratory, having their breeding grounds as far away as Siberia. Sites where one can observe waders during their autumn migration include Berbak and Sembilang along the coast of Sumatra, and Selangor on the west coast of Peninsular Malaysia.

east Asia whereas the Wallacean sub-region holds 36 species including the beautiful Yellow-crested Cockatoo. The Blue-crowned Hanging Parrot, which is smaller than a sparrow, is fairly common, being easily found, for example, at Taman Negara in Peninsular Malaysia.

With more than 30 species in this region, cuckoos are a particularly prolific family in tropical habitats. A majority of species of this interesting group are known to have used a foster parent plan, laying their eggs in other birds' nests. However, not the Coucals, thick-billed birds with rufous backs. Especially attractive are 10 species of malkohas, including the Chestnut-bellied Malkoha,

A great number of pigeons and doves can be seen in the vicinity of villages and parks. Familiar birds are the Spotted and the Peaceful Dove. More than 20 species of green pigeon occur in the region. They are particularly attractive to visitors from northern latitudes who are accustomed to grey and brown members of this group of birds. Most require large forests although the pink-necked Pigeon has adjusted to the gardens of downtown Singapore.

The parrot family has some 315 species worldwide with centres of diversity in tropical America, Africa and Australia. Only nine species are recorded on continental South-

which are endemic to tropical Asia. They run in a squirrel-like manner, rather than fly, through the forest canopy.

Owls are nocturnal predators that are rarely seen but are more often heard by their haunting night calls. The Collared Scops Owl, one of the smallest owls, is frequently heard in parks.

Nightjars, like owls, are nocturnal but feed on insects. The best time to see them is around dusk, when species such as the Large-tailed Nightjar can be observed hawking while repeating its monotonous "chonk" calls. It is next to impossible to find one of their related species, the frogmouths, unless introduced to their calls by a tape or by an accompanying expert.

Trogons are among the most colourful birds but are seldom seen as they inhabit the middle storeys of primary forest. Although groups of tropical birds are restricted either to the Old or New World tropics trogons occur in Asia, Africa and America. Nine species are known in Southeast Asia.

Twenty species of swifts, swallow-like but non-passerine birds, include the world's fastest flying bird, the Brown Needletail. Southeast Asia is famous for its breeding colonies of Edible-nest Swifts. Excessive harvesting of the nests, which are the basis of a highly prized soup, endangers this species.

Bee-eaters evolved as birds of the desert and savannas and occur on forest fringes or in disturbed habitat. Two species, the Blue-bearded and the Red-bearded Bee-eater occur in closed forests. Bee-eaters such as the Blue-tailed Bee-eater are a very colourful group of birds of predominantly green plumage, slender posture and with prolonged tail feathers. They feed on insects in mid-air.

Rollers are a family of beautiful blue, crow-sized birds. The Dollarbird is widely distributed in secondary vegetation while the Indian Roller is restricted to the north of the region.

Hornbills are a most extraordinary group of birds, which have the habit of sealing off

the incubating females in their nest holes, to prevent the nest from being robbed by predators. Hornbills are all large in size, with a large protuberant casque on top of their bill. Common representatives among this strictly arboreal group are the Black Hornbill and Southern Pied Hornbill.

Kingfishers, represented by a single species each in Europe and North America, have developed into a baffling diversity in Southeast Asia. This group is easy to recognise with their large heads and long pointed bills. The 34 species, of which 18 are restricted to the Wallaceana, are all spectacularly coloured. Not restricted to open water but inhabiting all kinds of habitat, a number of species, such as the Collared Kingfisher, are birds of open bush land.

Left, a Thick-billed Pigeon feasts on figs. Above, a Blue-crowned Hanging Parrot sees things differently.

Barbets like trogons are distributed pantropically. Although hardly seen, since they are strictly arboreal and live in the canopies of dense forests, their incessant, far-reaching calls are quite distinct, and form much of the acoustic backdrop in primary and secondary forest. Twenty species occur in the region.

There are 45 species of woodpeckers, in

all sizes and colours, the brightest coloured being the Greater Goldenbacks. Three species of diminutive piculets are, with a size of only 8 cm, among the smallest birds of the forest.

Ten species of broadbills represent a family of birds restricted to the tropics of Asia and Africa. While rarely seen, their songs in chorus with the calls of barbets are the main element of the acoustic backdrop which visitors from northern countries find "exotic".

The commonest swallow species is undoutedly the graceful and slenderly built Pacific Swallow, a resident of the region. Other species winter in the region.

habit the tropics of the Old World. Lesser and Greater Racket-tailed Drongos with their distinct prolonged outer tail feathers are respectively birds of montane and lowland forest.

The oriole family is widely distributed throughout Southeast Asia. The Black-naped Oriole with yellow appearance and melodious call similar to its European relative is one of the region's most common garden birds. Surprisingly, other species have developed a completely black and chestnut plumage. The beautiful black and blue Asian Fairy-bluebird, common in rich forests throughout the region, is also distantly related to this family.

Several species of the beautiful black and red minivets belong to the Cuckoo-Shrikes family. They are some of the most common passerines in lowland and montane forest.

Bulbuls are a large family of often nondescript olive or brown birds of the middle storeys of tropical forest in Asia and Africa. Some, such as the Sooty-headed Bulbul, are birds of the open country and more likely to be encountered. The Yellow-vented Bulbul is extremely successful in urban and rural man-made habitats and is one of the most common birds of the region.

Drongos are medium-sized, all blackbirds with characteristically shaped tails which in-

Crows are scavengers that can readily be recognised by their all-black plumage and their large harsh call. In the tropics, however, the family has evolved into many beautiful forms in the shape of colourful treepies and magpies.

No less than 140 babbler species are found throughout Southeast Asia. Unfortunately, it is a rather difficult group to identify because most members are rather dull brown and small. However, their songs or calls, such as that of the Striped Tit-babbler, are quite distinctive. The family includes seemingly unrelated species like different laughing thrushes and some beautiful birds of the montane forest, such as Cutia and Silver-eared Mesia.

The Magpie Robin is the most commonly seen species of the large thrush family which includes thrushes, chats and forktails. It is a common bird of open woodlands and gardens of the lowlands. Both the Rufous-tailed and White-rumped Shama are birds of closed forest. Their beautiful songs have resulted in the sad fact that in some regions more are in captivity than in the forest.

A member of the warbler family is the Common Tailorbird that is frequently seen in bushes and gardens. They are small, with a rufous coloured crown and a cocking tail. They owe their name to their ability to sew together a large leaf to serve as support for their nest. The Yellow-bellied Prinia is a common scrubland bird often seen clambering about in bushes.

About 50 flycatcher species are known from Southeast Asia. Most of the males are real emeralds, brightly coloured blue and red, while the females are dull brown or grey. Common members of this family are the Black-naped Monarch and the Asian Paradise Flycatcher, which are both forest birds.

Heard, but not often seen. Left, a Black-and-yellow Broadbill. Above, Silver-breasted Broadbill.

Most starlings and mynas, such as the Common Myna and the Philippines Glossy Starling, abundant town birds of Singapore and Kuala Lumpur, have a close affinity with man. The Hill Myna (*Beo* in the German language) is famous for its ability to imitate the human voice.

Sunbirds have adapted, like neo-tropical hummingbirds, to living off flower nectar but the two groups are in fact unrelated. Many sunbird species such as the common Olive-backed Sunbird favour gardens which provide a haven of flowering plants. Others such as the Crimson Sunbird require undisturbed forest.

Spiderhunters are also nectarine birds, despite their name, and have extremely long, curved and slender bills.

Flowerpeckers are nectarine birds that differ from sunbirds in having a stout short bill. All species of flowerpeckers boast very colourful plumage. The Scarlet-backed Flowerpecker, a delightful red-coloured bird, is a common garden bird of Bangkok and Singapore: most other species require primary forests.

Sparrows, weavers and munias include the cosmopolitan Eurasian Tree Sparrow and a variety of rice birds such as the White-headed Munia. They are fairly colourful and popular cagebirds.

HORNBILLS

Hornbills are bizarre and impressive birds. They have three peculiar characteristics: (1) a disproportionate form and shape; (2) The jerking of head and bill when making the loud, distinctive call; (3) Nesting in the natural cavities of trees.

Appearance: Hornbills are large, awkward birds. Their plumage is most often black and white, with coloured patches of bare skin at the head, throat and eye. They resemble the toucans of tropical America, although they are not related.

The bills of most hornbills are disproportionately large when compared to the size of their heads. Their bills are downward curved, or nearly straight, and are decorated with an extra part on the top called a "casque", varying in shape and size. In some species, the casque is a ridge or a raised crest. All casques are hollow, except for the casque of the Helmeted Hornbill, which is dense and ivory-like. The function of the casque is unknown, but may involve species and sexual recognition.

Because of the absence of underwing coverts, air passing through the base of the wing feathers creates an amazing whooshing sound resembling a locomotive in the forest. This sound is audible, before the hornbills are seen. The birds fly with a pattern of flaps and glides. The number of flaps and the length of a glide depends upon the species. Little wonder that the hornbills are not migratory but move only locally within a large intact forest.

Nesting Habits: Asian hornbills live only in tropical forests which provide nest trees and food sources. Hence, a forest containing hornbills must be large and unaltered by man. Since hornbills, unlike woodpeckers and barbets, cannot excavate their own nests, they must use existing suitable natural cavities in trees as nest sites. Dipterocarps are favourite nesting trees for all hornbills. Scientists speculate that fungi which cause heart and butt rot in dipterocarps eventually create the ideal nest cavity for the hornbill.

Left, portrait of a Wrinkled Hornbill. Right, it's not difficult to identify a hornbill in flight.

During the breeding season, which varies according to geographical location, the female seals herself into the cavity by plastering the entrance. She may be assisted by her mate. The plaster consists of mud, tree bark, wood dust and food debris. The proportions of the materials vary from species to species. The female mixes the material with her own faeces or regurgitated food. When dry, the mixture becomes very hard. A narrow, vertical opening is left in the plaster, through which the male feeds and the female

and the brood defecates. From the time the nest is sealed until the chicks become fledglings, the female and the brood are wholly dependent upon the male. The breeding cycle varies from species to species.

Diet: Hornbills are generally fruit eaters but can be omnivorous. Their favourite food is figs. When raising the young, they prey on insects and small animals, putting the hornbill number one in the food chain. The population and breeding status of hornbills makes them excellent indicators of the health of the forests they inhabit. Any tropical forest with a large, secure population of hornbills is ecologically intact.

Hornbills in Thailand: Thirteen species of hornbills are found in the forests of Thailand. Hornbill populations have shrunk or become extinct in certain areas because of human encroachment, deforestation and overhunting. The government's recent policy of protecting 15 percent of the country's forests and the banning of logging throughout Thailand make the hornbill's future secure for the time being. This policy, however, needs proper management and a conservation education campaign.

Seven species of hornbills are included in the list of endangered species in Thailand: Wrinkled Hornbill, Blyth's Hornbill, Rufous necked Hornbill, Rhinoceros Hornbill, Helmeted Hornbill, Black Hornbill and White-crowned Hornbill.

Khao Yai National Park in Thailand is perhaps one of the best places in Southeast Asia to observe large concentrations of hornbills.

Great Hornbill: If ever there was royalty in the bird kingdom, then the Great Hornbill, *Nok Gok*, *Nok Gahang*, or *Nok Gawa* in Thai would be king. It may not be as graceful as the eagle, but it is certainly no less impressive with its huge bill and horned casque. Its Latin name *bicornis* is derived from these distinctive features. In Thai, the Great Hornbill is named for its call. The call consists of the sound, *Gok, Gok, Gok...*, followed by few rapidly repeated cries of *Gahang, Gahang...*, or *Gawa Gawa...* In some areas, the Great Hornbill is respected as "Lord Buddha's bird". This stems from the belief that its call awakens monks and summons them to morning prayers.

This black-and-white bird is easily recognised by its size (between 110 cm and 130 cm in length) its large yellow and orange bill and its large yellow casque. Males are larger than females and have red eyes and black underparts on the front of their casques. Females have white eyes and no black on their casques. Both sexes have white tails crossed by a black band. The Great Hornbills paint their bills, casques, heads and wings yellow. The source of the paint is an oil gland located at the base of the tail and it is said that the species uses "cosmetics".

The breeding season of the Great Hornbill begins in January and ends in May. It has a better chance than other hornbills, even the Wreathed Hornbill which begins to breed at the same time, in selecting a nest hole. There is no doubt that interspecific competition for nest holes occurs. Nest trees used by both species may reach 150 cm in diameter at breast height, and nest holes are sometimes more than 25 metres above the ground. Two important genera, *Dipterocarpus* and *Eu-*

genia, provide suitable nest cavities. The Great Hornbill tends to select a cavity with an elongated entrance which is just big enough for the female's head to pass through. The selection of as small an entrance as possible reflects the conservation of energy in the sealing process. First, the female cleans the inside of the nest, removing all old nest debris and plaster, and then she begins to seal the nest. The male sometimes assists her by bringing her tree bark and fruits. Earth is little used for sealing. Mating occurs near the nest, either before the female begins the day's sealing work, or after. The sealing process take three to 10 days and is done only by the female.

After the female imprisons herself, the male feeds her primarily with fruits. Feedings increase during the seventh week, which is either in late March or early April. By this time the chicks have usually hatched and the food brought to the nest is diverse.

Fruits, especially the fig, are the main food of the Great Hornbill. Important non-fig species consumed include the *Cinnamomum sp.*, *Eugenia spp.*, and *Strombosia spp.* Animals provide additional nourishment for the growing chicks. The Great Hornbill takes full advantage of the diversity of the forest and preys on a fantastic variety of animals. Squirrels, rats, snakes, owls and nightjars are among its quarry. Its awkward appearance belies the grace and speed with which it hunts.

A few weeks after the single chick is hatched, the female breaks the plaster and assists the male in the feeding. The chick reseals the entrance using materials brought by the parents.

The chick begins to fledge about 40 days after the female has left the nest, usually by late May. At this time the feedings begin to decrease. The fledgling follows its parents and is under parental care until the next breeding season approaches.

Distinctive profiles of a Great Hornbill (left) and Brown Hornbills (above).

Wreathed Hornbill: The Wreathed Hornbill *Nok Ngauk Grarm Charng* (elephant teeth or *Nok Goo Gee* in Thai) derives its name from its wreathed casque, which is small and which consists of a series of ridges. These ridges can be used to determine the age of the immature, as a single ridge denotes a year-old bird. The species is about 110 cm long. In flight it appears black with a white tail. This species has a pouch which is yellow in the male and blue in the female. The male is also distinguished by its chestnut crown and nape and its white and buff face and throat. In contrast, the entire body of the female is black save for her white tail. The bills of both sexes are characterised by wrinkles which increase as the bird reaches maturity.

The Wreathed Hornbill makes the most spectacular flying sound of all of the four species and is the strongest flier of the group. The flight pattern is a long, continuous series of flaps interrupted by short glides. Its amazing whooshing sound makes it easy to identify as it approaches. The Wreathed Hornbill's call is similar to that of a puppy.

The nesting and breeding behaviour and cycle of the Wreathed Hornbill is similar to the Great Hornbill, although the female remains in the nest until the single chick fledges.

This species is the most gregarious of the hornbills. It prefers to feed and roost in flocks of 1,000 individuals or more. This species also travels the farthest, journeying more than 10 km from the nest site in the non-breeding season. The sight of a huge flock flying into a valley to roost is amazing and the sound of these flocks is akin to an approaching storm.

Indian Pied Hornbill: This black-and-white hornbill, *Nok Gaek* or *Nok Gaeng* in Thai, is about 75 cm in length and its white abdomen and wing tips are conspicuous in flight. The male has an ivory-coloured bill with a bigger cylinder lying along the upper part of the bill and a black patch across the front, and the female has a smaller bill and casque covered with irregular black patches. Its call gives it its name in Thai: *Gaek, Gaek,* . . . and its flight pattern is an alternating series of flaps and glides. The Pied Hornbill is not as loud in flight as its two larger cousins, but can still be very noisy.

The Indian Pied Hornbill begins to nest late in February or in early March. It is the most tolerant of the hornbill species and can even nest within a remnant forest. The species uses earth collected by the male from the roots of fallen trees. The female seals the nest entrance.

The female lays two or three eggs, but only one or two chicks hatch after an incubation period of 25 to 27 days. The female remains with the chicks throughout the breeding cycle, which ends in May and lasts about 80 days.

This species is more carnivorous than its two larger cousins. It eats rats, birds, crabs, frogs, molluscs, fish and a wide variety of insects, and it is not uncommon to see the Indian Pied Hornbill feeding on the ground. Animal food comprises 20 percent of its diet.

During the non-breeding season, the Pied Hornbill gathers in small flocks of between 50 and 100 individuals. These flocks are very noisy when travelling in or along the forest's edge to feed. They are seen more often than the other hornbills in wide open areas.

Brown Hornbill: This species *Nok Ngauk Si Nam Tarn*: brown in Thai, the only hornbill with brown plumage, is easy to identify. Males differ from the female in having lighter brown feathers and white sides and under parts up to rufous. Their wings and tail feathers are also tipped with white and their bills are yellowish white and their orbital skin is blue. The female has darker grayish brown plumage over her entire body and her bill and casque are smaller.

This hornbill is the rarest of the four species in the Khao Yai National Park. It prefers to nest at higher altitudes of about 800 metres above sea level. It is extremely noisy and remains in flocks of up to 50 individuals all year around. The Brown Hornbill is confined to the central, north and northwest of Thailand, where large, intact forest still exists. The Brown Hornbill is the most carnivorous of the four species. Animal food makes up about 40 percent of their diet, with insects being a speciality. The remainder of the diet consists mainly of figs and other fruits.

Interestingly, this bird has a cooperative breeding habit. In addition to the mate, the brood of one to three chicks is fed and protected by one to five males of varying ages. These males are called "nest-helpers". Non-breeding females are not permitted to feed the chicks. The advantages of having nest helpers may involve the security of the nest or the bird's own safety. Helpers also reduce the workload of the mate.

Interaction: Since the hornbill's main food source is the fig, the known favourite of fruit-eating animals, competition for food is intense. A fruity "fig tree" is like a large restaurant. Its main customers include hornbills, barbets, bulbuls, mynas, pigeons and orioles. Mammals in the restaurant area are macaques, gibbons, squirrels, binturong, and the palm civet. Customers compete not only for food, but for space. Hornbills are forced to interact with other animals, particularly gibbons, which also feed on other fruit sources favoured by the hornbill.

The hornbill's life history can be a successful tool and model for conservation and education programmes for the preservation of the tropical forest ecosystem. The motivation for the well-being of the members of this family can help to assure the protection of the region's natural heritage.

Left, three of a kind: Wreathed Hornbills in flight. Above, the aptly-named Rhinoceros Hornbill.

PITTAS

Among the denizens of Southeast Asia's forests, the pittas — brightly-coloured, ground-feeding birds — receive more than their fair share of attention from birdwatchers. With 29 species in total, this family has its headquarters in Southeast Asia. Though one species is found in equatorial Africa, the remainder are chiefly distributed from the southern and eastern flanks of the Himalaya, across into Indochina, Peninsular Malaysia and throughout the Indonesian archipelago to Northern Australia. Structurally, all appear remarkably similar, being plump-bodied and short-tailed, with strong bills and long, strong legs to leap across the forest floor in springing hops. Many species are brilliantly coloured, with iridescent blue or green, deep red, flame-orange or even purple hues often combined with striking black and white markings.

Pittas feed mainly on soft-bodied insects, worms and snails plucked from the forest floor and all species therefore frequent shady and moist, but usually well-drained, areas where such food is abundant. They breed mainly in the rainy season. While most species are sedentary and confined to the moister evergreen forests, the migratory habits of the Blue-winged Pitta, one of the most widespread and successful of the group, has enabled it to exploit the seasonally driest forests, bamboo and even scrub in the hottest and driest parts of continental Southeast Asia. Towards the end of the southwest monsoon, and at the beginning of the continental Southeast Asian dry season, in September or October, the Blue-winged Pitta migrates to the ever-wet rain forests of Sumatra, where it spends the non-breeding season. In May, when the monsoon rains bring new life to the parched deciduous forests in the lowlands of Thailand and Indochina, the Blue-winged Pitta returns to nest, feeding its young on the earthworms which now abound in the freshly-moistened leaf litter.

Notwithstanding their evident beauty, much of the allure of pittas for birdwatchers stems from their secretive nature. Like many other ground-feeding forest birds they can be extremely shy and are difficult to observe. At the start of the breeding season their calls, often short, fluty, but explosive whistles, are a good clue to their general whereabouts and taunt the observer struggling to catch even the briefest glimpse of them. More often, one stumbles across a pitta by accident while walking along a forest trail. In such cases, a startled bird may fly on to a branch or log, perching briefly to take a look at the intruder before flying or bounding off into the forest.

Birdwatchers love the challenge posed by these birds: perhaps it is the sublimation of the hunting instinct. At any rate, every sighting of a pitta is a little bit special: something to be savoured and carefully recorded in one's notebook. Thailand is a particularly good country for pittas, supporting 11 species. Doi Inthanon and Doi Suthep in the north are good places to search for the Rusty-naped Pitta, while Khao Yai supports both Blue and Eared Pitta. In the peninsula, Thaleban supports both Hooded and Banded Pittas while the Mangrove Pitta may be found in the mangroves of the west coast.

Because of the association of pittas with moist forest habitats they can be useful "indicator species", alerting conservationists to the health of forest ecosystems. This is especially true in the seasonal tropics where fires, started by man, may encroach from dry woodland or cultivated land into the edges of the evergreen forests. Fire has long been used by man as a tool to clear land for cultivation but, even where forest is not cleared outright, repeated fires from adjacent cultivated areas may gradually degrade moist evergreen forest patches and lead to their replacement by drier, more open woodland which supports a less rich wildlife community.

Rare Bird: One species in particular, the Gurney's Pitta, has captured attention because its world range is restricted to a small area of southern Thailand and extreme southern Burma where the rain forests are often referred to as "semi-evergreen" to distinguish them from the wetter, less seasonal "evergreen" rain forests of Malaysia.

It is not known why lowland forests are so important for Gurney's Pitta. However, they and about 40 other species of forest birds, ranging from the small babblers right up to some pheasants and other gamebirds, are missing from the hill slopes of southern Thailand and Peninsular Malaysia. Almost as soon as the boundary between lowland forest and the foothills is crossed, a noticeable reduction in bird diversity occurs. Rain forests of the level lowlands in Southeast Asia are richer in tree species than those anywhere else in the world and the bird commu-

Pitta patterns: left, a Mangrove Pitta and above, Gurney's Pitta, considered extinct for decades before rediscovery.

nity which has evolved to exploit this richness and complexity is presumably less well suited to exploit the botanically poorer hill slope forests.

At the only site where a potentially viable population of Gurney's Pittas remains, there is perhaps no more than 10 square km of lowland forest in total, most of which is secondary growth and is fragmented, situated among rubber gardens and other cultivation, albeit connected to a core area of roughly 100 square km of forest on nearby hill slopes.

Research suggests that the species is well adapted to survive in certain kinds of regenerating moist forest and secondary growth, particularly where there is an abundance of understorey spiny palms in which the species can nest. Quite possibly, Gurney's Pitta is particularly adapted to certain stages of forest regrowth such as might have occurred naturally in clearings caused by tree falls or along riverbanks. Future management which is aimed at enhancing the regeneration of secondary forest on already cleared areas could increase the areas available to Gurney's Pitta. Narrow corridors of protected secondary growth habitat, radiating through cultivated areas along streams, and which link fragmented forest patches, might also enhance survival by allowing Gurney's Pittas to disperse between them. Surprisingly, another rare and little known species, the Giant Pitta has been found in similar areas to the Gurney's and would probably also benefit. Since two more widespread and abundant species, the Blue-winged Pitta and the smaller Hooded Pitta also occur, this means that under some conditions, no fewer than four species of pitta may be found co-existing in areas which, until a few years ago, would have been considered far too disturbed to warrant serious attention.

One of the most promising aspects of the situation is the active involvement of villagers in conserving the birds. Since many of the birds occupy small habitat patches which are scattered among farmland, birds can only be conserved by enlisting the active cooperation of villagers.

Villagers have always harvested forest products in order to supplement their meagre income, most of which is now derived from growing rubber. In the past, the Gurney's Pitta site was a thriving supply centre for many rare birds, especially pittas, Hill Mynas and the Crested Wood-partridge and, at certain times of year, provided a good source of income for a small nucleus of bird trappers. As forest cover and bird populations were reduced, so trapping birds became more labour intensive and less worthwhile. The most skilled of these former bird trappers, is now employed as a forest guard. Instead of catching birds, he receives an alternative source of income by helping field researchers.

Income generated by birdwatchers who have visited the site has been channelled directly to the village committee for use in improving amenities. A modest level of properly regulated "wildlife tourism" can provide useful financial incentives for conservation. The Gurney's Pitta and lowland forest project is still in its infancy. Nevertheless, it has already yielded promising initiatives which may be applicable around protected areas elsewhere in Thailand and, indeed, throughout the tropics. Instead of pitting wildlife protection officials against villagers, as has so often happened in the past, the protection of Gurney's Pitta has provided them with a common cause.

Left, the brightly-marked Garnet Pitta. Right, a reptilian representative.

REPTILES AND AMPHIBIANS

About 10,000 species of reptiles and amphibians are known worldwide. The Indo-Malayan region has its fair share, including species unique to the region, such as the Komodo Dragon, gliding lizards, man-eating crocodiles, and species of turtles and tortoises.

Reptiles, a group of vertebrate animals which breathe air during the whole of their lives, have scales and the majority lay eggs. They are classified in five main groups, of which four are well represented in this region.

Of the world's 22 crocodile species, four, including the Estuarine Crocodile, live in the Indo-Malayan region. This awesome creature, reaching lengths of six metres, is quite capable of killing people. Its main habitats are coastal estuaries and even open sea but also extend well inland at brackish water swamps. Over the last decades, crocodiles have been hunted for their valuable skin, and these reptiles now occur only in protected areas. Irian Jaya, Indonesia's most easterly province, is believed to still hold 500,000 of such beasts.

At least 35 species of tortoises and turtles are found in the Indo-Malayan coastal waters. The Green Turtle is the most common of the marine turtles. It is a rather small turtle, rarely exceeding one metre and owes its name to its green coloured fat. The green turtle is believed to have the stamina to swim over very long distances. Although carnivorous when young, adults feed off sea-grasses and mangrove leaves.

Loggerhead turtles, named because of their massive stout head, prey on fish and invertebrates such as jellyfish and crustaceans. Hawksbill Turtles, each less than one metre long, are the smallest turtles in the region. They feed mainly on crabs and prawns.

Leatherback Turtles are the largest turtle species, reaching lengths of 2½ metres and weighing up to 900 kg. Like the other species, the leatherback lays its eggs from May to September. A great leatherback may lay as many as 100 eggs in a hole it digs in the sand with its flippers. Important nesting beaches on Malaysia are the islands off the coast of Sabah and Sarawak and Rantau Abang Beach in Trengganu, Peninsular Malaysia. Indonesia has many good sites for viewing nesting turtles which include Meru Betiri, Komodo Island and Ujung Kulon. River turtles include the Labi-Labi which can have shells as large as 80 cm.

The order of lizards includes the famous Komodo Dragon, found only on the island of Komodo and its satellite islands, and *chichaks* or the common gecko. The latter is often found in houses and grows to a length of 9 cm. This species is mainly nocturnal.

Considerably larger than the gecko, but also found in houses, although in fewer numbers, is the Tokay. It can grow to a length of 20 cm. Being of such size, it preys not only on insects such as dragonflies but also devours geckos, mice and even birds with its very powerful jaws. Although seldom seen, its far-reaching and distinctive "tokay" call can frequently be heard.

Other lizards are the brilliantly coloured skinks and flying lizards which, like flying squirrels, do not really master free flight, but rather glide. Such species can be seen in gardens.

A bigger lizard, growing up to two metres in length, is the Monitor Lizard which can often be seen basking on riverbanks. They are expert swimmers and hunt fish and small mammals.

Well over 100 snakes have been recorded in the Indo-Malayan region, the majority of which, such as the Reticulated Python, the world's largest snake, are non-poisonous. They feed on small mammals and birds and can reach a length of nine metres. Most sea snakes and estuarine snakes are venomous. The flat oar-like tail of the sea snake allows it to swim in the water. Only a few terrestrial snakes, including the cobras and pit-vipers, are poisonous.

Amphibians include toads and frogs, of which a large variety are found in the region. The majority are forest dwellers, some of which, including the flying frog, are capable of inhabiting trees.

INSECTS

Insects are probably the commonest of all living things. Currently there are around a million known species which constitute about three-fourths of all animal species. It is believed that there may actually be two to four million species of insects. They exhibit a great diversity of habits and inhabit practically every terrestrial and freshwater environment. Indeed, with the majority of insects, the larva and adult stages occupy different habitats and lead an entirely different mode of life.

Just as insects come in a great array of form and colour, human beings may perceive them quite differently. Nature lovers admire butterflies for their bright and attractive colours. Housewives are dismayed at the sight of cockroaches, ants and house flies. Fruit and vegetable farmers dread the presence of caterpillars and bugs. And travellers to the tropics regard mosquitoes as the scourge of malaria, dengue, filariasis and other diseases.

Beneficial and harmful insects exist. In the following pages we attempt to highlight the biodiversity of insect life and to introduce some of the more interesting and extraordinary insects likely to be encountered.

Ants, bees, wasps, and others: Although all kinds of insects are encountered in forest and wayside habitats, ants seem to be the most common. There are ants' nests on the ground, on rotten logs, high in the treetops, and even hidden inside the hollow stems of certain plants.

The most conspicuous and unmistakable ant in the jungle is the Giant Ant (*Camponotus gigas*). It usually wanders singly on forest paths. Measuring about three cm long, the worker ants are dark reddish brown in colour. They are not normally aggressive, although their powerful jaws make them look very formidable.

One of the most familiar and amazing ants is the weaver ant or Kerengga (*Oecophylla smaragdina*) which weaves leaf castles in the air by joining living leaves together with fine silk threads produced by their larvae. Weaver ants are very aggressive and the workers savagely attack any intruders biting with their jaws and injecting an acrid fluid over the wound.

Related to the ants are many types of bees, some of which are solitary while others are social insects that live in colonies. Some bees play an important role as pollinators of flowering plants. Others are valuable sources of useful products such as honey and beeswax.

The most conspicuous solitary bees are the carpenter bees whose females have formidable stings. Two of them — *Xylocopa latipes* which is very large and black in colour, and *Koptorthosoma confusa* which is smaller with black abdomen and yellow thorax in the female and completely yellow in the male — damage timbers by boring large holes.

Of the social bees, the true honey bees are represented by four species. The biggest is the Giant Honey Bee (*Apis* dor*sata*), the smallest is the Dwarf Honey Bee (*Apis andreniformis*, formerly referred to as *Apis florea*). Both build open hanging nests on tree branches, ledges and cliffs. Two other species — the Small Indian Bee (*Apis cerana*) and the Red Honey Bee (*Apis koschevikovi*) of Borneo — construct their nests in sheltered cavities. These wild bees are hunted extensively for their honey.

The majority of social bees are stingless honey bees of which more than 30 species occur in Southeast Asia. These bees are encountered in most habitats and their vestigial stings are harmless. They are especially attracted to human perspiration. Their presence is readily detected by the landmark entrance tube to their nests which are built in cavities in trees and buildings or in the ground.

Like bees, wasps are represented by solitary and social forms. Wasps possess a formidable weapon in the sting which is the egg-laying organ that has been modified into a hypodermic needle connected with a poison gland. Adult wasps feed on nectar, fruit and plant sap.

Preceding pages: the uncanny camouflage of a Leaf Insect. Left, not an exotic plant, but an exotic insect — the Flower Mantis.

Solitary wasps capture small creatures, such as spiders, caterpillars and other insects, and store these in their nests as food for their developing young.

They construct nests, usually of mud, in a variety of situations in cavities or on open surfaces. Potter Wasps (*Eumenes*) make the most elegant nest which resembles a little urn-shaped pot with a neat funnel-shaped opening.

Social wasps, like ants and social bees, build communal nests. Some species, particularly hornets, are quite dangerous. The two largest species are the Banded Hornet (*Vespa tropica*) and the Lesser Banded Hornet (*Vespa affinis*), both of which are black with an orange band on the abdomen. Unlike bees, hornets do not lose their sting in the process of stinging and hence can repeatly attack the victim.

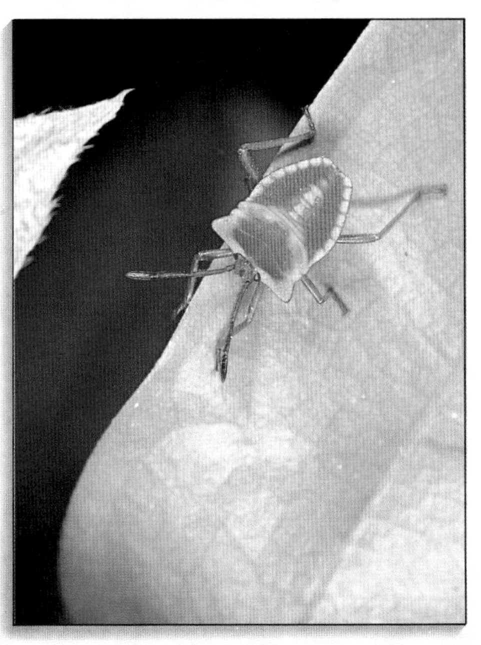

Of the smaller species, Hover Wasps are tiny and elegant insects. These little wasps can be seen hovering between the threads of spider webs to pick out the spiders' prey and use them as food for their young.

Termites are commonly referred to as "white ants". They are, however, not related to ants but are probably closer to the cockroaches. A termite community is made up of four basic castes. The king and queen termites live together for several years in a "royal chamber" deep down in the nest. The other two castes are sterile. They are the female workers who tend the nest and look after the royal pair and the young, and the male soldiers with large heads and jaws who defend the colony against intruders.

Most termites stay underground or inside wood. Some construct globular nests among tree branches and others build curiously shaped pillars of earth. The nests are built by the workers.

The most common termite is the small *Macrotermes gilvus* which builds mound-like nests and grows fungus gardens inside them. In the jungle, the most common are members of the genus *Hospitalitermes* which form dense, seemingly endless, moving columns on the forest floor or tree trunks.

Of the handful of cockroaches, the American Cockroach (*Periplaneta americana*) is perhaps the most common household pest. Against this, many species of cockroaches in the wild possibly play an important positive role in the ecosystem. Some are of giant size while others are quite small. Some are drably coloured but blend well with the environment. Others are brightly coloured and with beautiful patterns. They are found among the leaf litter on the forest floor, underneath tree bark, resting on leaf surfaces and even within bat guano.

Camouflage and Chorus: A close relative of the cockroach, mantids can be distinguished from other insects by their highly specialised forelegs and their characteristic praying attitude. Mantids use their forelegs as "hands" and move only on four legs. All are ferocious predators, living entirely on smaller insects and other small creatures. The voracious habits of mantids sometimes make mating a hazardous operation for the male: he may be devoured by the female as legitimate prey — even while copulation is in progress.

The Flower Mantis or Orchid Mantis (*Hymenopus coronatus*) is perhaps the most extraordinary mantid. It differs from other mantids in having the middle joints of the two hind most pairs of walking legs widely expanded and resembling the petals of flowers. With the hind part of the body arched over its back and the petal-like thighs surrounding it, the mantis resembles a large four-petalled flower. This and the camouflage colouration possibly serve to protect the mantis against predators, such as birds and lizards, which hunt by sight. They also serve to lure prey, such as bees and butterflies which visit the flowers for nectar.

Equally as famous as the Flower Mantis, if not more so, are the Stick Insects. These insects have an exceptionally long body with slender, delicate legs and look exactly like sticks. They are usually green or brown in colour and almost always remain motionless. All Stick Insects live among the foliage of trees and bushes and are plant eaters. Most of these insects feed and move about only at night.

Many species of Stick Insects do not have wings, thus enhancing their resemblance to twigs. In many of the winged species, the hind wings are brightly coloured but are carefully concealed when the insects are at rest. When they take to flight, a sudden flash of bright colour ensues. This bright colour disappears just as suddenly when the insects settle down again. Such "flash colouration" is confusing to a searching predator and is an effective protective device.

A close relative of the Stick Insects, the Leaf Insect (*Phyllium*) resembles a different part of the plant with great perfection. It has a flattened body with reliefs like leaf venation. Coupled with leaf-like legs and greenish or yellowish colour, it is a perfect copy of a leaf in every detail.

Grasshoppers and their relatives (bush crickets and crickets) characteristically possess greatly enlarged hind legs for powerful

Left to right, plants constitute a large portion of the domain for many forest insects.

jumping. Another feature is the ability of most species to produce sounds by stridulation for communication and courtship. The chirping chorus in the fields is the result of this stridulation. Usually, only the males can sing and each species has its own song.

Bush Crickets are cousins of the grasshopper and possess greatly enlarged hind legs for leaping and long, thread-like antennae. They are nocturnal creatures. Most are vegetarians but some are predatory, feeding on other insects.

Many bush crickets are superb leaf mimics, matching them in colour, texture and venation. Some are green and resemble living leaves while others are brown and look like dead leaves. Some even have "disease blotches" in their repertoire of deception.

Crickets, which are another relative of the grasshoppers, are famed for their cheerful singing. Unlike grasshoppers, most crickets sing at night. Their songs are produced by rubbing together specialised areas at the base of the wings. Crickets, unlike grasshoppers and bush crickets, are ground dwellers. They are omnivorous and eat a wide range of plant and animal foods.

Bugs: Cicadas, which are among the best known bugs, are the most accomplished insect singers. Their songs, which are far louder than those of any other insects, are produced by a "click" mechanism on each side of the front end of the abdomen. Like grasshoppers, only male cicadas possess sound-producing abilities.

Adult cicadas are usually associated with trees and their mouthparts can penetrate the bark to suck up plant sap. The nymphs are subterranean and suck plant sap from roots. Some spend very long periods in this underground phase.

Most cicadas have transparent wings as exemplified by the Giant Empress Cicada (*Pomponia imperatoria*) which has a wing span of some 20 cm and the smaller green-bodied species of the genus *Dundubia* with a wing span of 10 cm. Examples of cicadas with coloured wings are species of the genus *Tacua*, which look rather like moths.

Other members of this group are aphids which provide honeydew to the ants that herd them, lantern flies with strange prolongations of the head; assassin bugs which are predatory and feed on other insects, shield bugs which give off a horrible smell; water skaters which live and move about on the surface film of water in pools and streams; aquatic giant water bugs which prey on other creatures including frogs and small fishes; and bedbugs which are blood-sucking parasites and nobody's favourite bug.

Dragonflies and damselflies are often seen

near the water. Of the two, the dragonflies have a more powerful flight and are among the fastest of all insects. The bodies of dragonflies and damselflies are often brightly coloured. Their wings are usually transparent, but may be tinted or patterned. Dragonflies hold their wings stiffly extended on each side when at rest, whereas damselflies fold them over the back.

These insects are predatory in all their life stages. The adults catch other insects on the wing but can also capture prey at rest. The aquatic nymphs capture their prey with the help of a mask (the greatly enlarged lower lip), which is unique in the insect world. In addition to insects, prey items include tadpoles and small fish.

Mating of dragonflies and damselflies is a spectacular sight. Because of the male's peculiar sexual anatomy, the mating pair assumes the so-called "wheel" position during copulation. Before and after mating, the pair may fly in tandem, a position may even be maintained while the eggs are being laid.

True Flies: Dragonflies and damselflies as well as many other insects bearing the name "fly" are not true flies. True flies possess only one pair of wings (except those few which are wingless); the hind wings are modified into halteres, or balancers.

The most familiar example of the true flies is the house fly. This insect is a potential carrier of gastrointestinal diseases such as dysentery which is spread through contaminated food.

Even more dreadful are blood-sucking and disease-bearing mosquitoes. These include members of the genera *Anopheles* (carriers of malaria), *Aedes* (carriers of dengue fever) and *Mansonia* (carriers of filariasis). Despite their blood-sucking habit, not all mosquitoes are harmful. There are mosquitoes which are beneficial to mankind. Unlike other mosquitoes, female *Toxorhynchites* do not suck blood. In addition, their larvae are predatory, feeding on other mosquito larvae and other aquatic creatures, including their own kind.

Far left, a Lantern Fly; left, Metallic Wood-boring Beetle; and right, a bright spark, the Firefly.

Another disease-carrying group of true flies is the sand fly. The bites of sand flies, as well as those of biting midges, result in considerable discomfort. Biting midges, which are common in mangrove swamp and occasionally in inland secondary jungle, are not known to be disease carriers.

Hoverflies are attractive insects, often brightly coloured like wasps, but are harmless. They are magnificent aerial acrobats and are able to maintain their position in the air despite the presence of drifts and eddies. Some species are beneficial as their larvae feed on aphids. Similarly, robber-flies all live as predators on other insects.

Stalk-eyed flies are unique and indeed among the most curious in the insect world. Their compound eyes are borne on long stalks which are prolongations of the head. The purpose of locating the eyes far apart and away from the head proper remains a mystery.

Beetles and Butterflies: More than 250,000 species of beetles have been described and named and many more await discovery.

They occur in almost all habitats, other than the sea.

Among the most spectacular beetles are the fireflies (with wings) and glow-worms (wingless) which produce a cold light through a luminous compound. After dusk, adult male fireflies, particularly in the mangrove, send out flash signals at regular intervals to attract mates. The continuous light of the glow-worms may serve as a warning signal to would-be predators.

Tiger Beetles are fierce predators and their larvae are also carnivores. Diving beetles which inhabit freshwater habitats are also predatory. Among the most useful of all insects are Ladybird Beetles. These pretty insects prey on aphids, scale insects and other pests.

Although not as uniformly beautiful as butterflies, many beetles are popular with insect collectors. Among them are the longicorns with very long antennae, the stage-beetles with extraordinary antler-like mandibles and the metallic wood-boring beetles many of which have iridescent green, bronze and blue colours.

Butterflies and moths are acclaimed worldwide as the most magnificent of all insects. Foremost among them in this part of the world is the Rajah Brooke's birdwing (*Trogonoptera brookiana*), the male of which is velvety black with a curved band of triangular spots of a brilliant metallic-green colour.

Most butterflies are day fliers. They are most easily observed in gardens and along roadsides, as well as along paths, clearings and streams in the forest, and on mountain and hilltops. The males of some species often congregate with other species, at moist spots on forest roads and on sand banks by rivers and streams.

Moths are generally nocturnal in habit. Some are as attractive as butterflies. The Atlas Moth (*Attacus atlas*), one of the largest in the world is the most beautiful. Its wings are attractively patterned in shades of rich brown with transparent "windows" on all four wings.

Butterflies exhibit a great variety of defence mechanisms. "Warning colouration" and "flash colouration" are perhaps the most common but the most famous is the leaf-like appearance of the leaf butterfly.

Other Insects: Some insects are characteristically wingless. These include the springtails, the bristletails of which the silverfish is the most familiar and lice and fleas which are significant in terms of public health. Some species of book lice are also wingless.

Other insects which may be encountered are the nerve-winged insects whose larvae are predatory and adult ant-lions which look somewhat like dragonflies but have long antennae and a weak flight. Their larvae construct sand pits to capture their prey. Another predatory insect is the scorpion fly whose male possesses a "sting" like that of a scorpion but is in fact harmless.

The young of mayflies, stoneflies and caddis-flies all live in water. Of these, mayflies are the only insects which moult after they have acquired their wings.

A peculiar-looking insect is the earwig. Its abdomen ends in a pair of pincers which may assume different forms in individuals. Just as strange are the thrips, whose wings are very narrow with long fringes.

Insect Allies: Insects are arthropods creatures with "jointed legs" which also includes crabs, prawns, spiders, centipedes and millipedes. Spiders, represented by a very large number of species, are commonly encountered. They are predatory, feeding mainly on insects. The most interesting spiders are members of the genus *Liphistius* which live in burrows provided with a hinged trap-door and several communication lines to inform the occupants of the landing of an insect prey.

Millipedes are perhaps as common as the spiders. They are vegetable feeders. Giant millipedes may reach a length of 25 cm. A remarkable and curious group are the pill millipedes which, when alarmed, roll themselves into a hard spherical ball.

Centipedes, a close relative of the millipedes, are predatory and can inflict a poisonous bite. The large, reddish *Scolopendra* may reach a length of about 25 cm. A curious looking group are the Scutegerids which have very long legs and a short and stout body. The uninitiated may not even recognise them as centipedes.

Left, a Papilio Butterfly from the mountain forests of Java. Above, the Atlas Moth.

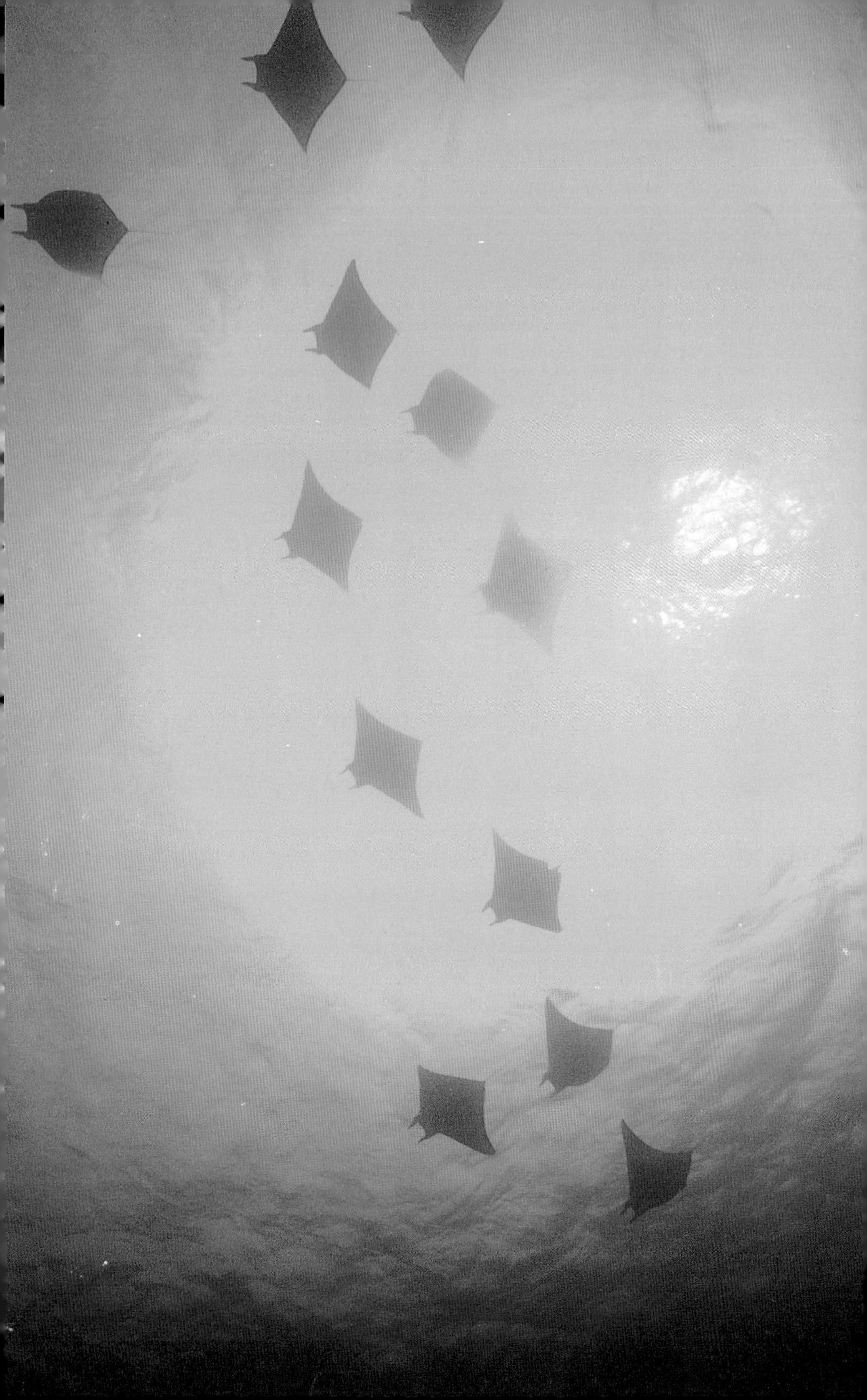

CORAL REEFS

Coral reefs are found in abundance throughout the seas of Southeast Asia. Of the estimated 600,000 square km of reefs worldwide, 25 to 30 percent are located in Southeast Asia. Within the region, the wide range in structure and profile of the reefs is matched by their high biodiversity. In recent times, many reefs, particularly the more easily accessible ones, have been heavily exploited or have become seriously degraded by man. At the same time, it is interesting to note that many more coastal resorts use coral reefs as the selling point to bring in the tourist dollar. Countries that have realised the potential and value of coral reefs have included reefs in designated marine reserves and marine nature parks, with management policies designed to protect them from total destruction.

Hard corals come in an astounding variety of shapes and sizes and give the reef substrata an interesting profile. Branching *Acropora* corals are common and often form extensive stands over large areas. Growth is profuse with branches commonly meeting and fusing with each other for greater strength and rigidity. Branching corals, including those of other genera, provide lots of protection and hiding places for fish and other invertebrates. Tabletop *Acropora*, also common, has a flat, horizontal disc-like upper portion supported by a narrow central column. These corals affect the neighbouring corals by reducing or totally blocking off sunlight to the area beneath.

Other forms of coral growth such as foliaceous, laminar, encrusting, massive, columnar, all contribute quite significantly to reef structure and development.

Hard corals are the main builders of a reef. Throughout their growth process, they continually secrete a calcium carbonate skeleton which forms the framework of a reef. This skeleton remains after the colony dies and serves as a useful substratum for the settlement and growth of other reef animals including coral species. The process of accretion is prevalent on a healthy reef. Other organisms which incorporate calcium salts within themselves also help in reef construction but not to the same extent as hard corals. Corals thrive best in clear waters as the unicellular algae present in their tissues require sunlight for photosynthesis, a process which enhances coral growth and skeletal deposition.

<u>Left</u>, a formation of Manta Rays glides beneath the surface.

Numerous species of reef-associated invertebrates make their home on a reef, living among the corals. Close relatives of the hard corals, such as soft-corals, sea-fans, sea-whips, sea-pens, hydroids and anemones, do not resemble hard corals in any way. Their interesting forms stand upright against the current or sway delicately with it. Clownfish share an interesting relationship with the anemones and are apparently immune to the anemone's stings which are lethal to other fish. The shallower zones of some reefs are sometimes dominated by soft-corals.

The structure of many lower reef invertebrates sometimes makes it difficult to accept them as animals. This is certainly true for the sponges, the simplest form of multicellular life. Some appear like irregularly-shaped boulders while others take the form of intricately branched colonies. Neptune's cup sponge, *Petrosia*, exhibits a distinct cup-shaped form and can grow to a large size.

The annelid worms that are most conspicuous on the reef are fanworms, such as *Sabellastarte* and *Spirobranchus*, the latter commonly called the christmas tree worm or bottlebrush. These worms make their burrows within live hard corals and are well protected. Through the opening of these burrows, their colourful tentacles remain fully extended to trap plankton, but are immediately withdrawn at the slightest hint of danger. Totally unrelated and belonging to a completely different animal group are the flatworms whose frail-looking but colourful soft bodies glide effortlessly over the reef substratum.

Reef crustaceans such as crabs, shrimps and spiny lobsters are known for their economic value although some species of crabs are extremely poisonous. An interesting group of shrimps is the sub-family *Pon-*

toniinae with the majority of species living as commensals off other reef organisms.

Reef molluscs are exploited for food as well as for their ornamental value. The shell trade has increased tremendously in some places, as evidenced by the heaps of shells piled in the trader's backyard. This contributes greatly to the decline of these molluscs on the reefs. Giant clams are also fast disappearing from many reefs. Even molluscs without shells are much sought after for the aquarium trade.

Common among the echinoderms on Southeast Asian reefs is the black long-spined sea urchin, *Diadema setosum*, often appearing in large numbers on disturbed reefs. Divers know all too well the effects of sea-urchin spines and learn to avoid them but, viewed from a safe distance, the different species are interesting in their structure and behaviour. Other echinoderms, such as starfish sea-cucumbers, featherstars and brittlestars, are all represented on these reefs by a great diversity of species.

Coral reef fish endow reefs with much visual impact, and contribute significantly to the kaleidoscope of colour. Pomacentrids and wrasses are usually abundant and the diversity of dainty butterflyfish often gives an indication of the condition of a reef. Fish occupy all niches of a reef, some swimming over a wide range, while others defend their established territories. The reef provides ample food and shelter for fish of all shapes and sizes to grow and to multiply. Turtles, like the hawksbill and the green, are commonly seen swimming about a reef.

The ecosystem maintains an equilibrium as it develops and this is best seen in pristine reefs. Human interference usually results in irreversible damage and this has been demonstrated on many reefs within the region. Destructive fishing methods such as the "muro-ami" or blasting have laid bare many reefs. Over-exploitation of reef resources has resulted in reefs being unable to recover and declining rapidly, thus leading to the

loss of a rich and useful natural heritage. Coral mining, where coral blocks are removed for construction purposes, has devastated many reefs. Run-off from land, where human activities are intense, has also been detrimental to the reef ecosystem.

Coral reefs play an important role as a buffer against strong waves which would otherwise erode beaches. Reefs also serve as a source of beach sand. When properly managed on a sustainable basis, coral reefs can support artisanal and commercial fisheries. They harbour a rich variety of species, many of which are valuable sources of protein, while others are of important commercial or ornamental value. To harness the full benefits of coral reefs, full emphasis must be placed on the practice of sustainability and effective management. This will allow them to continue to serve as well as to attract tourists. The preservation of genetic diversity has become a more important issue in recent years as international attention is being focussed on naturally-occurring substances that have great medical importance. The potential of coral reefs in this respect is considered to be enormous and no effort should be spared to maintain this important genetic bank.

Throughout the region, many coral reefs are within easy reach and are close to resorts which offer good diving facilities. More remote reefs can be reached with greater effort. Excellent reef life can be seen in some national marine parks where collection is strictly forbidden. The reefs are distributed over a wide geographical range from the northern tip of the Philippines (latitude 19°N) to Timor which is the southernmost island of Indonesia (latitude 11°S), and across from the western coast of Sumatra, which is influenced by the Indian Ocean, to the eastern coasts of the Philippines and northern coasts of Irian Jaya, which are influenced by the Pacific Ocean. A wide variation in coral reef life can therefore be expected throughout the region under the influence of climate and other geographical variables. While fringing reefs are dominant, all other kinds of reef structure occur in the region, including patch, barrier and atolls.

Left, the coral reef — a tropical ecosystem as rich as the rain forest. Above, the Starfish is a common reef dweller.

Indonesia and the Philippines have the greatest number of coral reefs and, together with Malaysia and Thailand, possess a number of marine parks and preserves. In Indonesia, particularly good reefs with prolific growth occur in the central seas such as Flores, Banda and Ceram which separate the Sunda shelf from the Sahul shelf. Atolls are also located in these deep waters with Taka Bone Rate, south of Sulawesi, being the third largest in the world. Patch and fringing reefs are scattered around most of the Indonesian islands.

The reefs of Peninsular Malaysia are better developed on the east coast than the west coast. Reefs of the east coast are mainly fringing, associated with offshore islands, and patch. On the west coast, the reefs further north towards Langkawi become more diverse. In East Malaysia, reefs extend along the western coast. The reefs on the southeast side of Sabah at Semporna and Pulau Sipadan are in excellent condition.

Scattered throughout almost the entire coastline of the Philippines are reefs and coral communities. The knowledge about, as well as the condition of, these different reefs vary widely. Little is known of the Philippines reefs facing the South China Sea while those in the southern part of the Sulu Sea are said to be the best Philippine reefs.

Singapore reefs have been subjected to high sedimentation levels but the upper reef slopes continue to support a high diversity of corals and reef-associated organisms.

In Thailand, reefs on the western side which face the Andaman Sea are the most developed. The majority are fringing reefs associated with the islands off the island of Phuket. On the eastern side, some fringing reefs are associated with the mainland but most are developed around the many offshore islands, with reef diversity improving towards the southern part of the Gulf of Thailand such as at Ko Samui and Ko Phangan.

Those who visit Southeast Asia to watch wildlife should not expect to photograph herds of ungulates galloping across plains or carnivores posing picturesquely with their prey. The natural vegetation of most of Southeast Asia was originally rain forest and the wildlife is therefore very different from that which inhabits the more open grasslands of East Africa. It also means that different techniques are needed in order to see the animals.

Getting There: Reaching the parks and reserves can be an adventure. The seas between Komodo and its neighbouring islands in Indonesia, for example, have some of the most dangerous currents in the world.

Access-roads into parks can be as difficult, if not as dangerous as the sea, particularly in the wet season. Bridges collapse, the "road" turns into a swamp, the hired jeep breaks down every 10 km and gets stuck in the mud, its four-wheel-drive system unable to cope. But, not to worry! Southeast Asian countries are heavily populated and willing hands are usually available to help dig out vehicles or to offer a folk-remedy for a mechanical ill, such as plugging a leaky radiator with an unripe banana...and it works!

Most parks can be reached by public transport which takes longer but is much less expensive than hiring a vehicle. The parks tend to become smaller and older the further one travels from urban centres and the visitor may end up entering the park on foot or on the pillion of a motorbike.

Once you get there: The rain forest presents particular problems for the naturalist. First, it is difficult to spot animals amongst so many trees, which are evergreen in most areas and which therefore do not offer the seasonal bonus of bare branches. Bali Barat in Indonesia is an exception to this, with its lower rainfall and deciduous vegetation. Second, most flowers and fruit of the rain forest are produced within the canopy since most of the sunlight is caught here with few rays reaching to the forest floor. This means that many of the birds, insects and other animals are completely arboreal, creating further difficulties for the earth-bound observer.

A third problem is that many forest-living species are solitary, secretive and very shy of mankind and speed away through the trees or simply freeze at the first sound, sight, or smell of people. Their natural fear is accentuated by the fact that hunting is still common, even in supposedly protected areas. Many parks and reserves have only been established relatively recently and often with inadequate resources and so villagers who have traditionally hunted there can be slow to adapt to the new conditions.

Then, many forest animals are nocturnal, which calls for further specialist viewing techniques. Finally there is the difficulty of actually getting around in the forest. In primary lowland rain forest, the forest floor should be fairly free of undergrowth because of the lack of sunlight, but many such forests have now been cleared for other purposes. Much of what is left is either swamp forest, subject to periodic flooding, or montane forest growing on steep hillsides which are difficult to traverse. Where the forest is more open, perhaps as a result of disturbance or due to the presence of long river banks, vegetation is much thicker and truly resembles what most people consider as "jungle", an impenetrable wall of greenery.

What to expect: Expectations must be adjusted so that the whole complex web of life in the rain forest — from a colony of weaver ants rushing out to defend their leaf-nests, to a pile of fresh elephant dung showing that Asia's largest land-mammal has recently walked along an improbably steep or narrow path — becomes the principal aim of observation. A flying speck of iridescence may turn out to be a minute, rotund beetle, while among the branches overhanging a dry river bed, an enormous black-and-yellow spider may have stretched its web. The fantastic variety of trees and other plants in the tropical rain forest is in itself marvellous after the paucity of the vegetation in the world's temperate zones. The experience of standing in the forest and gazing up at the massive trunks festooned with lianas, orchids and ferns has been compared to the elation inspired by the holiness and majesty of a cathedral. Once perceptions have been changed in this way, the sighting of any large animal becomes, not so much the focus of a forest walk, but a bonus.

Preceding pages: boat ride to Komodo. Left, the best way to view Mulu. Above, "hiding out" in Taman Negara.

Outside the rain forest, conditions on grasslands in places such as Khao Yai in Thailand, are obviously quite different. It is possible to see further and, provided animals remain unaware of the visitor's presence, to watch them for longer. The mangrove forests, which fringe much of the coastline of Southeast Asia, are an extraordinarily rich habitat, providing food and shelter for both land and sea animals.

When to look: The best time to see wildlife in the forest or on the savanna is undoubtedly in the early morning. This is the coolest time of day and many animals are at their most active. The gibbons in the area, if there are any, will start their whooping territorial calls as the sky begins to lighten and will soon be joined by the birds with their dawn chorus. Rising as it gets light and going out for a few hours' animal-watching will ensure a healthy tally of species seen — and the hotter part of the day can then be spent relaxing.

As the day cools in the late afternoon, deer, wild cattle and pigs emerge from the shade to drink at water-holes, and flying squirrels apparently play a game as they glide from tree to tree. At dusk, fruitbats with their metre-wide wing-span leave their roosts to feed while smaller insectivorous bats stream out from caves to take the airborne place of swiftlets. Nightjars sit on pathways and roads and fly off at the approach of walkers or vehicles, and in wetter areas the incessant singing of frogs is a constant night-time lullaby.

Where to look: Because most activity in the forest is in the canopy, any opportunity to get a better view of this should be taken. This may involve a climb to a vantage point on a hill, the ascent of a purpose-built watch-tower, or a stroll on an aerial walkway through the trees. The last is undoubtedly the best, as it enables the observer to walk through the forest at the level of the branches and to see a variety of birds and other animals which would be impossible to spot from below. However, walkways are difficult to build and to maintain in a safe condition and, in some ways, watch-towers are more effective as they act as a hideaway from which animals can be observed in relative comfort.

Although roads into parks are not always desirable, walking along them is a good way to see lots of birds, to obtain a clearer view of the surrounding forest and to enjoy an easy passsage. Where there are good roads, such as in Bukit Timah in Singapore and Quezon in the Philippines, access to the forest is possible for people who are unable to walk. Artificial waterholes to attract animals to a convenient viewing site are only effective when there is a clearly-defined dry season, as in Baluran, East Java. Rivers and ponds do, however, provide a useful focus for wildlife viewing. Some knowledge of the animals' habits is of course important, knowing where and when they are likely to feed and the type of habitat they prefer.

How to look: The best way to see Southeast Asian wildlife is to adopt some of the habits of the animals themselves: be secretive, shy and solitary! The chief element of success is silence: most animals have an acute sense of hearing. A slow, quiet walk alone through the forest will result in some memorable experiences: an orang-utan swinging peacably on its way above your head, possibly even unaware of your presence, or a family of green peafowl strutting across the path in front of you. Although walking long distances alone is not really recommended — in many parks it is not even permitted — even a stroll around the area close to the camp will reveal a surprising number of animals.

Being accompanied on a forest walk by a good ranger should be instructive, since the ranger should know what animals to expect and where to see them. Many of the animals are well-camouflaged against their background, and the rangers will normally spot and identify them long before visitors do. The more sophisticated parks normally have visitors' centres with information on the typical animals of the areas, and in some places, such as Kinabalu in Sabah, there are nature trails of different lengths on which important features of the forest are indicated.

Keen birdwatchers should attempt to visit a good zoo or bird-park in order to familiarise themselves with some of the birds before going into the forest. Jurong Bird Park in Singapore is an excellent place to watch the birds in well-planned and sympathetic surroundings. Similarly with plants, the Botanic Gardens in Singapore and in Bogor and Cibodas in West Java are well laid out, with labels on many of the plants and knowledgeable guides able to give tours of the gardens in English.

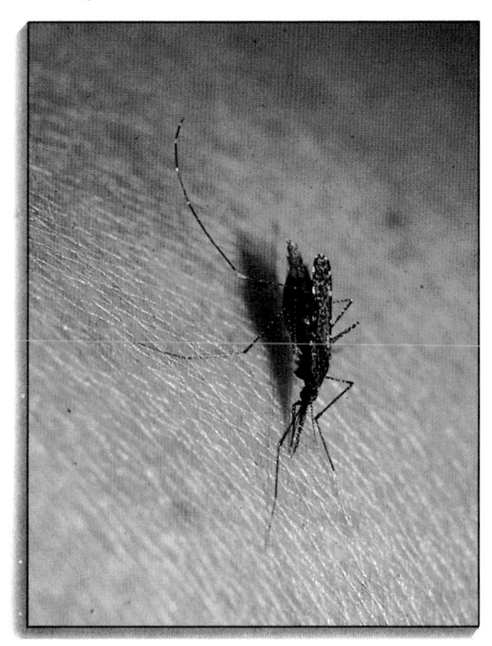

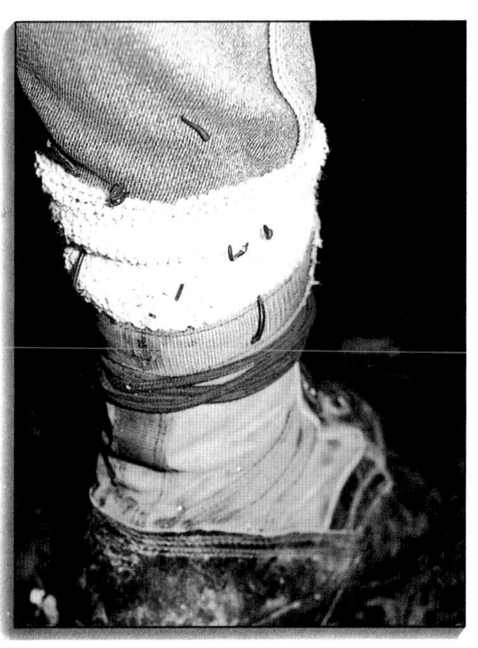

An important way to determine which animals are present and common in the parks is to look out for the signs of their presence, such as their tracks or droppings. Footprints show up well in the soft mud along river banks and other damp places, and a basic chart of animal tracks will make identification easier. Some animals leave their droppings in obvious places as territorial markers: the droppings of the *Musang* (civet-cat), for instance, mostly consist of large fruit-seeds and are nearly always left on a boulder by a river or stream. Crouching down on the forest floor so as to bring your eyes to the level of a small animal can be revealing: you may notice a path beneath the undergrowth quite clear at 50 cm from the ground, but invisible from a normal standing position.

Listening intently as well as looking hard is vital in order to identify a bird by its song, or to catch the tell-tale sounds that advertise an animal's presence. The Wreathed Hornbill, for instance, has an unmistakably loud wing-beat and a crashing through the branches usually means a troop of macaques is on its way. However, sometimes the noises are so strange that you will have no idea what is making it: it could be a bird, an insect or a tree-frog!

A good way of getting around in the forest is by water. In a boat with paddles or a quiet engine, progress will be slow and silent enough not to frighten the animals. It also means that the naturalist can sit still and enjoy the sights and sounds of whatever is going on around without having the stress of dealing with the humidity and the difficult terrain of the forest. The only way into Malaysia's Taman Negara is by river, and rafting tours are run by the tour company Pacto down the Alas River in Sumatra.

Left, access to parks can be difficult, especially during the wet season; undesirable companions: mosquito (above left) and dreaded leeches (above right).

Visiting the forest at night reveals an entirely different fauna. Where there is a road into the park, such as at Doi Suthep in Thailand, strong lights on a jeep or truck will reflect red in the eyes of watching animals and show up owls listening for their prey.

A Dangerous Place to Be?: Adventure films set in "the jungle" have given the forest rather a bad reputation. Fortunately, tigers do not pounce from behind every tree nor do cobras rise up at every turn of the path to strike the unwary leg. It is the smallest animals and insects which will cause most irritation, particularly mosquitoes, which make their presence felt almost everywhere. Using insect repellents and wearing long sleeves can keep the worst of them off. Anti-malaria tablets should also be taken regularly.

The thought of leeches arouses horror in some people, and if the sight of blood makes you want to faint then it is best not to risk getting bitten by one. They inject an anti-coagulant into the wound so that the blood flows freely and even after the leech has been removed blood will flow for some time. The bites are not painful. Patent remedies for getting leeches to fall off are putting salt on them or burning them off with a cigarette. However, locals simply scrape them off with a knife or fingernail. Spraying of shoes and socks with an insecticide such as Baygon is praised as the best protective device. The taste of the chemical spoils the leeches appetite for your blood.

Southeast Asia is home to some extremely venomous species of snake. A bite from a krait or pit-viper can be fatal within minutes. Never walk around barefoot, especially at night, and do not investigate fallen trees or piles of leaves with your hands or feet. Bring along a first-aid kit for reassurance.

For those from temperate zones, the humidity of the rain forest can be oppressive. Yet, with the excitement of being in a new place it is easy to attempt to do too much in the first few days. Paths are often steep, slippery, and inadequately maintained and heat exhaustion can occur. It is important not to be over-ambitious, to rest often, and to carry plenty of water. A sun-hat is absolutely essential in all Southeast Asian parks, and in savanna and coastal and marine environments it is obviously sensible to cover the rest of your body as well in order to avoid over-exposure to the sun.

The forest is a confusing place and it is easy to get lost. It is important to take a ranger along if paths are not well marked. A compass is always useful. Remember that in some parks where the rangers are inexperienced they too can get lost. Without a machete (called a *parang* in Indonesia and Malaysia) it is impossible to penetrate much of the forest where there are no paths, especially where rattan palms with their strong sharp thorns are common.

What to Wear and Carry: Throughout Southeast Asia it is hot by day and only the mountains are cool at night. Light clothes are therefore best, although shorts and sleeveless shirts are not really suitable for the parks because of the increased amount of skin exposed to mosquitoes, leeches and sunlight. It should not be necessary to burden yourself with heavy walking boots and leather can be a nuisance as it takes a long time to dry and goes mouldy quickly in the humidity. Trainers, which dry readily, are really the best footwear for all but the hardest treks, especially where there are streams and rivers to cross.

Sleeping at high altitude in mountain parks such as Doi Inthanon in Thailand or Gunung Gede/Pangrango in Java can be surprisingly cold, and a warm jumper or jacket and a sleeping bag are necessary. When it rains it is impossible to keep dry. Cheap plastic ponchos are available in most towns.

Binoculars are essential for any serious wildlife watching, and because of low light levels in the rain forest it is best to use fast camera film. A good-quality cassette recorder can evoke the atmosphere of the forest in a way that photographs never can, whether it is the call of a bird or gibbons, the sound of angry orang-utan babies demanding their milk, or the high-pitched shrill of an unidentifiable insect.

Although most marine parks have masks and snorkels for hire it is worth taking your own equipment when visiting coral reefs. Training-shoes will again come in useful for wading out to the reef to protect your feet against the unpleasant sharp spines of sea-urchins or the even nastier stonefish and stingray.

Right, all eyes towards the sky: birdwatching is a collective effort.

PLACES

Southeast Asia is not sparsely populated as is Amazonia or central Africa: rather, it is home to 400 million people. This large human population leaves room for only a few areas of large unbroken natural landscapes, yet hundreds of national parks, wildlife reserves and forest reserves have been set aside for the complete or partial protection of their natural assets. Some, such as Taman Negara in Malaysia, are large and virtually untouched, while others such as Bukit Timah in Singapore are tiny but still of considerable local value. Forest reserves are in principle set aside as a source of timber but can provide complete protection for fauna and flora when they are needed as a watershed, such as in the mountains of Malaysia. Some parks offer virtually perfect protection while others which still contain rich nature, document the problems of enforcing existing laws against a poor population which has not yet conceived of nature tourism as an alternative way to make a living, but which continues to poach wildlife and timber.

This section covers about 50 parks that serve as an appropriate introduction to an understanding of the region's nature. They should not be superlatives. Southeast Asia's tropics do not have the world's highest mountains, deepest canyons, or herds of uncountable ungulates. However, here are the world's oldest and most diverse tropical forests, a natural environment accessible until recently only by expedition but opened by modern technology to convenient travel.

It takes a subtle approach to enjoy the huge, dark forests, the quietness interrupted by the distant call of a gibbon, the ever-changing chorus of the cicadas, the passing of a mixed flock of colourful birds, the knowledge of the hidden presence of tapir and tiger. An observer prepared for a nature experience will be left awe-struck rather than exhilarated.

Preceding pages: an angry Anak Krakatau; typical swamp forest vegetation; light and shadow at work — to wondrous effect — in the rain forest of Indonesia. Left, even if there's not a wild animal in sight, the tropical nature experience is highly rewarding.

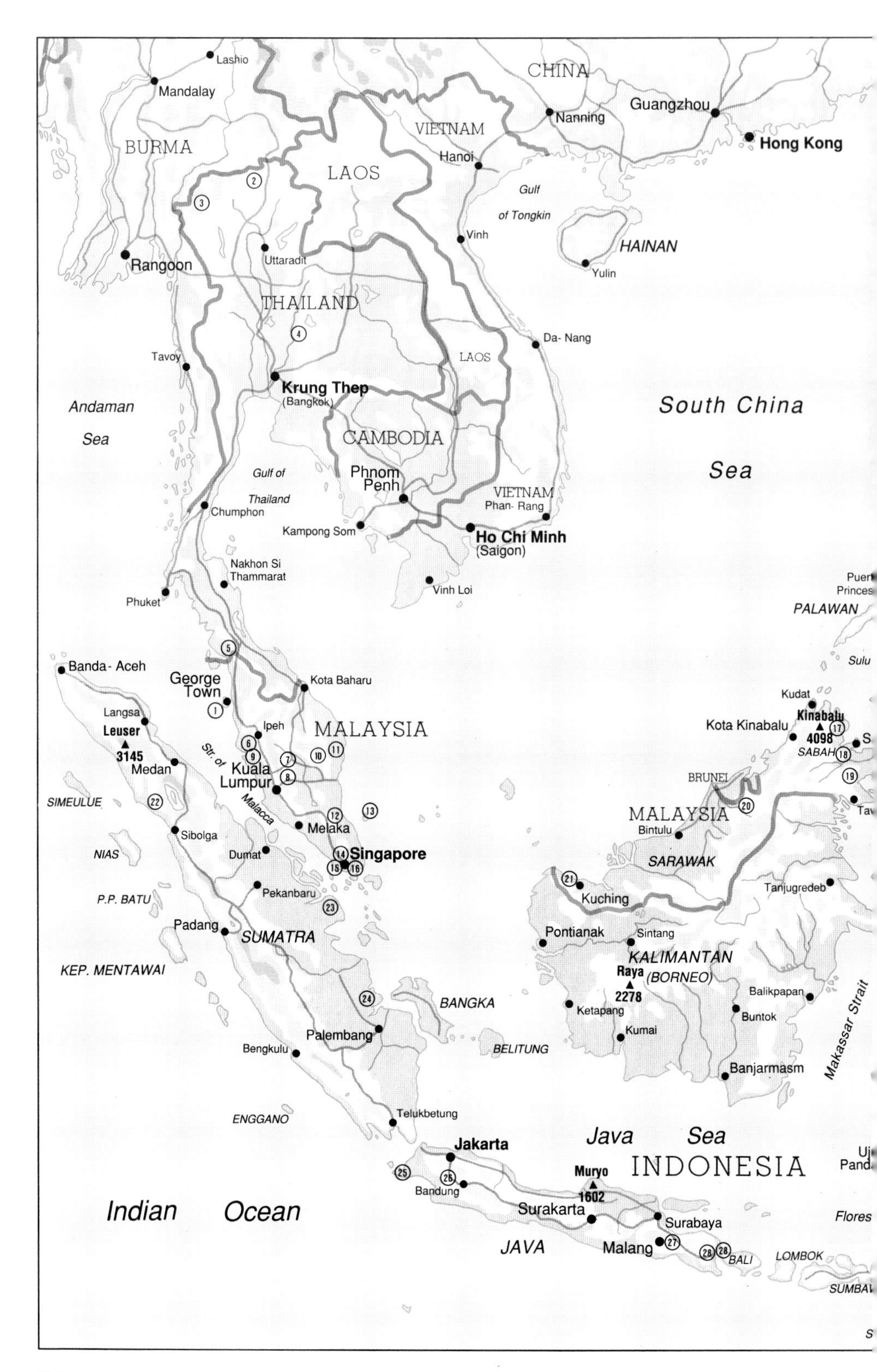
Lashio
Mandalay
CHINA
Nanning
Guangzhou
Hong Kong
VIETNAM
BURMA
Hanoi
LAOS
Gulf
of Tongkin
Vinh
HAINAN
Yulin
Rangoon
Uttaradit
THAILAND
Da- Nang
Tavoy
LAOS
Krung Thep
(Bangkok)
Andaman
Sea
South China
Sea
CAMBODIA
Gulf of
Thailand
Phnom
Penh
VIETNAM
Phan- Rang
Chumphon
Kampong Som
Ho Chi Minh
(Saigon)
Nakhon Si
Thammarat
Vinh Loi
Phuket
PALAWAN
Sulu
Banda- Aceh
George
Town
Kota Baharu
Kudat
Langsa
Ipeh
MALAYSIA
Kota Kinabalu
Kinabalu
4098
SABAH
Leuser
3145
Medan
Str. of
Malacca
Kuala
Lumpur
BRUNEI
SIMEULUE
MALAYSIA
Bintulu
Sibolga
Melaka
SARAWAK
NIAS
Dumat
Singapore
Pekanbaru
Kuching
Tanjugredeb
P.P. BATU
Padang
SUMATRA
Pontianak
Sintang
KALIMANTAN
Raya
(BORNEO)
2278
KEP. MENTAWAI
BANGKA
Ketapang
Balikpapan
Buntok
Makassar Strait
Palembang
Kumai
Bengkulu
BELITUNG
Banjarmasm
ENGGANO
Telukbetung
Java
Sea
Jakarta
INDONESIA
Muryo
1602
Bandung
Indian
Ocean
Surakarta
Surabaya
Flores
JAVA
Malang
BALI
LOMBOK

S. E. Asian
Wildlife Parks
480 km/ 300 miles
1 Penang
2 Doi Suthep - Pui
3 Doi Inthanon
4 Khao Yai
5 Thaleban
6 Cameron Highlands
7 Fraser's Hill
8 Templer Park
9 Kuala Selangor
10 Taman Negara
11 Rantau Abang
12 Endau Rompin
13 Pulau Tioman
14 Bukit Timah
15 Sungei Buloh
16 Singapure Botanic Gardens
17 Mount Kinabalu
18 Sepilok
19 Danum Valley
20 Mount Mulu
21 Bako
22 Mount Leuser
23 Berbak
24 Sembilang
25 Ujung Kulon
26 Mount Gede Pangrano
27 Mount Bromo Semeru
28 Bali Barat (4 Parks)
29 Komodo
30 Dumoga Bone / Bunaken
31 Mount Apo
32 St. Paul's Subterranean River
33 Iglit Baco
34 Mount Pulog
Pacific
Ocean
PHILIPPINES
Philippines Sea
LUZON
Manila
SAMAR
Cebu
Iloilo
Suriago
MINDANAO
Manukan
Davao
Celebes
Sea
Manado
Tobelo
HALMAHERA
Molucca
Sea
MOLUCCAS
Seram
Sea
Manokwari
Jayapura
(Sukarnapura)
Fakfak
IRIAN
JAVA
Jaya
5030
Kokenau
NEW GUINEA
Kubor
4359
PAPUA
NEW GUINEA
Kikori
BURU
SERAM
BUTON
Banda
Sea
INDONESIA
KEP.
ARU
Flores Sea
KEP.
TANIMBAR
P.DOLAK
Merauke
Gulf
of Papua
ALOR
Oekusi
Arafura Sea
Kupang
TIMOR
Torres Strait
AUSTRALIA
Somerset

THAILAND

Thailand is located in the northwest of Southeast Asia. In the eyes of the naturalist, its long north-south extension of more than 1,500 km makes it climatically the most diverse country of the region. The north lies deep in the monsoon belt and has a dry period which often extends for more than six months while the south is in the moist tropics where rain falls throughout the year. This is reflected in a fauna and flora which, in the north, is mostly of Indochinese origin and, in the moister south, is of Sundaic affinity. Thailand is densely inhabited and its vast plains have been intensely cultivated for more than 1,000 years. The husbandry of its forests over the last decades has been mostly profit oriented, and with little regulation. On the positive side, a complete ban on the logging of virgin forest since 1988 is unique in the region and provides much hope for the future of Thailand's nature. More than 40 national parks have been gazetted over the last years. Many of these, however, protect topographic features such as waterfalls rather than unspoilt vegetation and wildlife. The following chapters deal with some of the most interesting and dissimilar places of protected nature in Thailand: the mountains of the north, such as Doi Inthanon and Doi Suthep; the large and well managed Khao Yai National Park in the centre, and Thaleban with its tropical rain forests in the south.

Preceding pages: hiking in tiger country at Khao Yai National Park. Left, Yellow Bittern weighs down a lotus leaf.

NORTHERN THAILAND

In spite of Thailand's headlong rush into the late 20th century, each of the regions of the country has so far managed to retain its own distinctive ambience. Nowhere is this more so than in the mountainous north, where the lowland people speak with a different and distinctive northern dialect and where the uplands are chiefly inhabited by tribal peoples of either Tibeto-Burman or Chinese stock. Around Bangkok, the rice padis stretch seemingly boundlessly in all directions: in the north, by contrast, the cultivated alluvium is confined to between the mountain ridges. Chiang Mai and adjacent provinces lie on the fringes of a huge area of uplands, ridge after ridge of mountains, stretching through the Shan States of Burma into southwest China and the southeastern margins of the Tibetan Plateau. Biologically and ethnically, this area constitutes a distinct entity. In biological terms, it is one of the richest and most exciting areas in Asia and is still relatively unknown. While the remote uplands of eastern and northeastern Burma are likely to remain lawless and inaccessible for the forseeable future, northern Thailand has changed markedly in the space of only one human generation and improved roads and communications have opened much of the area for tourism.

Many of the tribal peoples of the mountains are shifting cultivators and grow opium as a cash crop. Traditionally, forest is cleared in order to plant opium, hill rice and vegetables. After a few seasons, soil fertility lessens and the cultivators move on to clear other areas. Secondary forest gradually returns to the old swiddens, so that at any one time, the hills may be covered with a mosaic of abandoned clearings, various stages of forest regrowth and some primary, undisturbed forest on the higher summits and on steep valley flanks. As long as human populations remained sparse such a system was an ecologically sustainable and efficient way of using available land. Much montane wildlife uses secondary habitats as well as primary forest and the cover of woody vegetation was always sufficient to maintain wildlife populations. However, as human populations have grown, so have forest cover and wildlife declined. This is particularly true in northern Thailand where increased pressure from the hill tribes has reduced the areas of montane forests and lowland peoples have also pushed upwards from the valley floor onto the hill slopes.

About 30 years ago, wild elephants and gaur still roamed the Chiang Mai hills. Today, most have gone and the last herd of elephants is confined to the remotest southwest corner of the province, in the Om Koi Wildlife Sanctuary. It would be wrong, however, to think of the Chiang Mai hills as a despoiled wilderness. Notwithstanding the many threats they face, many Chiang Mai mountains still have an extremely rich wildlife. The forested eastern flanks of Doi Suthep-Pui National Park tower over the city of Chiang Mai only a few kilometres from the town centre. Chiang Mai is the only city in Thailand with a forest in such close proximity.

Doi Suthep has long been revered by the people of Chiang Mai. The foundations of the Buddhist monastery of Wat Prathat, situated at 1,000 metres elevation, were laid 600 years ago. In 1935, a tarmac road was constructed, linking the monastery with the city and this was extended as a dirt road as far as Doi Pui, the highest summit of the mountain (1,685 metres) in 1962. Two villages of the Hmong tribe are on the mountain and, in addition, a number of Thai households were established before the National Park was declared in 1981. The royal family maintains a palace on the mountain, at Phuphing.

Doi Inthanon, which lies roughly 60 km to the southwest of Chiang Mai, is the highest mountain in the country. Also known as Doi Angka, its present name commemorates a former prince of Chiang Mai whose ashes are buried on the summit. Formerly accessible only by a long and arduous hike, the summit

green
ene: upper
ontane
rest on the
mmit of Doi
thanon.

is now reached by a tarmac road which was constructed in the 1970s to give access to a military installation. The mountain is home both to Hmong and Karen tribesmen, and there are also a number of Thai settlements.

Because both sites have many people resident within them and are subject to many different forms of land use, neither conforms to the usual definition of a national park. Nonetheless, both possess great biological richness. Because of this and because of their ease of access, Doi Suthep and Doi Inthanon receive a great many visitors and are two of the most favoured sites for birdwatching and nature study in the entire country, if not in all of Asia. The mountains have many similarities, as well as striking differences. Doi Inthanon, the higher of the two, supports more species of birds (over 380 species compared with roughly 330 species on Doi Suthep). Yet, Doi Suthep, because it possesses evergreen forest at a somewhat lower elevation than Inthanon, between 1,000–1,500 metres (a zone which on Doi Inthanon has been largely deforested) is very rich and supports a few species of birds which are scarce on or absent from Doi Inthanon.

Each mountain top may differ slightly from its neighbours in the precise make-up of its resident bird or mammal community. Mountains are like islands: in this case, "islands" of moist mountain forest surrounded by a sea of drier lowland forests. At the end of the last glaciation, about 20,000 years ago, the distribution of forest types would have been different, a cooler, drier environment probably favouring a pine-dominated community. Still more recently, around 8,000 years ago, the climate was wetter and warmer than at present. At any rate, areas of species-rich, moist, montane forest have certainly fluctuated over millennia and some mountain summits, which are today isolated would, in the past, have been connected with others. Dispersal between adjacent but unconnected mountains is limited, some species being relatively mobile and able to

Mountain Imperial Pigeons find haven in a fig tree...

traverse unsuitable intervening areas while others, including many of the denizens of the moist forest interior, are more strictly resident with less dispersal capabilities. The precise composition of the resident bird and mammal community of each mountain summit has probably been determined by post- isolation selective pressures.

On both mountains, the transition from the dry, deciduous dipterocarp woodlands on the lower slopes to broadleaved, moist evergreen formations at higher elevations is very marked. Dry dipterocarp woodlands are usually dominated by about four members of the Oriental tree family *Dipterocarpaceae*: two species of *Shorea* and two of *Dipterocarpus*. This woodland is low stature and open, with a grassy understorey. From January onwards, the leaves on the trees start to display brilliant yellows and bronze-reds before being shed at the height of the dry season, when the whole forest takes on a stark, dead appearance. The sudden proliferation of new greenery is all the more striking and usually occurs even before the first showers, in April. Dry dipterocarp woodlands are still quite rich in larger birds. In addition to the common Lineated Barbet they support a number of woodpeckers, together with the Black-hooded Oriole, Eurasian Jay, Blue Magpie and Rufous Treepie. However, perhaps because it is more disturbed and the total area is smaller, the dry dipterocarp on Doi Suthep is less rich in birds than that on Doi Inthanon.

.as does
is Black
iant
quirrel.

On both mountains, one can observe how the dry dipterocarp woodland gives way to evergreen formations along the courses of streams. This is particularly noticeable on Doi Inthanon, where the access road ascends the mountain along the course of the Nam Mae Klang, a large stream, sustained year round by the mountain's watershed. Here, one can see a well-developed evergreen "gallery" forest with many towering, silver-boled dipterocarps. Spray-laden air from the impressive Vajirathan waterfall, which has a vertical drop of roughly 100 metres, soaks the surrounding vegetation.

Native pine forests are also found on both mountains. They are chiefly confined to the drier ridges, and to the ecotone between the dry dipterocarp and the evergreen forests where annual fires, started in the dry dipterocarp forest, burn into the edges of the evergreen.

The best and richest evergreen forest on Doi Suthep is found on the eastern flank of the mountain, in the vicinity of Wat Prathat, at 1,000 metres and up to roughly 1,400 metres. To reach this the visitor needs to leave the main road, striking off on one of the many tracks which lead into the woods. One such track leads from the National Park headquarters and another leads from Km 16. In addition to the many montane birds such as Great and Blue-throated Barbets, Orange-bellied Leafbird and Maroon Oriole, there are also many species more characteristic of lowland evergreen, including Bar-winged Flycatcher-shrike, Puff-throated and Grey-eyed Bulbuls and Striped Tit-babbler. These lowland species fade out as one ascends the mountain. Also in this zone are found the scarce Red-billed Scimitar-babbler and White-hooded Babbler. In the wet season, the Orange-headed Thrush returns to breed on the mountain and its rich song is frequently heard.

Areas of high elevation (above 1,500 metres) forest on Doi Suthep are very small and rather disturbed. In addition, because they are relatively dry, they only support open low stature forest where pines and evergreen oaks are mixed and denser forest is confined to shady ravines and valleys. Nonetheless, they are still very rich. Some of the typical resident birds of these higher ridges include Yellow-cheeked Tit, White-browed Shrike-babbler and Little Pied Flycatcher.

On Doi Inthanon, the evergreen forest which formerly clothed the zone between 1,000–1,500 metres, has been almost completely cleared by the Hmong peoples. At Ban Maeo Khun Klang (1,300 metres), the park headquarters is situated along a quiet stream in a huge, open area, where houses perched on small hillocks overlook strawberry fields and vegetable gardens. Such crops, established under the Highland Agricultural Project of His Majesty the King, have replaced the fields of opium poppies which formerly dominated the landscape. Two huge waterfalls flow down the rock face at the head of the valley, dominating the vista. Elsewhere, plantations of native pines are beginning to clothe the formerly denuded slopes.

As the road ascends above 1,500 metres, one enters a rich and luxuriant moist evergreen forest with many huge trees 30–40 metres tall. This zone of the mountain, extending up to 1,900 metres, supports a very rich avifauna. Rufous-throated Partridges, Rusty-naped Pittas and both Lesser and White-browed Shortwings are among the many skulking species on the forest floor. Where dense banks of understorey and ground herbage occur, in small gulleys and along streamlets, may be found the White-tailed Robin, White-gorgetted and Snowy-browed Flycatchers. The Slaty-bellied Tesia, a

Birds of the montane forest: Black-throated Sunbird...

kind of tail-less bush-warbler with a glistening golden-olive crown and slaty underparts, is also found here. It has a loud, jumbled and explosive song and is often quite inquisitive, sometimes approaching quite close and giving its rattling alarm call. Its peculiar habit of bouncing from side to side on low twigs and globular, tail-less appearance make it appear quite comical. Doi Inthanon is a mountain which will amply repay repeated visits.

For five to six months of the year, during the wet season, the higher slopes of Doi Inthanon, extending to the summit at 2,565 metres, are swathed in low cloud and mist for much of the time. This favours a prolific growth of epiphyte-hanging lichens, mosses and orchids. The forest on the summit, which is almost 300 metres higher than any other mountain, is unique in Thailand, being dominated by native rhododendrons and magnolias. In late February and early March, when the flowering rhododendrons are at their peak, the visitor is rewarded by the sight of many brilliantly-coloured, jewel-like sunbirds feeding on the nectar of the magnificent, blood-red flowered *Rhododendron delavayi* which, in Thailand, is only found at this one station. In addition to the Gould's Sunbird, a winter visitor, there is also an endemic race of the Green-tailed Sunbird (*Aethopyga nipalensis angkanensis*) — a subspecies whose entire world range is limited to this small area on the summit of Doi Inthanon.

The birdwatcher quickly becomes accustomed to the extreme wariness of many birds in Thailand which is a by-product of the great hunting pressure even within parks and sanctuaries. However, on the summit of Doi Inthanon, many of the small, brightly-coloured birds defy expectations and are extremely tame and approachable. In addition to sunbirds, tiny Ashy-throated and Orange-barred Leaf-warblers hover; Yellow and Orange Chestnut-tailed Minlas descend from the canopy to pick berries from the roadside herbage Rufous-winged Fulvettas cling

..the Silver-ared Mesia.

to mossy trunks and larger limbs. Two of the most characteristic and evocative sounds piercing the mists are the yelping *ki-tiyook* of the Golden-throated Barbet and the tremulous, sibilant whistle of the Black-headed Sibia.

Mammals on both Doi Suthep and Doi Inthanon have long been reduced because of hunting. Among the large mammals, though, the Asiatic Black bear still survives on Doi Inthanon. This was also the first site where behavioural observations were ever attempted on gibbons in the wild, in 1937. In fact, at that time, the Karen tribes on the mountain believed that the crops in their cultivated fields benefited from being able to "hear" gibbon calls in surrounding forests. Though gibbons have long since been hunted from most northern mountains, three or four other primates, Slow Loris, Pig-tailed and probably Assamese Macaque and Phayre's Leaf-monkeys, though scarce, are still found on Doi Inthanon. Among the medium-sized mammals, large Indian Civet and Barking Deer still occur. The patient observer who goes out after dark with a flashlight can sometimes pick out the pink, reflected eyes of the Giant Flying Squirrel among the lofty boughs, and watch as it launches itself to glide across the road. However, the summit area of the mountain is out of bounds between 6 p.m. and dawn, so after dark, one's activities must be restricted to the lower margins of the forest, below the military checkpoint at the 37.5 km marker.

Doi Inthanon is the only place in Thailand where the visitor will see a number of small mammals of the Sino-Himalayan region. These include the Szechwan Burrowing Shrew, Chinese Pangolin and Père David's Vole. A total of at least 77 species of mammals have been recorded on Doi Inthanon, of which 31 are bat species.

The shift from opium cultivation to more stable agriculture, in which vegetables, fruits and flowers are grown, has both advantages and drawbacks as far as conservationists are concerned. On the one hand, it should lead to a reduction in the further clearance of forest.

An endemic orchid.

But, on the other hand, such intensive cultivation can lead to increased soil erosion, the silting of streams and increased pesticide pollution. Another noticable trend is the reforestation of denuded watershed areas with conifers, carried out under the auspices of the Royal Forest Department. Even though a native species, *Pinus kesiya*, is chiefly planted, such monocultures of pines appear to support little wildlife (and certainly far less than equivalent areas of native broadleaved forest). It could be argued that, where possible, native forest should be allowed to recolonize denuded areas as, given a chance it usually will. The principal difficulty is that cleared areas need to be protected from dry season fires if they are to regenerate naturally.

A further threat stems from uncontrolled tourist development. Holiday bungalows and resorts are springing up all over northern Thailand, aimed at both foreign and domestic tourists, and lead to further forest clearance and the pollution of waterways. Even though the access road to Doi Suthep is already perfectly adequate, a Bangkok-based company has sought to build an alternative transportation system in the form of a cable car leading up the eastern slopes of the mountain to the temple of Wat Prathat. Happily, the people of Chiang Mai, Buddhist monks, teachers, taxi drivers and others have united to reject such a development and to help preserve the sanctity of the mountain.

To visit Doi Inthanon and Doi Suthep is to recognise the enormity of the problems facing wildlife throughout Asia but, at the same time, to appreciate the continued richness and resilience of nature. The people of the lowlands, like the Highlanders, depend upon forests for their supplies of clean, flowing water. Given the rapidly growing environmental consciousness in Thailand, there is no reason why the mountains of northern Thailand will not continue to support a great wealth of species, providing that the means are found to rationalise the aspirations of the hilltribes and other poor, rural farmers.

uite a ngus: opical ictyophora.

THE ASIAN ELEPHANT

The Asian Elephant is a member of the order *Proboscidea*, a vast assemblage of 350 elephants and elephant-like animals that formerly roamed the earth. Asian and African elephants represent the sole remaining branches of a vast tree that stretches back to the Pliocene. The Asian Elephant has in the past been divided into eight, or even more sub-species but contemporary opinion suggests that only three now exist: the elephant of Sri Lanka (*Elephas maximus maximus*), which is the type species, the Sumatran Elephant (*Elephas maximus summatranus*), and all the elephants of the continental mainland (*Elephas maximus indicus*).

Today, the wild Asian Elephant is found in 13 different countries in Southeast Asia and the Indian subcontinent although its historical range was much greater, reaching west almost to the Mediterranean and north into China as far as Beijing. Present numbers of wild elephants total about 40,000 animals, although given the elephant's stealth and secrecy, any number is an estimate. India has by far the most (almost 20,000), followed by Burma (about 6,000), Indonesia (about 4,000), Thailand (2,000-3,000), and Sri Lanka (about 2,000). Vietnam, Laos and Kampuchea each has a population of about 1,000. Sabah has about 2,000 and Peninsular Malaysia about 1,000. Bangladesh, China, and Nepal have small populations of a few hundred or less.

Elephants are severely threatened over their entire distribution range and the Asian Elephant has been officially declared an endangered species. Although poaching and live capture have both taken their toll, the primary threat is the destruction of the elephant's forest habitat. For centuries — but increasingly over the last few decades — elephants have been forced from the fertile, grass-rich, alluvial river beds that are their preferred home into sub-optimal, hilly terrain where they must live in increasingly smaller groups, an isolation which has brought about the "pocketed herd phenomenon", small numbers of elephants cut off from other groups with which they might breed. Thus, they must face a new long-term threat, the danger of inbreeding. Probably nowhere in India or Southeast Asia exists what wildlife biologists call a "minimal viable population", that is, a population which can breed in perpetuity without harmful genetic effects.

Wild elephants live in small herds (more accurately called "family units") of about 20 to 30 animals led, normally, by the oldest cow and not by the "harem bull" of many historical accounts. In fact, young bulls are forced from the herd when they are about 10–13 years old, which is the period of their sexual awakening. Then, they become quite rambunctious, pestering the younger calves and also the cows who are still much bigger and stronger than the young males and able to eject them from the group. Thenceforth they wander with two or three companions, scarcely meeting family units, until they become full-sized bulls at the age of about 25. Even then the bulls will normally enter the cow-calf groups only when there is a cow in oestrus. The cow-calf group is thus matriarchal, and female elephants spend their whole lives together, forming deep and close social bonds.

Elephants are very social animals and are surrounded by loving and caring relatives. Indeed, a birth itself is often attended by an older cow, often termed an "Auntie", who helps ward off predators, take care of the afterbirth and aid and instruct inexperienced mothers.

The calf is nursed until it is three to five years old. The young elephant suckles with its mouth and only after several years during which it receives some instruction from the mother will the young animal be able to skilfully use its trunk. As the calf matures it will begin to use the trunk to suck up as much as four litres (one gallon) of water and transfer this to its mouth. The trunk is also used to gather the approximately 150 kg of grass and browse that an adult elephant consumes daily. Adult elephants spend as many as 18 hours a day feeding, gathering vast amounts of nutrient — poor plants which are subjected to a rather inefficient digestive system. Elephants, like horses and rhinoceroses as well as rodents and rabbits, are post-gastric digestives; that is, unlike most other ruminants, digestion occurs in the caecum and the colon, aided by fermentation and millions of protozoans and bacteria.

Probably the best place to feel fairly confident of seeing a wild elephant in Southeast Asia is Khao Yai National Park in Thailand. Located only a couple of hours drive from Bangkok, Khao Yai ("Big Mountain"), is an ASEAN National Heritage Site and deemed

Left, wildlife from the air: elephants at Khao Yai. Above, Baby elephant walk: heading for a morning bath.

one of the world's top five national parks. It still shelters about 200 to 300 wild elephants. However, even here the elephants are very shy, spending most of the daylight hours hidden in deep forest and emerging into open grassy areas only at dusk. By far the best chance of seeing them is from vehicles along the road at night (park officials can arrange transportation) where their eyes will be reflected by the searchlight and their ghostly shapes exposed. Some people find seeing elephants under such circumstances fascinating; others will find this a very artificial experience.

Possibly the next best place in Southeast Asia for seeing wild elephants is Taman Negara National Park in central Malaysia which is one of the region's best parks, stretching from an altitude of 120 metres to 2,150 metres and encompassing splendid tropical forest of many types.

Over much of its distribution the Asian Elephant is accorded an extraordinary respect which derives largely from its prominent place in two of the world's great religions. In Hinduism the elephant appears in many sacred forms especially that of Ganesha, a benevolent deity who is the god of wisdom, the guardian of books, learning and the arts. The elephant plays a parallel role in Buddhism where in a dream of the Buddha's mother, Queen Maya, a six-tusked white elephant named Chadanta entered her side, thenceforth to be born anew as the Lord Gautama. Thus the Buddha's penultimate reincarnation was as an elephant, though a very special one. The elephant also figures highly — usually as a compassionate being though occasionally as a fierce beast — in many of the Jataka tales, the delightful and informal stories which revolve around the birth of the Buddha.

Asian Elephants have been domesticated for at least 4,500 years, the earliest recorded evidence being from the great Indus River Valley civilizations of Mohenjodaro and Harrapa. Radiating from this centre, the art of owning and working elephants permeated through all of Southeast Asia, including much of what is now insular Indonesia. Carried along by Indian traders, the bearers of Hindu culture, elephant-keeping reached such importance that one index to the power of a king was the number of elephants in his stables. The Khmer empire might have had as many as 200,000 trained elephants and as many as 20,000 may have been used to build Angkor Wat.

Today, there are about 15,000 domesticated elephants in 13 countries in Southeast Asia and the Indian subcontinent with Burma and Thailand having by far the most, about 5,000 each. India has perhaps 3,000 and Sri Lanka about 400. Laos has about 1,000, Vietnam and Kampuchea about 600 each. Bangladesh and Nepal each has fewer than a hundred, and Indonesia has some newly introduced domesticated elephants.

Most surviving domesticated elephants work in forestry, hauling logs through thick forest, a job for which they are ideally suited. Able to drag about half their own body weight, elephants possess the advantage of doing minimal damage to the environment: they can thread their way through crowded forests and across steep hillsides without the necessity of building the roads and staging posts which are required by mechanised logging techniques. (Such service areas can consume up to 40 percent of the land area in a logging operation.) Additionally, the elephant's wide, soft feet do little damage to soil and grass. Thus, the elephant offers great potential in selective logging, the careful extraction of valuable hardwoods while leaving surrounding forest intact, a quality invaluable in exploiting essential resources while still conserving the region's rapidly dwindling forests.

Left, elephant power: workhorse of the forest. Above, a training session meets with some resistance.

Probably the best place in Southeast Asia to see domesticated elephants at work under more-or-less authentic conditions is at the Young Elephant Training Centre, just outside of Lampang in northern Thailand. (Lampang is less than two hour's drive from Chiang Mai and the Centre is about 30 minutes from Lampang, just beyond Km 656 on Highway 1.) This institution, run by the Forest Industry Organisation, is a *bona fide* school at which elephant calves are taught the skills they will need to drag logs. The calves enter the school at the age of five or six to meet their mahout, who might remain their companion for life, and are then taught the 40 or so spoken command words to which they must respond. They learn to haul logs singly and in tandem, to work with a partner, and to stack logs. There are normally 40 to 50 elephants at the Centre, though only about 20 usually take part in the demonstration which is held in March through May every morning except on public holidays and during the hot season. Entrance is free.

A less complete and more tourist-oriented show is held at Chiang Dao just north of Chiang Mai. Visits are easily arranged and the beauty of the riverside setting and the fact that many of the mahouts are Karen tribesmen compensate for the inadequacies of the presentation.

HUAI KHA KHAENG

Rainfall over most of Thailand is strongly seasonal, 80 percent or more falling during the southwest monsoon, from May to October. In contrast with Malaysia, therefore, where the major vegetation is rain forest, the lowlands over most of continental Thailand were once covered with predominantly deciduous forest, much of which has now been cleared and converted to agricultural land. Of the little lowland deciduous forest which remains, some of the richest lies in the **Huai Kha Khaeng Wildlife Sanctuary**, a 2,575 square km vastness only some 200 km northwest of Bangkok. The sanctuary, which takes its name from the *huai*, a large, permanent stream which flows north-south and bisecting it, is only one segment of a contiguous forested area of roughly 15,000 square km along the catchment of the Kwae River system. To the west of Huai Kha Khaeng lies an even larger sanctuary, Thung Yai (about 3,200 square km), which extends to the Burmese border.

In addition to the many plant and animal species characteristic of the dry, lowland forests of the Indo-Burmese region, a small number of Sundaic species, as well as Sino-Himalayan species, are found. For example, the Malayan Tapir occurs here close to its northern distributional limit in precisely the same areas as the Phayre's Leaf-monkey, a northern montane species found here near its southern limit.

In fact, as many as 10 species of primates may be present. Also, an incredible 21 species of woodpecker, ranging from the huge Great Slaty to the White-bellied Woodpecker are found in the sanctuary. This number is probably greater than that found in any area of comparable size anywhere in the world.

Much attention is now focused on the world's rain forests but, by comparison, forests of the seasonal tropics are even more poorly known. Any insect collecting trip is almost certain to reveal several species new to science. One reason for the richness of the sanctuary is that, until very recently, it was relatively remote, and, apart from small numbers of Karen villages to the south of the area, unsettled. Hunting was confined to a few wandering Karen and Thai villagers, and Huai Kha Khaeng has remained relatively rich in large mammals. It is the last stronghold of the banteng in Thailand and elephants, gaur and even a few wild water buffalo still occur. The area is rich in carnivores: eight species of wild cat, including tiger, leopard and Clouded Leopard, down to the diminutive Leopard Cat, are found, as are eight species of civet.

The deciduous forests of the sanctuary may be divided into two principal types: mixed deciduous, which predominates, and dry dipterocarp, a poorer, lower stature formation which can be found on small areas of lateritic, stony soil, chiefly in the east of the sanctuary.

While there is some overlap in the animal communities of the evergreen and deciduous formations, the former support more bird species, and probably more mammals than do the deciduous habitats. The deciduous formations abound in woodpeckers and support a high biomass of herbivorous mammals, especially banteng and bamboo rats. Many of the larger mammals range is influenced chiefly by the availability of water and by the distribution of mineral licks. Tigers and leopards, for example, are somewhat segregated from one another, the former apparently needing permanent water while the more ecologically tolerant leopards are able to go for weeks without drinking. Walking along the length of the permanently-flowing Huai Kha Khaeng River, in the heart of the sanctuary, one is able to see tiger pug marks on almost every patch of soft sand or mud: leopard signs are less frequent. By contrast, in the hills around the sanctuary's drier eastern margins, where most waterholes and streams dry up during the hottest months, February to April, leopard tracks and scats predominate. Some of the civets, particularly the binturong and the almost exclusively frugivorous

deciduous
rest in
arch, at the
d of the
y season.

Small-toothed Palm Civet, are confined to moister evergreen habitats.

Long before the existence of roads, rivers provided man's main communication links and consequently riverbanks were usually among the first lowland areas to be cleared or settled. Today, even the most remote rivers have been developed, chiefly through the construction of hydro-electric dams. Lowland rivers have been used to delineate the boundaries of parks and sanctuaries rather than being enclosed with them. Consequently, riverine habitats have suffered very badly in Thailand. Huai Kha Khaeng, together with the adjacent Thung Yai sanctuary to the west, are the only two protected areas in the country to have little-disturbed riverine habitats enclosed within their boundaries. The Huai Kha Khaeng River is, therefore, another key to the Sanctuary's richness.

In addition, the river may be regarded as a kind of lowland corridor, enabling a number of scarce species which have disappeared, because of human persecution and habitat destruction, from the extensive plains outside the sanctuary to survive in otherwise hilly country. Thus the sanctuary supports the only viable population of the Green Peafowl in Thailand, perhaps 300 or more birds, all of which are concentrated along the river. The male birds use level riverine sandbanks and adjacent natural grassy clearings as display areas. Other scarce species found there include White-winged Wood-duck, Lesser Fish-eagle, and the last surviving population of Red-headed Vultures in the country. The latter probably survive by scavenging the carcasses of large mammals killed by tigers, but there are probably fewer than 20 individuals remaining, so the species may be in danger of extinction. Many pairs of huge and vividly coloured Stork-billed Kingfishers are also found along the riverbanks, and serve to remind us that larger, open forested streams were probably the ancestral habitat of this species which may still be found in temple grounds and wooded copses in the well-watered

Stump-taile[d] Macaques prefer limestone outcrops.

rice-growing areas around Bangkok and Ayutthaya.

Follow the valley of the Huai Kha Khaeng and envisage something of the former richness of Thailand's plains. Waiting for peafowl by a grassy mineral lick in the early morning, one watches the flocks of pigeons, including the scarce Yellow-footed Pigeon, flying in to perch in nearby trees, pumping their tails up and down. The soporific purring of the Green Imperial Pigeon can also be heard and the ground is pitted by the tracks of the elephant, wild pig and Sambar Deer. Some of the smaller streams bear the name *huai raet*, the stream of the Javan Rhinoceros, a memory to the fact that even here, in the richest and remotest part of the country, some of the former inhabitants have already been hunted to extinction.

The present richness of Huai Kha Khaeng is severely threatened. At one time, forested areas outside its eastern boundary formed a *de facto* buffer zone between the sanctuary and the deforested and heavily cultivated plains to the east. The banteng and other herbivores of the dry dipterocarp woodland were able to move at will throughout. However, in the years before the government banned logging, it granted a timber concession in this area to a plywood company. New settlers, many of whom were immigrants from Thailand's arid northeast, followed the loggers and flooded into the area in order to clear land for agriculture. Now, the only cattle in the buffer zone are domestic stock: the effective habitat area for banteng has been halved and, in addition, poaching of both timber and wildlife has increased. On the positive side, Hmong tribespeople from the north who, over 20 years ago, migrated into the upland watershed dividing Huai Kha Khaeng from Thung Yai, have been resettled outside the sanctuary, thus removing a major source of hunting pressure. Another pressing concern is fire. Fires set in the drier forests frequently burn into the edges of adjacent moist evergreen and, over a period of many years, may cause it to gradually retreat. Along the Huai Kha Khaeng River are many places where the fire-blackened skeletons of tall dipterocarp trees stand as memorials to where a piece of evergreen forest once stood: of regeneration by their moist-forest successors, there is no sign. Even forest guards occasionally burn forest, as they were taught to do when they were still village children: old habits die hard.

e Green eafowl is re in utheast ia.

A management plan has been prepared for the sanctuary and should address these and other concerns. A number of other projects have been proposed, including the construction of a training centre for wildlife conservation personnel. There is a good chance that Huai Kha Khaeng will receive much additional funding aimed at maintaining its integrity in the years to come.

Huai Kha Khaeng and adjacent Thung Yai together constitute the single most important remaining natural area in Thailand and one of the most important anywhere in Southeast Asia. In recognition of this, the Thai government has proposed they be listed under the World Heritage Convention.

KHAO YAI NATIONAL PARK

Khao Yai National Park (*Khao*: mountain; *Yai*: big) is the most-visited park in all of Thailand. The park is situated about 160 km from Bangkok in the central north and covers an area of 2,168 square km and portions of four provinces. The four provinces also form the rough boundaries of the park: Nakhon Ratchasima in the north, Prachin Buri in the east, Saraburi in the west, and Nakhon Nayok in the south. Today, Khao Yai suffers badly from human encroachment, making the park an island of forest in a sea of farmland.

History: Khao Yai is part of the Phanom Dongrek mountain range which is situated at the southwestern tip of the Korat Plateau. Sixty years ago, 30 families from Nakhon Nayok moved into the Khao Yai area to farm, hunt and collect forest products. Due to its isolation, the area became a notorious hide-out for outlaws and criminals and the government decided to expel the villagers of the Ban Khao Yai settlement before the park was designated. The abandoned farmland has become the *lalang* grassland that one sees today.

The park was established in 1959 at the behest of General Sarit Thanarat, who was then prime minister of Thailand. The first road to the park from the northern side, highway no. 2090, is called Thanarat Road in honour of the general.

The park headquarters are located in the east, on Thanarat Road at Km 37.

Park Management: The objectives of the park management are to preserve the genetic resources, flora and fauna, and ecology of the central Thai rain forest for future generations; to protect the watersheds of at least five river systems which are essential for agriculture and city life; to fill the educational needs of the public and to have a place for environmental and research studies; to support local communities and to provide facilities for recreation and tourism.

Terrain, Climate and Vegetation: The terrain is primarily mountainous and varies in height from 250 metres to 1,350 metres above sea level; most lies above 700 metres. The lower elevations are found along the park boundaries. The three highest peaks are Khao Rom (1,350 metres) in the southeast, Khao Lam (1,330 metres) in the east and Khao Khieo (1,290 metres) in the south. The terrain gradually slopes downwards towards the Friendship Highway along the northern and eastern boundaries of the park. In the south and west the terrain consists of steep slopes which abruptly drop into an agricultural area.

Two monsoons play a key role in determining the climate of the park. The steep slopes are influenced by the southwest monsoon which creates a milder climate along the mountain ridges, resulting in heavy rainfall from July to October. The average rainfall is about 3,000 mm and is heavier along the higher ridges. During November, the park is influenced by the northeast monsoon, which results in the yearly shift to the dry season, during which all

eft, Khao
ai is the
est place to
pot a tiger.
ight, water
elief.

but the major streams dry up. The average annual temperature is about 23°C; the highest temperatures may reach up to 30°C during the daytime in the summer (March to May) and the mercury may drop to 6°C in the winter (December to January).

Khao Yai is one of the largest remaining intact tropical forests on mainland Asia and has a rich diversity of flora. It would be wrong, however, to think of the park as simply one large, steamy rain forest. Variations in terrain and climate combine to cause variation in vegetation. Visitors who take the time to explore Khao Yai thoroughly, will find not one, but five distinct types of forest.

Most of the park is covered by dry evergreen and semi-evergreen rain forests which account for 26 percent and 60 percent of the park area. The dry evergreen forest is found at elevations of 100 metres to 400 metres, and the semi-evergreen rain forest occurs at altitudes of 400 metres to 900 metres. Members of the dipterocarp family such as *Dipterocarpus alatus* and *D. baudii* are typical of both forest types. These disappear at higher elevations and are replaced by *D. costatus* and *D. macrocarpus*. Oaks and chestnuts are also found at these altitudes, particularly *Lithocarpus eucalyptifolius*, *L. rodgerianus*, *Quercus semiserrata*, *L. annamensis*, and *Q. myrsinaefolius*. The last two species are confined to elevations of 400 metres to 600 metres.

The forest changes to hill evergreen forest at altitudes above 1,000 metres. Dipterocarps vanish and are replaced by gymnosperms of the genus *Podocarpus* which are excellent indicators of high altitudes. Many of the same species of oaks and chestnuts in the rain forest also flourish at these elevations. The flat top ridge with wet open areas is well covered with *Khaao tok ruesi* (*Sphagnum spp.*) and other mosses mixed with herbaceous species such as *Ya Khaao kam* (*Burmannia disticha*)

Dry, mixed deciduous forest is found along the northern slopes, ranging from 400 metres to 600 metres elevation.

Early mornin[g] at Khao Yai.

Typical tree species which are of economic value include the *Makhaa* (*Afzelia xylocarpa*), *Pradu* (*Pterocarpus macrocarpus*) and *Ta baek* (*Lagerstroemia calyculata*). During the dry season this region experiences forest fires.

The last forest types are savanna and secondary growth forest. The savanna was formed by shifting cultivation and the secondary growth is the result of road construction and effective fire control of the grassland. The dominant species in the savanna is the *Yaa Khaa* (*Imperata cylindrica*). Because of the effective fire control, seedlings and saplings of pioneer species such as *Macaranga denticulata* survive. The well protected grassland will eventually become a secondary growth forest.

Plants which are endemic to Khao Yai are *Bulbophyllum khaoyaiensis, Palaquim koratense, Manauthes prachinburiensis, Neolourya thailandica, Thisma mirabilis, Combretum quadratum* and *Embelia kerrii*. Epiphytes and orchids abundantly cover tree trunks and branches. Some of the orchid varieties include *Ione, Bulbophyllum, Thelasis,* and *Dendrobium*.

Trails and Points of Interest: The park is a paradise for the nature lover, especially for bird and animal watchers. Over 300 species of resident and migratory birds are found within its borders. Among the most attractive and easily spotted birds are Scarlet Minivet, Vernal Hanging Parrot, Green Magpie, Long-tailed Broadbill, Asian Fairy-bluebird, Blue-winged Leafbird, 12 species of woodpeckers, two species of trogons, and the Hill Myna. In addition, the park supports more than 20 species of large mammals, including elephants, gaurs, Sambar Deer, Barking Deer, wild pigs, Asiatic Black Bears, Malayan Sun Bears, Hog Badgers, binturongs, tigers, Leopard Cats, serows, gibbons, macaques and many other small animals. A short stroll near the park headquarters, either along the highway or on one of the trails, can bring the visitor close to some of these animals.

urmannia isticha is ndemic to hao Yai.

Khao Yai has about 40 km of hiking trails, ranging from those designed for the casual walker to those meant for the tough hiker. All trails were developed from paths originally made by elephants and other animals and all are still used by elephants, gaurs, tigers and wild pigs. This can add a great deal of excitement even to a casual stroll. Chances of seeing animals are increased if you wear inconspicious clothing, do not use perfume, walk quietly and stop frequently to let your eyes adjust to the environment. The following is a selection of interesting trails.

The **Moh Sing Toh Trail** is ideal for the visitor who wishes to take only a brief stroll through a tropical rain forest. The trail head begins across the road from the park headquarters. Interesting and impressive plants that grow along this trail are the tree fern *Mahaa sadam* (*Cyathea borneesis*), which is a typical species found along stream banks at this elevation of tropical rain forest and the *Tetrameles nudiflora* tree with its huge buttresses. The hanging liana is also impressive. The visitor should be able to see forktails, Silver Pheasants, Green Magpies and hornbills. Elephants, Barking Deer, and wild pigs may also be encountered.

The trail climbs gently uphill to an intersection. A left turn brings the hiker out into the grassland. Although the trail is unmarked, it is easy to follow, and offers a lovely view to the east and southeast. The trail now goes downhill to a small reservoir by the highway where one can see Little Grebes. Other birds that favour the trail area are the Red-whiskered Bulbul, Spotted Dove, Indian Pied Hornbill and bee-eaters. The visitor can spot animal scratches and droppings in the gravel of the trail. Animals which share this trail with hikers include tigers, civets, Leopard Cats and deer. The shrub of *Melastoma sp*. transforms the trail into a blossoming avenue in February and March. The trail is two km long, and between 1- 1½ hours to walk.

The **Kong Kaew Falls Nature Trail** starts from a suspension bridge at the

Below left and below, there is muc to observe among the trees in Khao Yai.

Visitors' Centre. The waterfall is small and makes a good picnic site. The forest along the trail is a good representative of the forest types found within the park, especially of a developing, healthy tropical forest, which consists of three canopies. The top layer is dominated by magnificent trees belonging to the family *Dipterocarpaceae*, reaching heights of 25 metres, or more. The ground cover consists of gingers, members of the *Zingiberaceae*, climbers and rattan (*Calamus sp.*). Rattan is a spiny stemmed palm which sometimes climbs to the high canopy. Animals which one may see include gibbons, deer, wild pigs and even tigers and elephants. Birds which frequent the area include forktails, herons, kingfishers, barbets, woodpeckers and hornbills. Walkers may be fortunate enough to catch a glimpse of the noisy Brown Hornbill.

Another easy trail starts from the road near Km 32, just opposite **Darn Chang** (**Elephant Trail**), and leads into another interesting forest area. The most spectacular attraction along this trail is a number of large strangling fig trees. These trees have been numbered for fruiting records and produce crops at different times of the year, though sometimes at overlapping times. The strangling fig appears to be the most important food source for frugivorous birds and mammals such as hornbills, pigeons, barbets, mynas, bulbuls, orioles, squirrels, gibbons, macaques, palm civets, binturongs, bats and others. When fruiting, a single tree produces a large quantity of food that may last as long as 10 days. The trail leads out into a vast area of grassland called Bueng Phai. The undulating landscape interspersed with strips of forest is lovely. During the rainy season, the winds create a wonderful, wavy movement in the grass. It is even more spectacular when flocks of Wreathed Hornbills fly over this open area to feed or to roost. Herds of deer can be seen grazing on the lower hill and one can even encounter elephants eating grass.

A tough hiker can continue further

arking Deer re common round park eadquarters.

along the trail to the southwest, and reach a guardpost in another large grassland area, the Klong E-tao. This is a suitable place to spot gaurs. There is a salt lick favoured by gaurs in the forest near the station across some low hills, and the animals frequently gather here. It is worthwhile spending the night at the station and taking extra time to explore the area. The best place for bird watching in this region is at a nearby stream.

The trail begins to loop eastwards after Klong E-tao and enters another beautiful forest. Here one can see Silver Pheasants, trogons, hornbills, forktails and many other species of birds. The trail ends at the Nong Pak Chee Animal Watching Tower and brings the hiker back to the main road. The trail is 11 km long, and can be completed in one full day.

The Nong Pak Chee Animal Watching Tower stands one km west of the main road, between Km 35 and 36. A vehicle track also leads to the tower which is an excellent place for bird-watching. Red-whiskered Bulbuls, Chestnut-headed Bee-eaters, woodpeckers, rollers, mynas, hornbills and grassland species such as prinias and Bright-capped Cisticola inhabit the area. A medium-sized man-made pond and two salt licks near the tower provide additional opportunities for wildlife observation.

Wintertime at the pond is a good time to see Common Moorhen and White-breasted Waterhens and even occasionally the Oriental Darter. The needletails put on a spectacular show, diving into the pond to drink and to bathe and various species of swifts and swallows are often seen over this open area. On hot days, the Crested Serpent-eagle often soars over this region, taking advantage of the thermals. The tower is also a good place to see hornbills during the rainy season. Sambar Deer and Short-clawed Otters sometimes swim in the pond, and the salt licks are visited by elephants, deer, gaurs, wild pigs, and tigers. Tigers are seen frequently in this area. To the south of the tower, visitors can see herds

Tiger cubs on the alert.

of deer and wild pigs. The tower is a good place to spend the night and to listen to the night sounds of the forest and grassland, such as the scream of the Flying Lemur. However, permission from the park authorities is needed.

The viewpoint beside the road at Km 30 is an excellent place for watching hornbills and other birds. The forest in this area is beautiful, especially to the southeast. Here one can see the distinctive canopies which are the trademark of tropical forests. Large stands of dipterocarps and fig trees, as well as other fruit trees, make this area a perfect nest site for the hornbills. Flocks of Hill Mynas and Golden-crested Mynas, Asian Fairy-bluebirds and Vernal Hanging Parrots share this area with hornbills. Two large elephant salt licks near which herds are frequently spotted are located near the road.

A magnificent stand of dipterocarp trees on the east and southeast can be viewed from a small hill by the road, at the last curve before it reaches the **Haew Suwat Waterfall**. A valley just southwest of this small hill is an important roosting place for many Wreathed Hornbills from June to August.

The final beauty spot is the **Haew Narok (Gorge of the Devil) Waterfall** which is by far the most spectacular of the park's many waterfalls. This waterfall is easily accessible through a well-marked trail from the new highway from Prachin Buri. Haew Narok is the highest waterfall in the park, and consists of three cliffs, totaling about 150 metres in height. The first, and also the tallest, is 60 metres high. During the dry season, the falls are dry but the sight and sound of the rushing water during the rainy season as it plunges into the deep, narrow valley is unbelievable. The entire valley is filled with mist and spray from the falls and mosses flourish.

The number of visitors to Khao Yai has increased more than 10 times in the last 20 years, to an estimated 500,000 in 1986. The most quiet time to visit the park is on weekdays during the rainy season (June to September), when animals are easiest to spot.

Bright eyes: Sambar Deer by the light of a flash.

WAT PHAI LOM

Wat Phai Lom (*Wat*: temple; *Phai Lom*: surrounded by bamboo) on the banks of the Chao Phraya River about 50 km north of Bangkok, is well known as the home of the largest breeding colony of Asian Open-billed Storks in Thailand. The Wat Phai Lom Non-Hunting Area is accessible by road or boat.

The breeding colony of storks at Wat Phai Lom has long been known and some claim that it is nearly a century old. Long before the reserve was designated in 1977 the colony was well protected by the temple monks. Although the storks sometimes damage paddy fields and annoy farmers they are accepted as part of daily life.

The Asian Open-billed Stork is a medium sized stork with a length of 65 cm. Its name is derived from an open space between the mandibles. The species is found in India, Bangladesh, Pakistan, Sri Lanka, Burma, Thailand, Cambodia and Vietnam. Since 1960 the storks have been listed in Category I under the Wildlife Protection Act.

Nesting Behaviour: The storks arrive at Wat Phai Lom in November. It is believed that their arrival is related to high tide or flooding of the central plain. The high tide is apparently a factor in the increase of the *Pila* snail population on which the birds feed.

The sugar palm tree seems to be the favourite tree for nesting. Bamboo and other trees with spreading limbs and numerous small, forked branches are also in demand. Nests are built with twigs and decorated with green leaves. The storks either gather nesting materials from the ground or pluck it directly from the trees.

Female storks lay two to five eggs which are incubated by both the male and female. It takes 27 to 29 days for the eggs to hatch, and the young are under parental care for about two months. They can fly within 40 days.

Feeding and Diet: The parents feed their young by regurgitating food into their nests for the young to retrieve. Each nestling requires nearly half a kilogram of food a day, or about 20 to 30 snails. It is estimated that a colony of storks can consume as much as a ton of snails per day.

Migration: The storks at Wat Phai Lom begin to migrate northwards at the beginning of the monsoon in April and May and by June and July only a few remain. Banding recovery indicates that storks from Wat Phai Lom migrate across Burma into the Brahamaputra and the Ganges deltas of Bangladesh. However, the migration route could extend to as far west as Pakistan.

Stork Population: A 1967 census counted 5,000 nests. Ten years later, the number of nests had increased to 13,000; in 1986, the number had declined to 8,600. The causes of this decrease are unclear, but it is perhaps due to natural phenomena such as storms or insufficient food. The decline may also be due to human encroachment and visits to the nest sites.

Other Birds and Wildlife: About 200 resident and migratory bird species can be seen at Wat Phai Lom. Large wading birds like the Painted Stork are occasionally sighted. Endangered species such as the Spot-billed Pelican and the Black-headed Ibis have been seen. The Stork-billed Kingfisher is one of the migratory birds which announces its presence in the area. Other common species are Pond Herons, Pied Fantails and Brown-throated Sunbirds.

Mammals found in the area are mostly bats and rats of different species. Mongooses and forest badgers are occasionally sighted.

Recommendations for Visitors: November to May is the best time to watch the stork breeding colony. January and February are suitable months for observing bird species such as herons, darters, ibises and other winter visitors.

Since the paths and trails pass under the breeding colony it is advisable for visitors to wear hats to protect themselves from stork droppings. Mites can be a great annoyance to visitors, and when returning from the area one should shower and soak all clothes.

The unique breeding colony at Wat Phai Lom.

BANGKOK WEEKEND MARKET

The proliferation of temples and Buddha images in Thailand constantly reminds the visitor of the country's Buddhist heritage: approximately 95 percent of the population is Buddhist. Many westerners mistakenly assume that, because the Buddha preached tolerance of other life-forms, Thais are respectful of wildlife. A visit to almost any market, whether in the provinces or in the capital, is usually sufficient to quickly dispel such an illusion. Only a tiny proportion of Thais are vegetarians. Markets which sell fresh meat, particularly those in the provinces, often sell wild birds and sometimes also wild pig, mouse deer, leaf monkeys and even pangolins for human consumption.

Interest in wildlife for many Thais is synonymous with cages. Wherever one travels in Thailand, caged or tethered wild animals and birds are seen. Although there has been much growth in environmental awareness in Thailand, this has been far exceeded by the growth in the purchasing power of the urbanised Thai citizen. In Bangkok, where this spending power is concentrated, exotic wildlife from Indonesia and South America rubs shoulders with an enormous number of indigenous birds and smaller mammals at the infamous "Weekend Market" at Chatuchak Park. This market was first set up over 30 years ago to enable produce from distant provinces to be sold in Bangkok. Here, much wildlife, the harvest from the forests, can be found alongside cheap fabrics, handicrafts, fruit, vegetables and other food.

As the visitor moves through hot, crowded aisles between cages, he is assailed by the loud shrieks of native Red-breasted Parakeets, one of the principal species sold. The Hill Myna, famed for its mimicry of the human voice, is one of the higher priced native birds, usually selling for 600 baht (approximately $25) or more though the predominant species now sold is the Zebra Dove, much valued throughout Thailand and Malaysia for its mellow, rhythmic hooting call. These species may be legally sold under permit, subject (theoretically) to a quota, so many being allotted per trader per year. In fact, the quota system is ignored and, in any case cannot be enforced since it is impossible to monitor the actual numbers being sold. For a common, open country

A Lesser Adjutant being sold fo[r] a few dollars the value of its meat.

species like the Zebra Dove, the volume of trade probably has little impact on the wild populations. However, the theft of nestlings, combined with forest destruction, has greatly reduced the numbers of such forest birds such as the Hill Myna.

Almost all native Thai birds and a great many mammals are fully protected by law in Thailand, but the legislation has little effect. As a recent survey showed, more than one-fifth of all native birds and over three-quarters of all species on sale were those nominally protected by law. Beneficial insectivorous birds as well as seed and fruit eaters are all sold. Thus, crowded cages, each containing 60 or more tiny sunbirds and flowerpeckers or other brightly-coloured but protected species, such as barbets or leafbirds, or the iridescent Fairy-bluebird are displayed with impunity. Only the larger, more threatened species, such as hornbills and larger birds of prey are usually kept "under the counter", in anonymous-looking cardboard boxes, to be shown when a prospective buyer appears. Mammals, such as slow loris, leaf-monkeys, gibbons or the smaller "spotted cats" are hawked by transient sellers, one or two animals per person.

The heat and the noise is overpowering; little wonder that many animals die before they are sold. A careful look around usually reveals boxes of dead hanging parrots, wagtails and other, more sensitive species thrown away among the garbage. Both sellers and prospective buyers appear indifferent to the inherent cruelty of the trade.

The wealthy of Bangkok are now also major consumers of imported wildlife, some of which comes from Laos and Kampuchea, which share many species in common with Thailand, and much of which comes from further afield. Even lion cubs are sometimes offered for sale. Of an estimated 13 million baht (US $520,000) worth of wildlife sold at the market annually, at least $200,000 is now contributed through the sale of exotics, particularly Indonesian and South American parrots. Internationally endangered species, in which trade is fully prohibited, and which leave their countries of origin illegally, may still enter the country to be sold here because Thailand still lacks the necessary domestic legislation to enable it to live up to its obligations under the Convention on International Trade in Endangered Species (CITES). An Indonesian Palm Cockatoo, for example, may be purchased for $400, or a South American Scarlet Macaw for $1,600.

The Weekend Market is only one of an enormous number of outlets for the sale of wildlife throughout the country. Much of the trade in the rarer and larger animals is now conducted behind closed doors. Price lists, distributed internationally by the main animal trading houses, claim to be able to air freight everything from elephants to tigers and clouded leopards.

It is usually argued that the wildlife trade cannot be eliminated because the forests cannot be effectively patrolled and rural villagers, being extremely poor, have always supplemented their incomes by selling what they find or catch. Yet there has never been a concerted effort to suppress the trade among the relatively wealthy and supposedly well-educated people of Bangkok, who are both its principal beneficiaries and customers. Wildlife officials should penalise both buyer and seller and the wildlife-buying public should be encouraged to examine the implications of their actions. Many Thai people believe that it is a meritorious act of compassion to buy a suffering baby gibbon in the market and to give it a home. They fail to consider instead that the only real effect of their action is to perpetuate the trade by ensuring that further nursing female gibbons will be shot and their infant young captured for sale.

nother sad ght — the ins trade.

THALEBAN NATIONAL PARK

Biologically, peninsular Thailand has a closer affinity with Malaysia than it does with the rest of Thailand. The majority of the animals and plants found here are of Sundaic affinity — that is, shared with Peninsular Malaysia and with the islands of Sumatra, Java and Borneo, the "Greater Sundas".

Both the lowlands and the mountains of peninsular Thailand were once chiefly covered with rain forest, usually referred to as semi-evergreen or white meranti rain forest, to distinguish it from the less seasonal, wetter evergreen or red meranti forest found throughout most of Malaysia. Meranti refers to the particular species of dipterocarp trees of the genus *Shorea* which occur in the forest community. To the layman, however, both types of forest look very similar.

Virtually the entire lowlands in peninsular Thailand has now been cleared, though good, tall forest remains on hill slopes. One of the best and most accessible areas lies at *Thaleban National Park*, Satun Province, on the Malaysian border. The semi-evergreen rain forest of which Thaleban provides a good example continues unbroken across the hills of the border ridge into the Malaysian state of Perlis, the only part of Malaysia to support the "white meranti" type of forest. Thaleban, though only 101 square km in size, possibly owes its continued richness to the fact that it is a part of this larger block. Although Thaleban is protected as a national park, the forest on the Malaysian side of the border is being rapidly cleared as part of agricultural development schemes.

Thaleban lies 37 km from the provincial capital, Satun. A tarmac road leads south through a steep-sided valley, past the park headquarters and on into Malaysia. At the head of the valley, sits the border post, which is open daily from 8 a.m. to 4 p.m. The Malaysian town of Alor Star is 81 km to the south. The park is popular, therefore, with both Thai and Malaysian visitors.

The park headquarters is situated in the valley floor at the side of a small lake on the east of the road. The lake, created when a small landslip blocked the outflow of a stream, is a focus for the parkland and bungalows around the headquarters area. To the east, forested hills with towering trees slope gradually upwards reaching their maximum elevation at Khao Chin (740 metres).

To the west of the road, precipitous limesone crags, covered with a much drier forest, rise to over 480 metres. It is possible to climb to the summit ridge and to enjoy a view looking almost vertically down on the park headquarters. The Yaroj and Chingrit waterfalls, which lie north of the headquarters, are good picnic sites and are popular with local residents. The former may be reached by car, and the latter after a short walk through padi fields and rubber estate plantations.

The open parkland, around the headquarters, is a good place to search for the more common birds, such as bulbuls, as well as migrant visitors from the north,

The attractive Diard's Trogon.

which favour garden and forest edge habitats. The rare Narcissus Flycatcher can sometimes be found here. It winters exclusively in lowland forest and forest edge, usually close to water, in a small area of southern Thailand and northern Malaysia.

The forest bird community is still very rich, though lacking some species which would normally be confined to the forests of level lowlands. The calls of the Great Argus and the Helmeted Hornbill may be heard, echoing from the tall forest on the hill slopes. The male argus chooses prominent sites such as ridge tops in which to carry out his elaborate display ritual and "dancing grounds" often lie on or close to trails. The dancing ground is usually a circular or ovoid area, about 3 metres across, from which the leaf-litter has been "swept" by the bird.

The limestone cliffs support nesting Dusky Crag Martins and at least one pair of a scarce, resident form of the Peregrine Falcon. The lake is sometimes a good place to observe the scarce and shy Masked Finfoot. This seems to be a non-breeding visitor to Thaleban and elsewhere in Thailand with most sightings coming during the period from January to May. The park is also a good place to observe bird migration. In the late autumn, diurnal migrants, including flocks of hawks such as the Black Baza and sometimes even the larger eagles, may be observed following the sides of the valley, passing south into Malaysia. Some return migration is also observed in spring and, consequently, the park has a large bird-list — almost 200 species have been recorded.

The park has never been fully surveyed for mammals. Dusky Leaf Monkeys and White-handed Gibbons and even Lesser Mousedeer may be seen without much difficulty, though the serow, a kind of large goat-antelope which frequents the precipitous crags, is now very shy and scarce.

A local guide might be advisable to explore remote parts of Thaleban national park, since the trails are not well marked.

aleban
m the
ights.

MALAYSIA

Malaysia consists of two geographical parts: Peninsular Malaysia (also called West Malaysia) covers the southern tip of the Malaya peninsula. East Malaysia consists of the state of Sabah and Sarawak in the western part of the island of Borneo.

Plant communities and wildlife in the two parts of Malaysia are very similar because of landbridges which existed between Malaya and Borneo during the ice-ages. Nevertheless, several large mammal and bird species, which are of great interest to naturalists, occur only in one part. Thus, gaur, tapir and tiger are found in the peninsula and Orang-utan and Proboscis Monkey in East Malaysia.

In the Malay peninsula and Borneo, plant and animal biodiversity reach a maximum, and Malaysia and the island of Borneo can be considered the core area of Southeast Asia's wet tropics. The area was thinly populated until the turn of the century. The introduction of oil palm and rubber plantations around the 1900s led to extensive forest clearings and, after World War II, the demand for tropical timber resulted in the massive destruction of lowland forest. In Peninsular Malaysia the large national park Taman Negara and the state park Endau-Rompin are the largest tracts of protected lowland forest, and the mountains still protect large expanses of primary forest, even within view of downtown Kuala Lumpur. While these forests are supposed to be reserved for tree growth and logging, they represent conservation potential in the face of changing ecological value systems.

Preceding pages: Limestone outcrops fringe much of the northern peninsula; Leaping Lizards! A close look at the "flight pattern" of a Flying Lizard. Left, the Asian Paradise Flycatcher.

Penang And Langkawi

Penang does not have national parks, but the beachcomber and amateur naturalist will find many forest reserves in the central and northwestern parts of the island located off the peninsula's northwestern coast. **Pantai Acheh Forest Reserve** covers two square km and encompasses small hills, granite outcroppings, rocky coastal slopes and small sandy beaches.

The area is hilly, and mostly covered by dipterocarp forest. The forest in some places was logged early this century but there are still large areas of virgin forest. There are small sandy beaches, and in some sheltered bays small streams empty to the sea. The coastal shore constitutes more than half of the reserve's boundary. A number of walking tracks pass small streams or springs. At Sungei Tukun, a shelter has been built in the tidal waters of a small bay into which a clear stream flows. There is a patch of mangrove at the mouth of stream and another section of the trail along this stream reaches a steep cascading waterfall.

Many tree species on the walking trails have been labelled. There are epiphytes such as orchids, bird nest ferns and the spectacular stag horn fern.

The only large mammal in the reserve is the wild pig, although many nocturnal species such as Leopard Cats, civets, Slow Loris and Flying Lemurs also roam the reserve. Lemurs can sometimes be seen during the day hanging from branches or clinging to tree trunks. The hero of Malay folklore, the mouse-deer, also dwells in the reserve. The only two primates, the Long-tailed Macaque and the Dusky Leaf Monkey, can be readily observed when they jump from tree to tree with a crashing sound as they grab hold of the branches. Crashing branches may also mean that the Black Giant Squirrel is about. At dusk and dawn, another night creature, the large Black Flying Squirrel can be observed. If one is extremely careful, one can spot otters at the beaches and at the river mouth, but they quickly disappear into the forest. Among the reptiles, monitor lizards can be seen scurrying in the undergrowth, flying lizards glide from tree trunk to tree trunk, and sea turtles are still believed to nest on some of the reserve's beaches.

At **Pantai Kerachut** a lagoon builds up during the months of March to June. The stream's mouth is gradually blocked by beach sand, and saline water slowly collects behind the beach, mixing with the fresh water from the stream. This causes the lake to have two layers of water. The lower level is saline sea water, which is warm and the upper level is fresh water from the stream, which is cold. When the lake is dry it leaves a grassy and muddy flat that attracts waders, kingfishers, bulbuls, little herons, eagles, hawks, amd kites. The reserve is inhabited by a number of attractive forest birds which include the beautiful Asian Paradise Flycatcher.

Some of the most colourful birds of the old-world tropics are the bee-eaters, of which six species exist in Southeast

A Blue-taile Bee-eater, one of the region's mo beautiful species.

Asia. Penang boasts one of the largest nesting grounds of this bird family and here, in one colony, can be found 200 to 500 pairs of three species. The Genting bee-eater nesting grounds are small flat parcels of porous white sand and are the most suitable substrate in which to excavate nesting burrows. They have attracted the Chestnut-headed, the Blue-throated, and the Blue-tailed Bee-eater.

Penang Hill, 650 metres above sea level, is a good spot for birdwatching although the hill tends to get more rain than the rest of the island. At least four species of sunbirds and two of spider-hunters can be seen feeding in flowering trees and shrubs.

The Langkawi Island group, off the very northwest coast of Peninsular Malaysia, consists of more than 100 beautiful islands. The islands are developed and promoted as a major beach resort but contain extensive forest reserves such as Machinchang and Gunong Raya whose peaks are 700 metres and over 880 metres respectively above sea level.

The two main forest reserves have a variety of mammals such as the wild pig, the Lesser Mousedeer, and the rare Large Mousedeer. Among the arboreal mammals are the cream-coloured Giant Squirrel, the Long-tailed Macaque and the Dusky Leaf Monkey. Green and Ridley's turtles come ashore on some beaches to nest.

Among the more attractive birds on Langkawi are Pied, Rhinoceros and Great Hornbills. The birdwatching specialist will find Langkawi a good place to search for the huge Brown-winged Kingfisher, a relative of the Stork-billed Kingfisher, that is restricted to this part of Southeast Asia.

Gua Langsir, a cave on the island of Pulau Dayang Bunting, is a roost for thousands of bats. Telaga Tujoh (seven wells) is a series of seven pools through which a freshwater stream cascades.

In Penang and Langkawi, the best season both for native observation and for swimming is from December to April, during the dry season, (while the wet season is from July to October).

ng-tailed acaques e easily countered.

KUALA SELANGOR NATURE PARK

A section of mangrove forest, 65 km west of Kuala Lumpur, has been preserved as Kuala Selangor Nature Park.

Visitors can drive to chalets at the edge of the park. The park can also be entered by walking down from the nearby Bukit Melawati hilltop, which features in itself a park, some historic fortifications — plus a great view across the mangrove.

From the entrance, short trails named after some of the typical birds lead into the Nature Park. This is probably the best place in Malaysia to see some of the region's characteristic mangrove birds: Mangrove Whistler, Mangrove Blue Flycatcher, Pied Fantail, Ashy Tailorbird should be sure "ticks" on an early morning walk and, with a bit of luck, the Mangrove Pitta. Woodpeckers are numerous — Brown-capped, Laced, Goldenback occur — and look closely to determine which Goldenback, Common or Greater, it is. Fortunate birders may also spot a Great Tit which is not a common species in this part of the world. Forest birds like Ruby-cheeked Sunbird, Thick-billed and Green Imperial Pigeon, Chestnut-bellied Malkoha are also present. In the clearings you may see a shy wood-swallow and often many koels. In general, the density and diversity of birds is impressive and, usually some of the birds may appear unafraid, almost tame.

At the artificial lake are four hides from where it is possible to look out over the cleared terrain. Many bee-eaters and kingfishers take advantage of the perches. This is a good place to view close-up raptors such as the Crested Serpent-eagle and Black-shouldered Kite as well as the more common Brahminy Kite. The White-bellied Sea-eagle will probably fly over since one pair builds its nest each year high up in the radio mast on nearby Bukit Melawati.

The scrub in front of the hides provides habitat for the Grey Heron which

Unobstructe view of Kual Selangor.

has its only breeding colony in Malaysia here. The large stork Lesser Adjutant is rarely seen now. And for large flocks of waders including rare species such as the Spoon-billed Sandpiper and Nordmann's Greenshank, the visitor must proceed six km north of Kuala Selangor to Tanjung Karang where huge numbers congregate during seasonal migration times. This area is not part of the reserve and does not enjoy any protection. However, with the habitat-improving operations being undertaken inside the Nature Park it is hoped more waders will be able to rest and feed in protected surroundings.

On the coastal side of the old bund the mangrove is doing well and is an important breeding ground for fish, crabs and prawns. As part of Phase II in the development of the park plan, a floating plankwalk will lead through here to the sea. This may provide visitors with a rare opportunity to study and to experience an otherwise inaccessible habitat.

The most important mammal to look out for in the park is the Silvered Leaf Monkey. The Kuala Selangor coastal forest has long been a well known habitat for this monkey.

Leaf monkeys are strictly vegetarians. To digest the cellulose in their diet their stomachs are even modified similar to those of ruminants. They live in the canopy of forest trees, moving about in small groups in search of leaves and fruits, jumping noisily from one tree to the next. There are only three species of leaf monkey in Peninsular Malaysia and even though they are diurnal like all monkeys they are usually difficult to see, mostly living in tall, primary forest and retreating as soon as they spot humans. The Silvered Leaf Monkeys at Kuala Selangor however have grown used to human presence and some occasionally venture out of the forest and come up onto Bukit Melawati and jump down to the ground in order to take hand-outs. They are handsome animals with their silvery fur, small crest and very long tail.

The more common monkey, the Long-tailed Macaque, an omnivorous feeder, also occurs in the park, just like the Short-tailed Mongoose. The Smooth Otter comes out into the open areas near the new lake, mostly early in the morning or in the evening, and can be viewed from the hides. The Leopard Cat, which is thought to live in Kuala Selangor, has not been seen for several years.

Kuala Selangor Nature Park is an easy-access location in which to see some good mangrove forest and wetland wildlife. More than that, it is a showcase of ecological land management, a pioneering example of co-operation between a nature group organisation, the government and the local people who participate in the running of the park through public discussions and who act as assistants and guides. Further enrichment of the habitats will take place through re-planting of mangrove vegetation and fruiting trees. Even re-introduction of wildlife is planned, with the Zoo Negara prepared to release milky storks into the area, which was previously a major breeding ground for this endangered species.

The Fiddler Crab, a common resident of the mangrove.

CAMERON HIGHLANDS AND FRASER'S HILL

The highlands of Malaysia enables visitors to get away from the tropical heat and to experience, close to the equator, a cool breeze, fog and cold drizzling rain. Ferns, moss and pitcher plants cover hill slopes while tree ferns and fish tail palms sway in the mist. Minivets twitter and brighten up the green expanse with flashes of red and yellow. The tranquility is often abruptly broken by the hooting of siamangs and the melodious, powerful warbles of the Lesser Shortwing ringing from some deep ravine.

The scenic settings and pristine forests of both Fraser's Hill and Cameron Highlands have made them popular with naturalists, especially plant enthusiasts and birdwatchers. Both hill stations, which are among the oldest in Peninsular Malaysia, are located in the state of Pahang. They are situated in the Main Range, the central spinal chain of young fold mountains, which extends from southern Thailand in the north to the state of Negeri Sembilan in the south. Fraser's Hill lies in the watershed between Pahang and Selangor and comprises a series of ridges with one small shallow valley at a mean altitude of over 1,200 metres. It begins at The Gap (820 metres), ascending to 1,310 metres at High Pines (the highest point) and descending to about 945 metres at Jeriau. Cameron Highlands, on the other hand, consists of broad shallow valleys criss-crossed by the Bertam River and its tributaries and surrounded on all sides by peaks. It consists of three settlements, namely Ringlet at 1,036 metres, Tanah Rata at 1,370 metres and Brinchang at 1,676 metres. Mount Brinchang at 2,032 metres is the highest point. In contrast to Cameron Highlands which has been extensively developed for vegetable and tea cultivation, comparatively little forest has been cleared at Fraser's Hill which retains much of its original pristine state and

Tea, a Cameron Highlands staple.

which amazes the visitor with its wealth of plant and bird life.

Climate: Although the climate is equatorial the altitude plays an important part in moderating it. Rainfall is both of the convectional and airstream boundary types. Both Fraser's Hill and Cameron Highlands receive over 3,000 mm of rain annually with the wettest period coinciding with the Northeast Monsoon from November to February. Temperatures are generally mild, around 24°C during the day and seldom dipping below 13°C at night. Mist usually blankets the forest from early evening till morning, leaving the foliage and the barks of trees damp.

Two types of montane forest: The montane forests covering both Fraser's Hill and Cameron Highlands are havens for botanists and plant lovers because they contain a number of interesting and colourful epiphytic orchids and myriad ferns. Malaysia's tallest palm, the 30 metre tall Giant Fish-tail Palm (*Caryota maxima*) is a conspicuous feature, standing majestically like a solitary sentinel with greenish yellow fruits hanging down in long beady strands.

The approach by car to gives an impression of the changing vegetation from lowland forest through lower montane forest to the upper montane forest, which starts above 1,500 metres. Trees become stunted with increasing altitude and are frequently only 10 metres tall or even less and characterised largely by slender and usually gnarled limbs. Areas with much smaller trees are frequently called elfin woodland. A stroll from Brinchang town along the metalled road to the Telecoms Station on the summit of Mount Brinchang reveals some of the characteristic and picturesque features of upper montane vegetation. Similar features may also be seen on walks to Cameron Highlands peaks, notably Mount Beremban (1,841 metres), Mount Jasar (1,696 metres) and Mount Perdah (1,576 metres).

Mammals: The mammalian fauna of montane forests consists of both truly montane species and of some which

Looking inquisitive, a Banded Leaf Monkey.

occur in lowlands. Many are shy and nocturnal in habits and are rarely seen. Large herbivores such as elephant and gaur are absent but tapirs may be seen. Wild pigs are very common and groups of up to 10 to 15 individuals, mostly sows and piglets, may be seen crossing roads or running along jungle trails.

Small animals are fairly common. The Short-tailed Mole is only known from Cameron Highlands where it frequents both forest and cultivated land from Ringlet to the summit of Gunung Brinchang. Montane squirrels such as the Mountain Red-bellied and the tiny Himalayan Striped are frequently seen. The Red-cheeked Ground Squirrel is rarer. Despite its name it also frequents trees. Because of its noctural habits, the Spotted Giant Flying Squirrel is rarely seen though it is by no means uncommon. The montane race of the Malayan Fruit Bat feeds on fruit and pollen and at Fraser's Hill and Cameron Highlands, may sometimes be seen feeding on the pollen of bananas. The Black-capped Fruit Bat and the Grey Fruit Bat are some of the other montane bats found here.

Among the large primates the siamang is fairly common. Its loud hooting call preceded by booming notes is a characteristic sound of montane forest. Strictly arboreal, siamangs move around in troops of five to seven and may be seen at both Fraser's Hill and Cameron Highlands. Dusky and Banded Leaf Monkeys are also fairly common.

The tiger also inhabits montane forest and its plug marks are sometimes seen imprinted in soft earth along jungle trails and roads. Sightings, however, are rare but visitors have on occasions seen this magnificent cat along The Gap/Fraser's Hill Road. The leopard, Clouded Leopard and Malayan Sun Bear, which are also known to inhabit montane forest, are very rarely seen. Other lesser carnivores include civets, the Yellow-throated Marten and the Malay Weasel.

Birds: Although only 73 of the 600 species of birds in Peninsular Malaysia

A Streaked Spiderhunter probes a flowering banana tree.

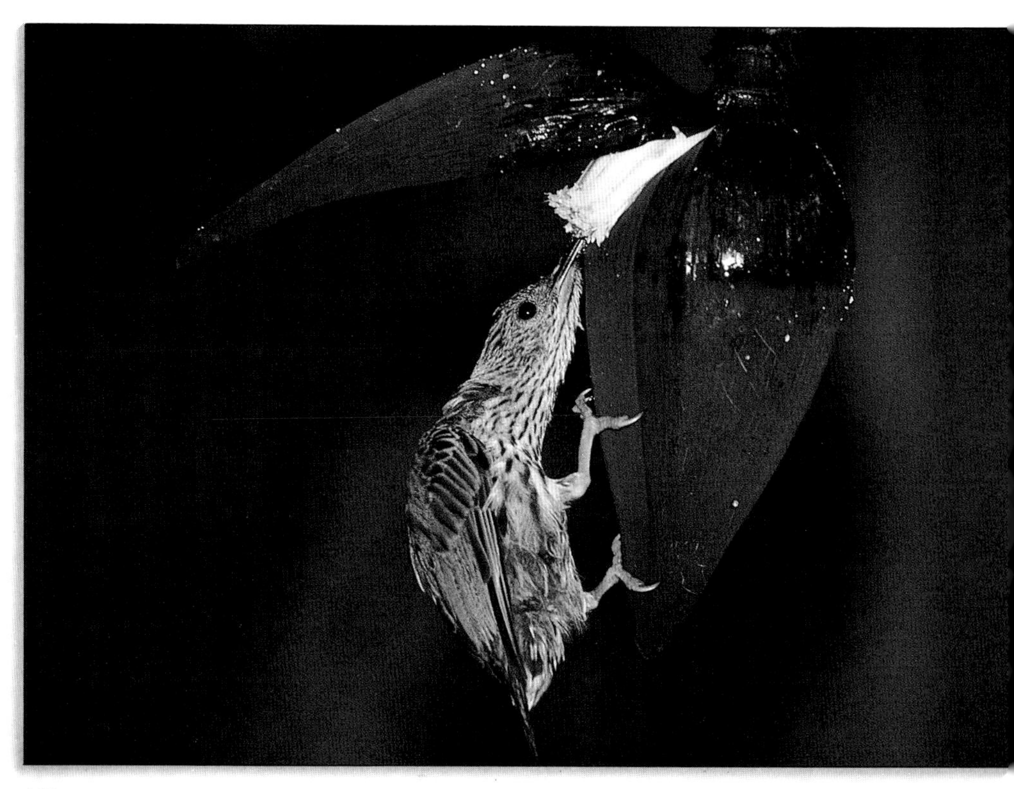

are montane residents more than 230 species have been recorded at either Fraser's Hill or Cameron Highlands. A large proportion of these are lowland species which extend into the lower elevations of lower montane forest and also autumn and spring migrants which winter in Peninsular Malaysia. Records of the latter are obtained largely from ringing and these include the Common Koel, Dollarbird and Ruddy Kingfisher. Most migrants do not winter at the hill stations but merely pass through on their way to lower elevations. Some, however, do spend winter here and among these are the fruit eating Siberian Thrush and Eye-browed Thrush and insectivores which include the Mugimaki flycatcher.

A walk along the eight-km Gap/Fraser's Hill Road will reveal a number of lowland and sub-montane resident species, especially around The Gap although they may sometimes be seen near the summit of Fraser's Hill. These include the Striped Tit-babbler, Large Wood-shrike, Greater Racket-tailed Drongo, Asian Fairy-bluebird, Blue-crowned Hanging Parrot, several bulbuls and Silver-breasted Broadbill. Hornbills are fairly common and one may see the Rhinoceros, Bushy-crested, Helmeted, Wreathed and Great Hornbills.

At 900 metres, truly montane species become more common and lowland species become increasingly rarer. Bird waves, a curious phenomena characteristic feature of montane forest, is where many species band together, and move through the forest in large numbers, stirring up insects and feeding on them. Regular participants include the Speckled Piculet, Lesser Racket-tailed Drongo, Mountain Fulvetta, Golden Babbler, White-throated Fantail, Little Pied Flycatcher and Blue Nuthatch. A number of attractive babblers are common, some of which include the Chestnut-capped, Chestnut-crowned and Black Laughing Thrushes, the brilliantly coloured Silver-eared Mesia, Long-tailed Sibia, White-browed and Black-eared Shrike-babblers and the rarer Cutia.

Large Niltava, Rufous-browed Flycatcher and White-tailed Robin may be seen along jungle trails. The Mountain Peacock Pheasant and Sumatran Hill Partridge are amongst some of the terrestrial birds found here.

In gardens, the Black-throated Sunbird and Streaked Spiderhunter feed among the cannas. The Slaty-backed Forktail may be observed along rocky streams at both Fraser's Hill and Cameron Highlands. The Malayan Whistling Thrush, endemic to the Main Range, frequents gullies in dark jungle and may sometimes be observed along the trails.

In the elfin forest above Brinchang, upper montane species may be observed and these include the Golden-throated Barbet, Malaysian Niltava, Snowy-browed Flycatcher, Chestnut-tailed Minla and Brown Bullfinch. The Brown Bullfinch also breeds at Fraser's Hill in the gardens of the High Pines bungalow where the planting of exotic conifers has created an artificial upper montane type habitat.

At the southeastern fringe of its distribution, the Himalayan Striped Squirrel.

TEMPLER PARK

Templer Park lies in the Kanching River basin, 21 km north of Kuala Lumpur along the north-south highway. The park encompasses about 12 square km and is set amidst the scenic Kanching, Serendah and Ulu Gombak Forest Reserves. Despite its small size, it offers the casual visitor a facile and relaxing retreat from the hustle and bustle of busy Kuala Lumpur as well as a chance to appreciate the beauty and harmony of nature.

The park was originally divided into three sections. The western section, comprising some 105 hectares (259 acres) is now part for the Kanching Forest Reserve and is managed by the Selangor Forest Department. A system of well-kept trails provides access through this forest and offer good views of the picturesque Kanching River, Tasik Barat and the undulating forested hills in the distance. In the northeast, there is a series of waterfalls and rock pools which can be reached by a footpath winding in a series of steps through the jungle, allowing the visitor opportunities to observe the rich and varied plant life in the rain forest.

The central section, about 252 hectares (622 acres) consists of a former tin mining area and **Anak Takun**, a limestone outcrop containing a network of dark caves. The spectacular **Bukit Takun**, though a traditional landmark of Templer Park, actually lies in the Serendah Forest Reserve. Nonetheless it has always been considered an integral part of Templer Park. This section, which is largely strewn with tin tailings, has been taken over by a private company to be developed into a recreational park which will include a golf course.

The eastern section, the vastest, comprising some 857 hectares (2,117 acres), consists largely of forested low hills which are the source of the Kanching, Gadoh and Udang rivers. Much of this area is mature secondary forest, having been exploited for timber in the

One of Malaysia's most common forest birds, the Greater Racket-tailed Drongo.

past. The former logging tracks and the new trail system constructed by The Friends of Templer Park Society allow visitors to enjoy and to appreciate scenic views of the forest. The trails lead through attractive jungle with streams, waterfalls and rocky gorges.

Climate and Flora: An equatorial type climate is experienced with rainfall all year round. Total annual rainfall is about 2,200 mm. There are two wet seasons: February to April and October to December. The former is inter-monsoonal while the later coincides with the northeast monsoon. Day temperatures are high, around 32.5°C.

The lowland forest covering most of the park is predominantly of the secondary type, much of which was logged in the past. Dipterocarps (*Dipterocarpus spp* and *Shorea spp*) together with *Campnosperma, Artocarpus, Ficus, Euguenia, Mallotus* and *Elaeocarpus* are some of the dominant tree species, especially in the eastern section. The bertam palm (*Eugeissona tristis*) is particularly abundant in the hilly areas. The *kapur* (*Dryobalanops aromatica*), a characteristic feature of the neighbouring Kanching and Serendah Forest Reserves, is also found in the park, mostly in the western section but also in parts of the eastern section. It is believed that the kapur was planted by aborigines as early as the 17th century; its crystalline camphor has been exported to Europe in the past. Along streams, Pandanus is abundant. The damp forest floor abounds in a variety of fungi, most of which are soft and fragile. These include bird's nest fungus (*Cyathus nidularis*), *Clavaria, Craterellus, Xylaria, Scleroderma* and the attractive *Trametes vesicolor*, a dark brown fungus with yellowish white outer arch bands with gills of fine pores. In the central section of the park, the tin tailings are colonized by plants such as the attractive *Spathoglottis plicata* and *Lantana aculeata* and also *lalang* (*Imperata cylindrica*) and *Saccharum arundinaceum*.

Bukit Takun, a 400-metre high limestone outcrop, is home to many unique and rare plants. Attractive flowering

. glimpse
f the
opical sky,
ırough the
ansparent
eetops.

plants growing on its slopes include *Monophyllaea horsefieldii* and *Epithema saxatile*. The latter bears clusters of minute flowers sessiled at the leaf axils. On the more rocky slopes, some short trees occur, including the *Aqlaia spp* which bear cream coloured fruit, *Atalantia roxburghiana*, a small tree bearing yellowish orange-like fruits and *Glycosinis calcicola*, an endemic and limestone-restricted plant bearing small reddish or dark purplish berries. Many plants growing on Bukit Takun are limestone restricted species. Among these is the palm *Maxburretia rupicola*, which is endemic to limestone in Selangor, having been recorded only from Bukit Takun, Anak Takun and Batu Caves. Others include the endemic Malayan boxwood (*Buxus malayana*), attractive shrubs like *Jasminum cordatum* and *Schefflera musangensis* and clusters of the pale and woolly *Boea paniculata*.

Mammals and Birds: Because it is contiguous with the Kanching, Serendah and Ulu Gombak Forest Reserves, a number of mammal and bird species are found in Templer Park. The most common mammal is the wild boar whose tracks are a common sight in the park. The tapir also occurs but sightings are extremely rare. Several species of squirrels including the Grey-bellied, Plantain and Black-banded are found. The serow inhabits the limestone cliffs of both Bukit Takun and Anak Takun. Its remarkable agility is seen as it descends steep slopes speedily and effortlessly. Other mammals seen include the Lesser Mousedeer, Malayan Porcupine, Long-tailed and Pig-tailed Macaque and civets. Tigers which inhabited the park in the past, have not been seen for a long time.

Birds are plentiful. On taking an early morning walk along the trails, particularly in the eastern section, the harsh cries of the Rhinoceros Hornbill, the monotonous "chonk", "chonk" of the Striped Tit-babbler and the distant hooting of the Gold-whiskered Barbet are often heard. Bulbuls are fairly common. The Yellow-vented Bulbul is found in

A Masked Palm Civet waits for a passing meal.

the open scrubby area in the central section while in forested areas, Black-headed, Cream-vented, Olive-winged and Red-eyed Bulbuls are more common. The White-rumped Shama with its rich melodious song is fairly common in the park. Fruit eating birds frequently congregate in fruiting fig trees and these include a number of green pigeons and barbets whose green plumage often blends well with the foliage and makes them difficult to observe. Other forest species often seen include malkohas, the Greater Coucal, a variety of Babblers, the Black-and-red Broadbill, White-bellied Woodpecker, Asian Paradise Flycatcher and Black-naped Monarch. In the more open central section, bee-eaters may be seen hawking for insects. The Yellow-bellied Prinia, Richard's Pipit, Lesser Coucal, Common and Jungle Mynahs and Scaly-breasted Munias are some of the open country birds seen in the park's central section. In the rocky habitats on Bukit Takun, the Blue Whistling Thrush and the Blue Rock Thrush may be seen.

Cave Fauna: The cave system of Anak Takun is home to a variety of cave fauna, notably bats, which depend directly or indirectly on the cave system for their survival. The most abundant bat species is the greater Roundleaf Horseshoe. Others also found roosting include the Diadem Roundleaf Horseshoe and the Common Roundleaf Horseshoe. The Glandular Frog and the Malayan Giant Toad are some of the cave dwelling amphibians. The Cave Racer occurs on the cavern walls and preys chiefly on roosting bats. The Anak Takun cave system also supports a rich variety of invertebrates, both in species and numbers. The smell of bat guano may often deter the visitor from exploring the Anak Takun caves but the variety of life forms seen together with some of nature's most artistic limestone sculptures more than compensate for the odour.

To complete the picture, Templer Park is also the home of thousands of insects and a great number of reptiles and amphibians.

**ie nimble-
awed
'evost's
quirrel.**

TAMAN NEGARA

Lying 300 km northeast of Kuala Lumpur and covered by the world's oldest tropical rain forests, is **Taman Negara**, Peninsular Malaysia's only national park and one of Asia's finest. Covering 4,343 square km, it contains the largest single expanse of protected lowland dipterocarp forest in Peninsular Malaysia and is a store house rich in both plant and animal life and also home to between 200 to 400 Negrito aborigines who live by hunting and gathering forest produce. The remarkable forest orienteering skills of the aborigines make their services invaluable to both visitors and researchers. This park has a historic past. Bronze relics dating back as far as the first century A.D. have been unearthed from the banks of the Tembeling River and it was here that the Pahang warriors, Dato Bahaman and Mat Kilau, made their way north along the "Great Warrior Trail" to Kelantan and Trengganu in the early 1890s during their struggle against the British. Early explorations of the park were carried out during the 1890s and 1900s, in the form of expeditions by British pioneers attempting to scale Mount Tahan, the highest mountain in Peninsular Malaysia.

River Journey: Taman Negara is unique because the approach to it and many journeys within the park are by river. Park boats meet visitors at Kuala Tembeling, 59 km downstream from Kuala Tahan, before ferrying them to Kuala Tahan, the park headquarters and first point of call for all visitors to the park. The river journey upstream takes about three hours (depending on the level of the river) and is a thrilling experience with numerous opportunities to see rural riverine villages and their inhabitants going about their daily chores. There are no roads within the park and so exploring the forest must be done on foot for those who do not wish to travel by boat. A good network of well defined jungle trails provides visi-

Taman Negara: unbroken canopy to the horizon.

tors the unusual experience of appreciating nature without being dependent on vehicles.

Geology and Topography: Most of Taman Negara consists of sedimentary rocks (mostly shales) from the Carboniferous to the Cretaceous-Jurassic eras. The Triassic rocks in the western section of the park are interbedded with limestone outcrops, a number of which have interesting cave systems such as Gua Peningtat which, at 730 metres, is the highest limestone outcrop in Peninsular Malaysia. Other noteworthy limestone caves are Gua Telinga in the Tahan Valley and Gua Daun Menari and Gua Besar in the Kenyam Valley. Land below 300 metres accounts for about 57 percent of the total area of the park while the remaining area consists largely of hilly terrain, varying between 300 metres and 2,100 metres. Land above 1,500 metres is entirely on the Tahan massif, a sandstone quartzite block with Gunung Tahan (2,187 metres) as the summit.

The park lies in the headwaters of three major river systems, the Relai-Aring-Lebir in the north, Terenggan in the east and Tembeling in the south. Only the Tembeling and its feeder tributaries, the Atok, Tahan, Trenggan, Kenyam, Sat and Spia are used by visitors for journeys within the park. The spectacular Tahan with its rocky bed and clear water shaded by leaning *neram* trees (*Dipterocarpus oblongifolius*) along both banks, forms a picturesque feature of the park. The Tahan and Kenyam rivers abound in fish and the latter is well patronised by anglers. Logging and clearing of forests on the upper Tembeling (outside the park) have caused heavy silting of the river in recent years, the effects being seen in the formation of sand islands in the river near Kuala Tahan.

Climate and Flora: The park experiences an equatorial type climate, characterised by rainfall throughout the year with no distinct dry season. Temperatures are generally high, up to 35°C during the day and about 20°C at night with a high relative humidity of up to 80

ew arrivals
the park.

percent. Rainfall is mainly of the convectional type with heavy thunderstorms in the late afternoon following a hot sunny morning and midday. The northeast monsoon greatly influences the precipitation, bringing heavy rainfall from November to February and causing floods in low-lying areas of the park. During this wet spell, the park is closed to visitors from (15 November–15 January) yearly.

The park's moist and humid rain forest ranges from lowland dipterocarp forest to montane and ericaceous forest at higher altitudes. The lowland forest is a highly diverse and complex ecosystem comprising tall, largely evergreen trees which include many tropical hardwood species of which *meranti* (*Shorea spp*) and *keruing* (*Dipterocarpus spp*) are especially abundant. Woody epiphytes as well as thick stemmed lianes are characteristic features. The 50-metre-tall *tualang* (*Koompassia excelsa*), Southeast Asia's tallest tree, occurs in association with the others, its buttressed roots and emergent crown being distinctive features of the lowland forest plant community in the park. This is also the home of a number of locally cultivated fruits, notably durian (*Durio spp*), mango (*Mangifera spp*), rambai (*Baccaurea spp*), rambutan (*Nephelium spp*), langsat (*Lansium spp*), cempedak (*Artocarpus spp*) and jambu (*Eugenia spp*). Common litter trapping epiphytes include bird's nest fern (*Asplenium nidus*) and stag's horn fern (*Platycerium coronarium*). The latter, with its unique shape and apple green colour, adds to the elegance of the host tree.

At higher elevations on the Tahan, Rabong and Gagau massifs, lowland dipterocarp forest is gradually replaced by montane forest, consisting largely of oak (*Fagaceae*) and native conifers (*Dacrydium spp, Podocarpus spp* and *Agathis spp*) with rattan (*Calamus spp*) and dwarf palms (*Arenga spp* and *Licuala spp*) occupying the shrub layer. Above 1,500 metres on the Tahan massif is the cloud forest where bryophytes, lichen and bracken thrive in the high humidity. The damp tree trunks and fallen timber are thickly carpeted with mounds of sphagnum moss and liveworts clothe tree trunks and fallen timber. An elegant sight to behold. At 1,700 metres, the aspiring mountaineer will find himself towering over miniaturised trees among ericaceous species such as *Rhododendron spp* and *Vaccinium spp*. This is the famous *padang* or plain of Mount Tahan. The montane and ericaceous vegetation of Mount Tahan house a number of endemic plant species. Among these are the beautiful tahan fan palm (*Livistona tahanensis*) and a native conifer (*Agathis flavescens*) which differs from damar minyak (*Agathis borneensis*) of lower elevations in having smaller, thicker leaves and much darker denser timber.

Wildlife Viewing: The dense foliage of Malaysia's rain forests does not permit easy viewing of wildlife, so seeing mammals in Taman Negara is a great test of one's patience and endurance and a bonus rather than a certainty. Salt licks (both natural and artificial) are visited by a variety of small and large herbivores and are the best places to

Tapir pause for liquid refreshment

spot mammals. Six observation hides, each holding six to eight persons, have been constructed overlooking salt licks in order to facilitate wildlife viewing. Kumbang, Yong, Tabing, Belau and Cegar Anjing Hides which are situated between six and 15 km from Kuala Tahan and which are reached by boat and then a brief walk have facilities for overnight stay and must be reserved. All, except Cegar Anjing overlook natural salt licks. Tahan Hide, only seven minutes walk from the park headquarters overlooks an artificial salt lick and grazing ground and does not have overnight facilities. Since many mammals are active at dawn and dusk, visitors should be in the hides by 3 p.m. and not leave until 9 a.m. the following morning. A powerful flashlight is essential for viewing mammals at night.

Among the large herbivores, the magnificent seladang or gaur occurs in the valleys of the major river systems in the park and has been observed from Kumbang and, on rare occasions, from Belau and Yong hides. Individuals may sometimes be seen from Tahan Hide in the late afternoon or at dusk.

The park supports a population of 160 elephants, distributed in Ulu Atok, Ulu Kenyam, Kuala Koh, Ulu Aring and in the area adjacent to the Kenyir Dam in Trengganu. Elephants have been observed most frequently at Kumbang and also Belau and Tabing. Sparse populations of the highly endangered Sumatran Rhinoceros occur but are very rarely sighted. Malayan Tapir are fairly common, especially at Kumbang and Belau, and are perhaps one of the most commonly seen and photographed large mammals.

Two species of deer, the sambar (*Cervus unicolor*) and the *Kijang* or Barking Deer are well distributed throughout the park and may be observed from most hides. The former is more often seen because of its larger size and because small groups often come close to the park headquarters at night. Wild pigs are extremely common and together with the two species of deer form an important item on the

aman egara ıvites xploration y boat.

tiger's menu. Both the Larger and the Lesser Mousedeer occur in the park; the latter is more common and frequently seen along the trails.

Primates abound. The call of the White-handed Gibbon heralds the early hours of daylight. Other primates include Long-tailed and Pig-tailed Macaques and Dusky and Banded Leaf Monkeys. Leaf monkeys feed largely on leaves and can be recognised by their long hanging tails while their bodies remain concealed in the foliage. The Common Treeshrew and the Slow Loris are two of the lower primates in the park.

A variety of squirrels is found with the Black and Common Giant Squirrels being among the largest. At dusk, the nocturnal Red Flying Squirrel may be seen gliding from tree to tree.

The tiger exists in fair numbers and may be found along all tributaries of the Tembeling. Its existence in the park is fairly secure because of abundant prey and a relatively vast protected area. Tigers have been sighted at Kumbang and Belau hides and swimming across the Tahan River. The leopard which exists mainly in the melanistic form, is much rarer with few sightings. The Clouded Leopard and the Malayan Sun Bear, though rarely seen, enjoy a wide distribution in the park. The latter leaves its calling card in the form of deep claw marks on tree trunks. Other smaller predators in the park include the Leopard Cat, Yellow-throated Marten and Civets. Otters, especially the Smooth Otter, may be seen along the banks of the Tembeling.

Avifauna: More than 250 species of resident and migratory birds have been recorded in the park. The best period for birdwatching is from September to March when migrant species such as Ashy Minivets, Arctic and Eastern-crowned Warblers, Japanese Paradise Flycatchers, Siberian Blue Robins and Eye-browed Thrushes, come to spend the winter months. Around the park headquarters, the birdwatcher may see up to 70 species especially when trees are fruiting. A variety of bulbuls and green pigeons are common.

The curious gaze of a Barred Eagle-owl.

Along river courses, the Lesser Fish Eagle may be seen gracefully in flight over the water. In winter, ospreys are often observed perched on riverside trees or plummetting into the water to fish. Kingfishers are plentiful.

The park shelters some of the world's most spectacular terrestrial birds. The Great Argus is the largest, whose loud penetrating double call is often heard day and night. The Malaysian Peacock Pheasant, Crested Fireback and Crestless Fireback are fairly common and can sometimes be seen along trails. The Mountain Peacock Pheasant has its haunts on Mount Tahan. Brilliantly coloured pittas are represented by resident Giant, Garnet and Banded Pittas and during winter, migrant Blue-winged and Hooded Pittas add colour to the forest's beauty.

The Crested Serpent Eagle and Changeable Hawk-eagle are some of the large raptors in the park. Of the six Malaysian trogons, the Red-naped, Diard's, Scarlet-rumped and Cinnamon-rumped Trogons are common.

In montane forest on Mount Tahan, Silver-eared Mesias, Chestnut-capped Laughing Thrushes, Blue-winged Minlas, White-browed and Black-eared Shrike-babblers may be seen foraging. Taman Negara is the home of two endemic montane birds, the Hill Prinia which occurs only on the *padang* and the Crested Argus which is found on Mount Tahan and also on Mount Gagau and Mount Rabong.

Reptiles and Amphibians: Monitor Lizards (*Varanus spp*), agamids and skinks can be seen around the park headquarters. Many species of snakes including the Reticulated Python (*Python reticulatus*), Common Cobra (*Naja naja*), King Cobra (*Oppiophagus hannah*), Grass-green Whip-snake (*Dryophis prasinus*) and the beautiful Paradise Tree Snake (*Chrysopelea paradisi*) are also found. Cave Racers (*Elaphe taeniura*), several species of bats and Malayan Giant Toads (*Bufo asper*) can be found in caves, especially in Gua Telinga, the cave nearest to Kuala Tahan.

he Lesser Iousedeer : about the ize of a ıbbit.

MARINE TURTLES

The east coast of Peninsular Malaysia is famous for its marine turtles which nest during the months of April through September. At the **Turtle Information Centre** in the village of **Rantau Abang** visitors can enjoy museum exhibits, watch a film about turtles, buy souvenir items, and obtain information about turtle watching. Adjacent to the village, 18 km of beach have been set aside as a sanctuary for nesting turtles. Here breeds the famous leatherback — the largest turtle in the world. Nesting females typically weigh between 250 and 550 kg; the largest leatherback on record is a 900 kg male captured in Wales. Although leatherbacks always lay their eggs in the warm sands of tropical beaches, they actually spend most of their lives foraging in temperate and polar seas, feeding exclusively on a diet of jellyfish. Leatherbacks are champion divers and can easily reach depths up to 1,000 metres.

In Malaysia, the leatherback breeds almost exclusively in the vicinity of the Rantau Abang Turtle Sanctuary, apparently because the offshore approach to that nesting beach is particularly deep and free of obstacles such as rocks and coral reefs that might injure these soft-skinned, highly pelagic animals. Three other species of sea turtle — the Green Turtle, the Hawksbill Turtle, and the Olive Ridley Turtle also nest in Malaysia and are encountered within the sanctuary. The Green Turtle, which typically reaches adult weights of 135 kg to 180 kg, is the only herbivorous sea turtle, and lives on a diet of sea grass and algae. Hawksbills are smaller (adults usually weight 35 kg to 75 kg, inhabit coral reefs, and feed largely on a type of sea sponge that most other marine animals find inedible. The smallest of the four species, the Olive Ridley, reaches a maximum size of about 36 kg to 50 kg and feeds mostly on shrimps and crabs.

Nowadays, the Green Turtle is the most common sea turtle nesting in

Leatherback hatchlings head straight for the sea.

Malaysia. It breeds most abundantly on the offshore islands of the state of Terengganu (many of these islands are now Marine National Parks), at the Turtle Islands of Sarawak, and within the Turtle Islands Park of Sabah. Green Turtles also nest on mainland beaches along the east coast of the peninsula and, on the west coast, in the states of Perak and Penang. The Hawksbill Turtle nests primarily at the Turtle Islands Park of Sabah, on beaches in the state of Melaka, and on the offshore islands of the states of Pahang and Johore. Olive Ridley Turtles nest primarily along both coasts of northern Peninsular Malaysia.

Marine turtles spend almost their entire lives in the sea. Only the adult female comes ashore and then only to lay eggs, generally under cover of darkness. After laboriously crawling up the beach to an area of dry sand, the female excavates a depression using her front flippers. Then, with her rear flippers, she carefully digs an urn-shaped egg chamber into which she deposits an average of 80 to 150 eggs (depending on the species of turtle). She covers them with sand using her rear flippers, and then camouflages the nest site by throwing on more sand with her front flippers. During the nesting season, the average female nests three or four times (the range is from one to 12 times) at intervals of about two weeks. Females rarely nest during two consecutive seasons.

The eggs incubate in the sand at a depth of 50 to 80 cm, depending on the species of turtle. The sex ratio of the hatchlings produced depends on the temperature of the sand: warmer temperatures produce more females and cooler temperatures, more males. After 50 to 60 days, the hatchlings break out of their shells and, as a group, make their way to the surface of the sand, leaving the eggshells in the bottom of the nest. The hatchlings, which usually emerge from the sand at night, scuttle towards the brightest point on the horizon. On a nesting beach, undisturbed by artificial lighting, this behaviour leads them to the sea. Sea turtles generally

Hawksbill
its
ement.

take a long time to reach maturity. In Green Turtles, 20 to 50 years pass between the time a hatchling emerges from its egg and when it returns to the nesting beach as a reproductive adult.

Threats to Survival: At Rantau Abang, during the 1950s, an estimated 2,000 female leatherbacks laid more than 10,000 egg clutches annually. Since that time, the leatherback nesting population has declined alarmingly. During the 1989 nesting season, only about 200 egg clutches were laid. This decline can be attributed to a combination of factors — most notably, the over-harvesting of eggs for human consumption during past decades and, more recently, accidental capture in fishing gear. The unruly behaviour of human visitors to the nesting beach has also discouraged successful nesting. In recent years, however, the Fisheries Department and the State Government of Terengganu have done an excellent job in protecting the remaining Leatherback Turtles and their eggs on the nesting beach. Enforcement personnel from the Fisheries Department ensure that visitors remain at least five metres from the nesting turtles, that they do not shine lights, use flash guns, or build campfires on the beach (light frightens the nesting animals and disorients the hatchlings), and that they do not make noise or play loud music on the beach. All the eggs laid by Leatherback Turtles are protected in hatcheries operated by the Fisheries Department. Unfortunately, accidental entanglement in fishing gear continues to take a heavy toll of the Leatherback population.

Accidental capture of turtles in fishing gear is a serious problem worldwide, and each year, well over 100,000 turtles die when entangled in trawl nets, drift nets and even in the lines of fish traps. In Southeast Asia, over-harvest of the animals and their eggs, along with habitat destruction caused by rapid coastal and offshore development and pollution, have also contributed to the rapid decline of marine turtle populations. Virtually every nesting beach in the region has a history of over-exploi-

A Green Turtle makes its way up the beach.

tation for turtle eggs, and at most sites, that exploitation has continued, unabated or at greater intensity than ever, into the present decade. Juvenile and adult turtles are slaughtered for their meat, leather and oil, or stuffed and mounted as wall hangings. Hawksbill Turtles are killed primarily for their shell which is usually either fashioned into curios or exported unworked to Japan. Each year Japan imports more than 20 tons of hawksbill shells — an amount that entails the slaughter of at least 30,000 hawksbills. Most turtles that supply the Japanese market come from Indonesia and the Philippines. In fact, in recent years, Indonesia has probably slaughtered more turtles and harvested more turtle eggs than any country in the world. This is because the more than 13,000 islands of Indonesia which encompass prime nesting and foraging habitat for marine turtles are also home to a burgeoning and relatively impoverished human population. Marine turtles are highly migratory, so that over-harvest in any one country threatens marine turtle populations throughout the region. In Malaysia, marine turtles enjoy relative freedom from purposeful slaughter because the majority of the population is Muslim and consider turtle meat to be *haram*. The same people, however, are avid eaters of turtle eggs. In 1989, the state government of Terengganu banned the sale of Leatherback eggs. Recently, the Malaysian Federal Fisheries Department, in conjunction with WWF Malaysia has mounted an educational campaign to discourage the consumption of other turtle eggs.

Sadly, sea turtle products — stuffed animals, items fashioned from turtle shell or turtle leather, turtle steak and turtle eggs are still sold in many parts of Southeast Asia. This despite the endangered status of marine turtles and the fact that importation of sea turtle products into most countries that have signed the CITES agreement (Convention on International Trade in Endangered Species) — including the U.S.A. and EEC countries — is now illegal.

lose-up: a awksbill is gged for cientific urposes.

TIOMAN

Pulau Tioman (Tioman Island), located about 30 km east of the town of Mersing on the coast of Peninsular Malaysia in the South China Sea, is the only large island in an archipelago of about a dozen small islands and many tiny rocky outcrops. Its entire surface area of roughly 100 square km is mountainous, with the north-south ridge rising from average heights of about 500 metres to several impressive, rocky summits, with Mount Kajang in the south, at 1,047 metres, the highest point on the island. Granite cliffs fringe most of the coast and separate numerous sandy beaches that have a beach vegetation rich in coconut palms and a backdrop of forest covered hillsides.

A merciful destiny resulted in Tioman escaping lightly from the logging activities that denuded most of Malaysia's coastline and so only a narrow belt of secondary scrub separates beach and coastal rocks from the primary forest that covers much of the island. The nature-oriented tourist can visit Tioman with the expectation of comfortable resort-style accommodation alongside beaches, coral reefs and virgin forest, a unique opportunity in Southeast Asia. However, do not expect an overwhelming diversity of wildlife.

Tioman has daily air-connections with Singapore, Kuala Lumpur and Kuantan, and regular boat and speedboat connections with Mersing and Singapore. If time permits take the slow boat which provides excellent opportunities to observe dolphins, flying fish, frigate birds and several species of tern.

The forest is difficult to reach and adventurous visitors who wish to climb a mountain should enquire about a guide. A convenient trail, however, starts in **Tekek** and leads eastward across the mountain ridge to **Juara**. While this trail is not sign posted, it is easy to identify, since its start is the only asphalted footpath that leads from the beach, a few hundred metres north of Tekek's short airstrip. For the first 20 minutes, the trail meanders through a deserted rubber plantation but soon enters virgin forest. It takes 2 to 4 hours to hike to Juara. Transport from Juara back to the west coast is by mailboat service, or by prearranged boat.

Additional opportunities for nature observation exist on the golf course south of the resort and along a trail that continues from there further to the south. In the early morning, before golfers arrive, Long-tailed Macaques and Monitor Lizards venture from the forest in search of fallen fruits. Green Imperial Pigeons can be heard all day, or be seen in large flocks in the early morning or late evening, when they migrate from sleeping roosts to feeding grounds. For the birdwatcher, the greatest attraction is the beautiful black-and-white Pied Imperial Pigeon.

At dusk, several hundred dots approach the tiny island 200 metres off the main tourist beach. Most of these are Lesser Frigatebirds but there are usually some Greater and even some Christmas Island Frigatebirds. The last species

Bumboats, the tradition[al] means of transport to Tioman.

breeds only on Christmas Island which is more than 2,000 km south of Tioman.

Tioman provides an interesting example of the biological rule that islands are always home to only a small fraction of the species of an equal sized area on a continent. During the Ice Age, when the surface of the ocean was up to 180 metres lower than today, Tioman was connected to the Malayan peninsula by land corridors that undoubtly permitted the migration of all forms of life including large mammals. After being isolated as an island, an extinction process began on Tioman that reduced — probably not even influenced by man — the fauna to its present impoverished composition. All large mammals have disappeared; the Long-tailed Macaque is the only surviving monkey: the mousedeer the only ungulate. The remaining mammal species are mostly rodents, shrews and bats but include Slow Loris, Flying Lemurs and several civets. Most forestbirds have disappeared, but some groups such as bulbuls, babblers, flowerpeckers, and sunbirds still are represented by several species. This restricted biodiversity of an ocean island like Tioman may be a natural model for the dangers that wildlife faces in today's small nature reserves, islands of nature that are surrounded by a "sea" of man-made landscape.

Tioman and its surrounding islands provide excellent opportunities for diving and snorkeling. Unfortunately, in front of the resort, the corals have been damaged. Although this situation exists all around, it is somewhat better at Pulau Tulai, northwest of Tioman, accessible by excursion boats. The rocky cliffs just south of the golf course are also good snorkeling sites, where one can expect regular encounters with Black-tipped Sharks and turtles. A comparable marine fauna can also be explored from most other islands in the archipelago, such as **Rawa**, **Babi Besar** and **Aur**, a few of which provide accommodation. However, their small size, and the absence of natural vegetation, make these islands fairly unattractive for nature explorations on land.

ore modern ansport rings isitors to ioman's ristine eaches.

ENDAU-ROMPIN

For years naturalists in Malaysia have been campaigning for the protection of **Endau-Rompin**, and now their efforts show signs of bearing fruit. The two states of Pahang and Johore, across whose border stretches a rugged forest area ringed by hills, are in the process of declaring adjacent state parks.

These protect roughly 800 square km around the headwaters of the river Endau and several smaller drainage systems. The area is one of mixed granite and volcanic intrusions with later deposits of sandstone existing as eroded remnants on the higher hills. The result is steep slopes and cliffs leading to flat-topped mesas. With acid soils, peat and few nutrients, these plateaux support patches of heath forest rare in Peninsular Malaysia. Acting like giant sponges, this peaty forest gradually releases water which forms some of the country's most spectacular waterfalls.

Since 1930 parts of Endau-Rompin have been included within the older Endau-Kluang Wildlife Sanctuary and within the Labis Forest Reserve. "Forest Reserve" means reserved for logging, and in the 1970s logging here became the focus of major publicity. At issue was the future of the Sumatran Rhinoceros for this is "home" to one of the peninsula's two main populations. Strong opinions voiced by the Department of Wildlife & National Parks, by scientists and by the public peaked in 1977. At times loggers worked day and night to remove timber in case licences would be revoked. A bone of contention has always been the lack of a legally defined boundary to the park thus enabling loggers and forestry officials to deny that logging was being done within the park; it all depended on where you thought the park was. Now, thanks to enlightened government action, some of these difficulties are being sorted out, and, at least for the present, logging seems to have stopped.

Access to Endau-Rompin is easiest by four-wheel drive vehicle, either from the small east coast town of Kuala Rompin in Pahang, or via the even smaller villages of Kajang and Kampung Peta in Johore. At least one tour agency runs frequent nature-based trekking and camping trips. The most common camping sites are near Kuala Jasin on the river Endau and on the river Kinchin. The rivers form important highways into the forest for the local Orang Hulu villagers, who fish and collect rattan, bamboo and other jungle produce. They also collect various medicinal plants for traditional cures and tonics, always re-planting a little "to give something back to the forest". Their boats are a study in themselves, and here is one of the very few places where it is still possible to see not only dug-outs with two mm of precarious draught, but also bark canoes sewn together with rattan twine.

In addition to fishing and camping, the area is important for its rare and endemic plants. As Endau-Rompin forms an isolated hilly unit round the headwaters of a major river system, and

The forest canopy is a source of food and shelter for some.

as it has distinctive geological features, it is not surprising that many plant species are unique here. Amongst the local specialities are Didy*mocarpus falcatus* and *D. craspedodromus* in the African violet family, both of which are found on rock faces and earth banks. In the same family are the silky-leaved *Loxocarpus tunkui*, also endemic, and *Didissandra kiewii* with magnificent rich purple flowers and dark crinkled leaves. In addition to these steep country specialists other endemic plants grow along the rivers such as the strongly rooted *Phyllanthus watsonii*, which is resistant to even the biggest flood. The list of plants unique to Endau-Rompin is continually growing, as more and more are discovered.

Most spectacular is the fan palm *Livistona endauensis*, a tall slender-trunked tree which grows mainly on the sandstone plateau where it can be the commonest tree, making up more than 30 percent of all trees. The tough fallen leaves, slow to decay, hinder growth of other tree seedlings, so this unique palm forest has an open understorey and is easy to walk through. Other trees, it is thought, possibly take root only where deeper soils exist in fissures through the sandstone. Short of nutrients, these trees have long meandering roots, some of which zigzag up the neighbouring palm trunks and obtain food from litter trapped in the palm's frond bases.

Where soil conditions are shallowest of all, open grassland and sedge can be found. Soil, waterlogging, and grubbing by wild pigs help to maintain this extreme form of stunted heath forest. Spectacular orchids, many of them not rare, can be seen here: *Bromheadia, Arundinaria* and *Spathoglottis*. Different plants such as the spectacular sealingwax palm *Cyrtostachys renda* can be found in different patches of open heath.

This heath forest apparently allows certain montane plants to grow at unusually low altitudes. Examples are the elfin tree *Leptospermum falvescens* and the fern *Dipteris conjugata*, living no more than 50 metres above sea level.

rafting xpedition Endau-ompin.

Leptospermum is one of several trees important as home to a range of ant-plants. The monkey's head plant *Hydnophytum* and several others grow in association with ants. In *Dischidia* species the ants live inside the flask-shaped leaves. *Hoya mitrata* is like an upside-down climbing cabbage with not only ants but also rare bats living inside. In *Hydnophytum* there is a fist-sized tuber, the monkey's head, which is filled by a labyrinth of ant-sized tunnels. The ants nesting inside may aid the plant by bringing nutrients and deterring herbivorous insects. Incidentally, they carry other plant's seeds back to their nests; some of these germinate and form ant-gardens sprouting from their new arboreal home. To see these pitcher plants, orchids and others, take a one-day return trip to **Gunung Keriong**, or a longer camping trip to the southern plateau, **Padang Temambung**.

Big game abounds, but is difficult to see. Sumatran Rhino still lives here, concentrated mainly in the western part of Endau-Rompin which is least accessible. Adventurous trekkers may find wallows or footprints but even that involves hard work. Sightings are unlikely. Along the river banks it is easier to pick up tracks of elephant, sambar deer and tiger. Tapir tracks are sometimes hard to distinguish from those of rhino, even for the experienced. Several visitors to the park have reported close encounters with tigers or leopards, but only to the extent of a mutual stare. Attacks by these beasts are very rare in Malaysia. Elephants and bears are other large mammals from which it is sensible to beat a retreat.

Amongst the smaller animals are a wide range of reptiles and amphibians, birds and mammals. One of the frogs, *Rhacophorus tunkui*, is a newly described endemic species. A crab, *Geosesarma malayanum*, is not only new but has the distinction of living inside the cups of pitcher plants, sometimes the only wet spot to be found during dry weather.

Virtually 200 bird species have been discovered, typical for rain forest in the

The Hill Myna, a forest dwell with many urban cousins.

region. Local specialities are Green Imperial Pigeon and the Giant Pitta, both of which depend on the fragments of extreme lowland forest. The Malaysian Rail-babbler seems particularly common at Endau-Rompin. Hornbills, pheasants, pigeons, woodpeckers, babblers, barbets and bulbuls all await the patient, sharp-eared birdwatcher. Look out for Lesser Fishing-eagles continually flushed from the riverside trees ahead of your boat; Pied Hornbills along the lower stretches of the river, and various spectacular kingfishers.

River trips are one of the most enjoyable activities in the park, but be prepared to get out and push. At low water sandbanks may hinder progress and there are always rapids which make some hauling on ropes necessary. Take care when swimming as well as boating; rapids and sudden floods are one possibility: undertow is another. It is easy to boat about 45 minutes from Kampung Peta to Kuala Jasin. From there it becomes more difficult. Trips up the Endau and the Kinchin rivers are best planned to last several days. Take camping gear and be prepared to get wet.

On the Johore side of the park two of the best known waterfalls are at Upeh Guling, with its curious jacuzzi-style bathtubs in the rock and at Buaya Sangkut. The latter, which means "trapped crocodile", is by far the highest and widest waterfall in the park and is spectacular when seen from the air and breathtaking when you are standing next to it. A more romantic atmosphere surrounds the Selindang waterfall on the Pahang side, where a constant fine spray fills the air with moisture and encourages luxuriant plant growth.

The biggest cliffs in the park are close to this last waterfall. Care and common sense win the day, and marked trails provide an energetic and interesting day's outing. Since the sandstone is relatively soft and since easily negotiable routes exist, these cliffs are not likely to attract the dedicated rock-climber but rather the enthusiastic scrambler.

eaving a b over a est trail.

The most interesting cliffs are approached from the Pahang side by camping on the banks of the Sungai Kinchin and tackling the western slopes of the hill Mount Keriong. This provides good views of the endemic plant specialities, including fan palm forest on the summit, and open heath forest for those prepared to use their compass.

For a good all-round experience of the different facets which Endau-Rompin has to offer, nothing has yet been found to beat the trail up **Mount Janing**. Leave by river from Kampung Peta and camp at Kuala Jasin.

From there it is a morning's strenuous uphill walk through the different kinds of forest, past umbrella palms and rock faces, to the impressive fan palm forest and to the hilltop swamp with its conifers and pitcher plants. In the distance the Upeh Guling waterfall can be glimpsed through the trees.

None of the mountains in Endau-Rompin is particularly high but all present a challenge because of their steepness. Generally, the higher one gets the tougher the going becomes, until you suddenly emerge onto the level top. In most cases a long trek leading through the forest precedes the actual climb . Well beaten trails exist up Mount Keriong and Mount Janing. More difficult is Mount Beremban which is almost at the centre of the park. Mount Besar, Pukin and Cabang Tiga are more remote and best approached from the west side of the park via Labis town. The use of guides is a sensible precaution.

As the Endau-Rompin park system becomes better established, more facilities and more possibilities should open up for outdoor enthusiasts. However, this will always be a place where Nature says "Here am I; it is up to you to discover me".

Advice on visiting Endau-Rompin can be obtained from the Malayan Nature Society, P.O. Box 10750, 50724 Kuala Lumpur (03–7912185) or from Wilderness Experience at 6B Jalan SS 21/39, Damansara Utama, 47400 Petaling Jaya, Malaysia (03–7178221).

Taking a shower — wilderness-style.

MOUNT PULAI

Scattered across southwest Malaysia are interesting patches of both primary and secondary forests which clothe the foothills and slopes of many mountains. These areas were logged, quarried and exploited in every way yet, over the years, somehow resisted the relentless spread of cultivation across the land. The greenery has reasserted itself and covered the scars made by man and machinery. Although not comparable to Taman Negara or Endau-Rompin, they are accessible by car, and continue to provide a viable habitat for the assortment of wildlife that has survived the destruction of the virgin jungle.

The forest around **Mount Pulai** (654 metres) is the most accessible. Largely unspoilt, it serves as the water-catchment for Pulai Reservoir. To visit, follow the tract at the Ladang Midlands signboard on the left of the Kuala Lumpur trunk road, three km north of Kulai, and after 14 km, turn left onto the road leading uphill to the carpark.

Linger awhile at the lovely waterfall which comes tumbling down the boulder-strewn slope in several stages under the shadows of the trees. Camping is permitted, but it is crowded on Sundays. Watch out for the Pig-tailed Macaque —so named because of its cute little tail that is usually arched over its hindquarter. A large troop with many young has been regularly seen foraging at the fringe of the jungle. The dominant male, a thick-set fellow with soft brown fur and blackish crown, prowls around the open ground, keeping guard over the group. When approached, it fronts the stranger and crouches low, baring its fangs and staring with gaping eyes. Fail to move away and , it goes into a series of aggressive stances that are simply fascinating to behold. Named "*berok*" by the Malays by virtue of its "brok" grunts, these monkeys can be trained to pluck coconuts and even botanic specimens from high up in the forest.

The two-hour walk to the summit of

e beautiful
ian Fairy-
ebird.

Pulai is an exhilarating exercise. White-handed Gibbons can be seen swinging along the high branches, their whooping calls echoing around the mountain slopes. Small groups of Dusky Leaf Monkeys and Long-tail Macaques run along tree branches and crash away into the deeper jungle as the visitor passes. The Dusky Leaf Monkey is unmistakable with its long, charcoal-coloured body and tail, and thick white rings around its eyes. Unlike the Long-tailed Macaque, it is strictly vegetarian. The Cream-coloured as well as the Black Giant Squirrel may be seen foraging in the higher branches. Keep a lookout on the tarmac too, at the grassy and shrubby verges, for the Brown Tortoise attempting to cross the road. Despite its nondescript name, it is a lovely specimen, with an orangish carapace.

The rich birdlife of the jungle here can be conveniently sampled from the road. Bulbuls and flowerpeckers of many species abound, busy flitting here and there along the roadside vegetation. The incessant "*tock-tock-tock*" rattling of the Yellow-crowned Barbet fills the air with excitement. Now and then, the mighty vomitive cough of the Rhinocerous Hornbill reverberates across the canopy. If you hear heavy wingbeats, reach for your binoculars, for hornbills are flying overhead. Usually, it is the Rhinocerous or the Wreathed, but others, such as the Black Hornbill, may also make an appearance. Further up the road, the shrill "*kweeee-kwee*" cry of the Crested Serpent Eagle can be heard, as the bird glides in circles among the clouds. When passing a stream, be alert for the darting Rufous-backed Kingfisher, a jewel of a bird, usually glimpsed flashing by rather than fully seen. Where the lianas festoon profusely around trees, peer into the shadows, and possibly catch sight of a trogon — handsome in its resplendent plumage. Scarlet-rumped and Red-naped species are good possibilities here.

Nearing the summit, the air becomes cool and nippy. Ferns of many varieties grow abundantly on the steep slopes, hanging down in lovely tongues of

Pig-tailed Macaque finds food for thought.

green. Palm trees also become a common sight. A species of large pandanus, with long dark green leaves, called "*Peropok*" by the Malays, stands out here and there along the slopes. Not far from the summit are two lookout points which provide a good view of the lowlands to the west.

Those who wish to sample a rougher but not too arduous uphill excursion, will make for **Mount Panti** (481 metres) whose attraction is swamp forest which is botanically interesting. To reach there, turn right into the junction at the Police Station of Kampung Batu Empat along the Kota Tinggi Waterfall Road, and head for **Kampung Bukit Melintang**, about three km away, where you can park your car and walk in. The climb to the summit, which has a small camping site, takes several hours. Water is not available and has to be fetched further back along the trail.

The jungle around Panti is heavily logged but enough vegetation has been left to harbour a small population of large mammals. Small herds of the Asiatic Elephant still roam the jungle, browsing on the new growth in the open patches during the misty mornings and retreating into the deeper jungle at dawn. Their fibrous droppings, smelling pungent, are scattered on the trails, but a glimpse of the elusive beast is rare. The wild pig seems to thrive, for now and then you bump into them scrambling out of your way. Tigers are very much in evidence, judging by their many pug-marks on the muddy grounds. The tapir too is around. Bird waves are common, and much rewarding time can be spent in the foothills observing gems such as the Olive-backed Woodpecker, Dusky Broadbill, Banded Kingfisher, Rail Babbler, and Bushy-crested Hornbill.

To cool down after Panti visit the **Kota Tingi Waterfall**, about 15 minutes up the road from Kampung Batu Empat. At the foot of the falls is a deep pool where bathing and swimming in the cold water can be fun. Chalets for rent are available. Try to plan your visit on a weekday, when it is not crowded.

eft, range-ellied lowerpecker oes its hing. ight, *angka*— elicious orest ruit tree.

TRIAL FOR ASS

The purpose of th
cultivars for garde
general appearanc

SSMENT OF ANN

trial is to assess se

in Singapore. Six cu

floriferousness, pest

BUKIT TIMAH NATURE RESERVE

Stamford Raffles, founder of Singapore, arrived in the region in 1819. At that time, the 620 square km island was inhabited by only a few hundred Malay fishermen. The island was covered by forest and mangrove and had an abundance of wildlife. Raffles and his British employer recognised the strategic and commercial potential of the island situated at the southeastern tip of continental Asia and wasted little time in establishing a crown colony. Today, Singapore is a successful city-state with a thriving population of 2.6 million people. Understandably, any nature tourist entering the country will wonder whether Singapore's overwhelmingly urban environment is only a great habitat for man or whether wildlife still exists.

His fears are not justified: pockets of natural forest and mangrove remain. Growing affluence has stimulated an awareness for the need for trees in the city and the country's parks make Singapore the greenest large city of Southeast Asia.

Bukit Timah Nature Reserve is in the geographic centre of Singapore, only a few kilometres from the bustling city, easily reached by taxi, once the driver has understood that the reserve is where he is supposed to go, not Bukit Timah, part of the town. After turning off from busy multi-lane Upper Bukit Timah Road, the reserve's heavy, moist and dark green quietness, overlaid by the uninterrupted buzzing of cicadas, is somewhat unreal, an isolated patch of land showing how the region would look like if man had not intervened.

The reserve includes Singapore's highest hill, **Bukit Timah** (163 metres), and protects, on 71 hectares, the nation's only virgin lowland rain forest. Towards the east, the reserve is contiguous with the protection forest of the water catchment area, 25 square km of ecologically valuable secondary forest. Housing and commercial develop-

Preceding pages: Singapore-style oasis; an experiment with nature? **Below left**, Long-tailed Parakeet. **Below**, Crimson Sunbird.

ments embrace it in other directions.

Subsequent to intensive logging of the forests of Singapore in the middle of the last century, Bukit Timah was declared a forest reserve as early as 1884 in response to research on climatic changes after deforestations. Already at that time people observed locally a danger that is feared globally today. Over the past 100 years, boundary changes have reduced the size of the reserve, and poaching of timber and animals reduced its ecological diversity. Today, most large mammals including tiger (last shot in 1924), leopard, Sambar Deer and Barking Deer are extinct in Singapore as are ecologically sensitive birds such as hornbills, trogons and broadbills. It is to be hoped that today's legal protection, the increasing age of the adjoining secondary forest and people's growing ecological consciousness will help stop the extinction process.

The reserve has a small parking lot, where a ranger sells from a little wooden outlet trail maps of the reserve. Larger entrance facilities are in construction. At the entrance, there is normally a large troop of Long-tailed Macaques demanding, rather than begging, to be fed. They pay no attention to the various notices informing that littering is an offense and happily empty garbage bins in search of food.

To explore the reserve follow the asphalted road from the parking lot to the hilltop. Many trees along this road are labelled with English and scientific names and give the newcomer to the tropics a feeling for the enormous diversity of plant species. Bukit Timah and the Singapore Botanic Gardens are two of the very few places in the region where the visitor can improve his skill in identifying at least a small fraction of the several thousand tree species found in Southeast Asian forests.

Singapore is home to a rich variety of beautiful tropical birds that occur in all parks and gardens. However, their native home is not the forest that occurred naturally in the region and consequently they are missing from Bukit Timah. The

he Chestnut-ellied Malkoha is elated to he cuckoos.

Olive-backed Sunbird is replaced by the Crimson Sunbird, the Scarlet-backed Flowerpecker by the Orange-bellied Flowerpecker, the Common Tailorbird by the Dark-necked Tailorbird, the Yellow-vented Bulbul by the Cream-vented Bulbul and the Pink-necked Pigeon by the Green-winged Pigeon. Some secretive birds of the undergrowth are best found by knowing or at least guessing the nature of their calls, such as the "*chongchongchong*" of the Striped Tit-Babbler, the fine whistling song of the Short-tailed Babbler or the scolding of the Little Spiderhunter. Also, Greater Racket-tailed Drongo and Asian Fairy-bluebird may be localized in the lower canopy through their cackling and whistling calls. Very few mammal species have survived, but the frequently heard hissing identifies members of large populations of Slender Squirrels and Plantain Squirrels. Another squirrel-like mammal with a long pointed nose is the unrelated Common Treeshrew.

The more adventurous hiker may leave the road and turn right onto the **Rock Path** which winds its way over slippery steps and rocky outcrops through a beautiful setting of large trees, which holds nests of the White-bellied Sea-eagle, and which are home to Singapore's only common forest woodpecker, the Banded Woodpecker.

Another attractive trail, leaves the main road to the left and leads to **Humpstead Hut**, one of about 10 rain shelters on the reserve. This is the best place in Singapore to find the Blue-rumped Parrot. Listen for its jingling call. Or try to spot the source of the deep cooing calls in the canopy. It is the Red-crowned Barbet, one of Singapore's two members of a beautiful family of tropical birds. And the shadow that races along a tree top branch may be a Chestnut-bellied Malkoha, a tropical cuckoo that prefers to run rather than to use its wings.

The hilltop is a perfect place to wait for White-bellied Sea-eagles or Brahminy Kites, and the besotted bird-watcher can meet the challenge of iden-

The Commo Treeshrew is distinguishe by its pointe snout.

tifying at least 12 species of swifts and swallows. The magnificent view over the protection forest of several water reservoirs makes the visitor forget that he is in one of the world's most densely populated countries.

Unfortunately, no easily identifiable trail connects Bukit Timah with that area, and, to explore it, the visitor must enter from the east, **MacRitchie Reservoir**, for example. For those who have time, it is worthwhile exploring this vast secondary forest, which holds even more animal species than Bukit Timah, in spite of its loss of plant diversity. And, because this forest is becoming older, rare resident and vagrant birds such as Thick-billed Pigeons may establish larger populations.

Like several nature reserves of the region, Singapore's forests can be considered an interesting experiment on the fate of ecological islands in the middle of a sea of man-made landscape. Is wildlife cut off or will it eventually recolonize the area once the habitat improves? Can any particular species maintain a viable population over long periods of time, or is extinction programmed due to the small size of the ecosystem?

In the evenings, the sun often disappears behind thunderstorm clouds that regularly build up over Sumatra, 100 km to the west. The chorus of cicadas changes as Brown Hawk-owls, Collased Scops Owls and Long-tailed Nightjars raise their voices and bats dive through the light of the street lamps hunting insects of the night. Just as in any rain forest, you will feel sweaty and probably be bitten by ants and mosquitos. But 20 minutes later, a taxi will have fetched you to the air-conditioned comfort of your Orchard Road hotel room. If you enjoyed the Bukit Timah experience, consider yourself fit for the national parks of Malaysia and Indonesia. However, if you found this forest excursion bothersome, this is your chance to change plans. Probably, resorts such as those on Pulau Tioman or Langkawi will provide sufficient comfort while allowing for some proximity to nature.

Plantain quirrel bites ff more than can chew.

BOTANIC GARDENS

The **Botanic Gardens** which lies just outside Orchard Road, one of the world's busiest shopping areas, is Singapore's oldest national park. Set up by the Singapore Agri-Horticultural Society in 1859, it is known worldwide as a living museum of tropical plants. Spacious and beautifully landscaped, with well-paved walkways winding around the luscious greenery, the Gardens are popular for family picnics, jogging or strolling. The cultured orchids and other tropical plants are famous, but the many varieties of birdlife are unknown to most. A walk at dawn, when it is cool, quiet and uncrowded, with eyes opened and ears attuned to the birds, is an enriching experience.

Long-tailed Macaques whose antics formerly provided amusement for visitors, have all gone, removed by the authorities when they became a problem to humans. The Gardens, however, are not devoid of animal life. The Plantain and Slender Squirrel as well as the Common Treeshrew can still be found. Reptiles such as the Green Crested and Changeable Lizard and the Common Flying Lizard are still common, while the Reticulated Python is occasionally sighted in the culverts and drains along the boundary.

The birdlife of the Gardens is interesting. Surprisingly, a wide variety of species is found in such a small area, which measures about 23 hectares (57 acres), excluding the newly-acquired land bordering the campus of the Institute of Education. Since the early decades of the 20th century, 80 species of birds have been recorded within the Gardens, of which 67 are Singapore residents. At least half of the resident species have nested regularly or, at one time or other in the Gardens, although 13 species are no longer found.

Most of the species that have gone required large tracts of undisturbed forest in order to survive. With the whittling down of the Botanic Jungle, it

The Botanic Gardens is famed for its orchid collection.

is inevitable that forest species suffered. Despite its reputation as the second patch of primary forest left in Singapore — Bukit Timah Nature Reserve is the other — the Botanic Jungle is unfortunately in a degraded state. It is becoming less and less dense as the old giants topple over. Reclusive forest birds are most vulnerable to this degradation.

Gone are the days of the Green Broadbill perching silently on a leafy branch and the Changeable Hawk-eagle nesting in one of the giants. Gone too is the liquid-like warbling of the White-rumped Shama and the Grey-breasted Spiderhunter flitting with noisy squeaks across your path. The Green-winged Dove, Olive-winged Bulbul and Short-tailed Babbler have neither been seen nor heard for a very long time. Orange-bellied Flowerpeckers, once the commonest flowerpecker in the Gardens, have been reduced to a remnant population, replaced completely in the open areas by their cousin, the Scarlet-Backed Flowerpecker.

Nevertheless, the Botanic Jungle still harbours small numbers of some forest species. The Banded Woodpecker, Striped Tit-babbler, Abbott's Babbler and the Greater Racket-tailed Drongo are still in evidence as they were decades ago. These birds have survived by moving beyond the confines of the Jungle to forage for food. Banded Woodpeckers can be seen pecking at the ants on old Tembusus (*Fragraea fragrans*) at the fringe of the jungle, while Racket-tailed Drongos rule the shadowy groves of the Albizias (*Albizzia falcataria*) along Tyersall Avenue. Abbott's and Striped Tit-Babblers venture out of the Jungle into the denser hedges and appear in the gardens of neighbouring bungalows. As in the days of Bucknill and Chasen, joint authors of the first book on Singapore birds published in 1927, the visitor can still see flocks of Long-tailed Parakeets and Pink-necked Pigeons flying in to land on fruiting trees. A couple of Hill Mynas can also be seen perched in the morning hours, as in colonial days, on the topmost points of the Jungle canopy

he "greening" f Singapore esulted many cenic spots.

and whistling away in their loud, clear voice.

After Independence, as Singapore embarked on its course towards modernisation, the Botanic Gardens was caught up in the relentless development initiated by the new government. A general campaign to clean and spruce up Singapore to fit its modern, progressive image was started. Wastelands and roadside zones were cleared of their wild native vegetation and replaced with neat grasses and instant stands of rain trees (*Samanea saman*), Angsana (*Pterocarpus indicus*) and exotic palms — all planted in an orderly pattern.

For the comfort of visitors, the edges of the main lake were cleared of reeds and shrubs and cemented all around and the swampy corner at the northern end was tamed. The sprucing campaign caught up with the corner at the northern end of the Gardens, a wild patch hedged in by Tyersall and Cluny roads. The marshy ground here was curtailed, and a second lake with a prim and proper look created. The thick hedges, consisting mainly of ferns, fishtail palms (*Caryota mitis*), wild cinnamons (*Cinnamomum iners*) and creepers such as passion fruits (*Passiflora laurifolia*) and wild water lemons (*Passiflora foetida*), growing in riotous profusion around ageing oil palms (*Elaeis guineensis*), were largely cleared. Species, such as the Yellow-vented Bulbul, that depend on these dense hedges for nesting, breeding and foraging decreased although this species still holds out as one of the more common garden birds in Singapore.

A sad and poorly understood episode in the avian life of the Botanic Gardens, as well as the entire main island, is the disappearance of the Magpie Robin. A handsome black-and-white thrush with sprightly habits and sweet melodious voice, the Magpie Robin is a popular cage-bird among the Chinese people. Several pairs of them can still be seen here, but they are the survivors of the population introduced several years ago by the authorities. In the days of the Straits Settlements, Magpie Robins

The Pink-necked Pigeon has adjusted to urban habitats.

were so ubiquitous that the British colonialist called them the "Straits Robin" in nostalgic memory of the little Robin of their homeland. As late as the 1960s, Magpie Robins could still be seen in great numbers hopping on the lawns of the Gardens as they foraged for worms and insects and retreating into the dark shadows of the hedges when people streamed in with the advancing day. The decline was dramatic for, by the late 1970s, they were practically gone from the Gardens and suburban areas. In the 1980s, they had disappeared from the main island. The cause for the demise of the Magpie Robin is still wrapped in controversy. The loss of suitable nesting and roosting sites, as a result of the clearing of wasteland vegetation and hedges, is probably a significant contributory factor, if not the decisive one.

In its present more civilised appearance, the Gardens favour species that are adapted to a parkland habitat. Most of all, the White-vented Myna proliferated — a species introduced to Singapore from Java probably in 1920. Unlike the Magpie Robin, it can nest and breed in almost any habitat. The Spotted Dove is generously represented and can now be seen ambling on every lawn and road. Richard's Pipit and the Scaly-breasted Munia appear on the grassy patches. The Black-naped Oriole, which invaded Singapore in the 1930s from Indonesia, and the Pied Triller, not present in the days of Bucknill and Chasen, are now numerous. The Dark-necked Tailorbird, so common at one time, has been replaced by the Common Tailorbird, whose vociferous chirping can now be heard in almost every Singapore garden.

The Gardens have proven attractive to some species such as the Collared Kingfisher, Brown-capped Woodpecker and the Brown-throated Sunbird that were originally more at home in a mangrove or coastal habitat. The Collared Kingfisher thrives by being more flexible in its diet — eating lizards, frogs, insects as well as fish — and has become as adaptable away from the coasts as has the White-throated King-

e Yellow-nted Bulbul found ywhere in e region.

fisher. The Brown-capped Woodpecker survives on ants crawling on trees. The Brown-throated Sunbird, like the more common Olive-backed, enjoys the nectar of the exotic plants such as canna (*Canna orientalis*), common hibiscus (*Hibiscus ross-sinensis*) and Indian coral tree (*Erythrina variegata*).

The most amazing adaptation to an increasingly humanised Gardens is that of the White-breasted Waterhen. A wetland species, this bird can be seen on the lakes, walking with long-toed feet on the leaves of the water-lily and cocking its tail up and down. It can also be seen in odd situations, trotting leisurely on neat lawns or even along roads and scampering away into the shrubs only when approached too closely. Feeding on insects, seeds and snails, it survives wherever there is some wet ground or ditch, no matter how insignificant. Its raucous calls have become as familiar to Singaporeans as the fluty whistlings of the Black-naped Oriole.

Despite the changes over the past decades, the Botanic Gardens still hold surprises for birdwatchers. The rare Red-legged Crake has been heard calling in the woods along Tyersall Avenue, where some Rufous-tailed Tailorbirds have also been spotted. The Grey-rumped Treeswift can be seen nowadays circling above the trees. A flurry of excitement swept through the local birdwatchers community in 1987 when a pair of Crested Goshawks nested in a conspicuous tree, moving in and out of the nest in clear view of visitors walking below. This is the first nesting record for the species in Singapore and the third for the whole of the Malay peninsula.

Especially in the evenings, the metallic "*tonk tonk*" notes of the Coppersmith Barbet can be heard. This is another colourful bird which, with the opening of the forest, invaded Singapore and the entire Malay peninsula only 50 years ago. While the visitor from northern countries looks on this bird and colourful kingfishers as desired exotic species, it should be remembered that they do not really represent the avifauna typical of the region's natural habitat.

Atop its flowery perch — a Brown-throated Sunbird.

Linger into the twilight hours and until complete darkness sets in. Large-tailed Nightjars will come out, their *"tock, tock"* calls issuing from almost every tree when the night is clear and fine. Some sit on the warm tarmac and when disturbed, take off in soft, silent flight. Although the Brown Hawk-owl can no longer be heard, the lilting hoots of the Collared Scops Owl is guaranteed. The Collared Scops, originally a forest species, is becoming common in suburban areas, wherever there are patches of trees. The most awe-inspiring call in the night comes from the rare Spotted Wood Owl whose call is a sequence of booming coughs, sounding like the deep barking of a giant dog. Reverberating throughout the Gardens, it sends chills down the spines of lovers in the night.

The Gardens received a big boost when a stretch of land on the other side of Cluny Road was acquired recently, extending the boundary right up to Bukit Timah Road. Unfortunately, the cleaning and sprucing habits surfaced once again. Marshy patches, where snipes and rails love to haunt, were filled up. Beautiful and intriguing old hedges that grew uncontrolled in luxuriant variety were totally wiped out, and in their place, orderly rows of several ornamental species appeared. The area now has a clinical look and is certainly dull because lack of the variety in plant and animal life that is typical of old natural hedges and marshy grounds.

The Botanic Gardens continues to provide a haven for many varieties of birds. Although many species, unable to adapt, have disappeared, some are clinging on tenaciously, while others have proliferated in the more humanised landscape. For the Gardens to continue to nourish its birdlife, it is best that as much as possible of the local vegetation be retained. Developments, especially the construction of buildings, should be minimised, not only within the Gardens; but also at periphery. Only then can a rich and multifaceted appreciation of nature be wholly sustained.

A pair of Black-naped Orioles.

SUNGEI BULOH BIRD SANCTUARY

Beyond the Bright Lights: Most visitors to Singapore, famous around the world for its clean roads, bright lights, elegant skyscrapers, shopping centres and efficient airport, are invariably trapped within the cosmopolitan city, leaving after a sojourn of a few days or weeks with the impression that they have seen all there is to this small island republic. There is, however, more to Singapore than meets the eye. Nature lovers and adventurous types will be pleased to know that a sort of hinterland exists in this small country — untrodden even by locals and waiting to be explored. This so-called hinterland is a stretch of rural and semi-rural landscapes just to the north of the city, beyond the Pan-Island Expressway. Most of these areas are ideal for visitors who need to get away from the hustle and bustle of city living for a half-day or one-day immersion in a quiet, soothing, green environment. Here is a world of mangroves, orchards, coconut and rubber plantations, fish ponds and farmlands — all were integral features of the old Singapore. For those keen on birdwatching, some areas provide ample satisfaction in search of rarities that would not be as easy to come by, even north of the Johore Causeway.

One for the Birds: The **Sungei Buloh Bird Sanctuary**, consisting of approximately 85 hectares (210 acres) of orchards, ponds and mangroves, is on the northwest coast of Singapore, flanked on the west by Lim Chu Kang Road and on the east by Sungei Buloh. The area was proposed as a bird sanctuary by the Malayan Nature Society in 1987 and was immediately agreed upon by the government.

The sanctuary, which consists of a variety of habitats within a small area, is remarkable for the abundance and diversity of its birdlife. It is estimated that more than 126 species of birds — resident, visiting or transient — can be seen. Woodland, mangrove and shore

Birders keeping vigil at Sungei Buloh.

birds abound in their respective niches.

A visitor centre, a carpark and a bridge across the Sungei Buloh are planned for the near future and will provide easy access into the sanctuary from the Kranji Dam, where there is a picnic area. Meanwhile, visitors are advised to enter the area along the track opposite the bus terminus canteen at the end of Lim Chu Kang Road.

Entering from the west, the visitor first encounters a series of small freshwater ponds, green with algae and surrounded by shrubs and mangroves. These ponds were formerly used for breeding aquarium fish. Some of the fish still thrive, and with the owners gone, birds have a heyday. Collared as well as White-throated Kingfishers sit on the low stakes by the bunds, waiting to swoop. Cinnamon and Yellow Bitterns stalk along the reedy edges, bursting into the air in alarm as visitors pass. During the migrating season, White-winged Terns in their white plumage, gather above, swooping down now and then to skim over the placid surface of the water for insects. The patches of woods and clumps of fruit trees are alive with resident birds — Yellow-vented and Olive-winged Bulbuls, Black-naped Orioles, Common Goldenbacks, Pied Fantails, Spotted Doves and Pink-necked Pigeons.

To enter the main part of the sanctuary, cross the ponds along the northern end of the bund and cut through the wall of mangroves on a wooden bridge crossing a charming little river. Beyond the bridge is a series of large ponds fringed by shrubs and tall grasses which provide excellent cover for close observation of wading birds feeding on the exposed mud-beds.

It is these migrating waders, of which more than 20 species have been sighted, which are the stars of Sungei Buloh Bird Sanctuary. From as early as September to as late as April, large flocks of plovers, sandpipers, stints, curlews, godwits, and egrets gather to feed on the mud exposed by the ebbing tide. Whimbrels, Curlew Sandpipers, Mongolian and Lesser Golden Plovers dominate

Many species of shorebirds arrive from Siberia to spend the winter months in Singapore.

the scene. Asiatic Dowitchers, a very rare wading species, have been regularly spotted. Great and Little Egrets grace the scene as they wade in the shallows or fly gently across the ponds to roost in the mangroves. Local herons, such as the Grey, Purple and Little, are also conspicuous. With the rapid reclamation of coastal areas in other parts of the Republic, the sanctuary now stands out as Singapore's last significantly sizable feeding ground for migrating wading birds. The ornithological importance of the site is underlined by the fact that Singapore is the last stop over in the migration path down the Malay peninsula before the thrust to regions further south.

The best time to watch the waders is when the tide moves into the ponds. During low tide, most of the waders feed at the coastal mudflats, especially at the estuary of Sungei Mandai, just east of Kranji Dam. When the incoming tide sweeps over these, the waders retreat inland and swell the crowds at those ponds where the mud is still exposed. When these mudbeds are in turn flooded, the waders retreat to the islets within the ponds and to the bunds criss-crossing the area, where they can be seen as a dense, agitated community. Their shrill cries as they circle in spectacular flocks searching for roosting sites make this an exciting place for birdwatchers and photographers. Landing in unison on any available piece of exposed mud, they stand shoulder to shoulder like sardines, jittery and twittering. The slightest disturbance causes an eruption like exploded confetti as they burst into the sky and circle the ponds in nervous agitation.

Those wishing a diversion from watching-wading birds will enjoy exploring the mangroves fringing the irregular coastline to the north. Large Monitor Lizards prowl on the mud and scramble and splash into the seawater when disturbed. When the ponds were in operation, the owners showed these creatures no mercy, killing and even eating them as a local delicacy, believing that they were the predators of the

A Monitor patrols its mangrove habitat.

fish and prawns in their ponds. Nowadays, with the owners gone, they seem to be slower in taking flight.

Large-billed Crows gather in large flocks in the trees. Dollarbirds sit, quiet and motionless, on the more prominent points. Watch out for mangrove birds, such as the Mangrove Whistler, Brown-capped Woodpecker, Ashy Tailorbird, Pied Triller, Common Iora, Laced Woodpecker, and Copper-throated Sunbird. The last named, a rare species in Singapore, are readily spotted as they flit about feeding on the nectar of mangrove flowers.

On the eastern side of Sungei Buloh, mangroves extend all the way along the coast to the Kranji Dam. This stretch of mangroves has scarcely been studied or explored and may yield interesting sightings to those who are venturesome. The mangroves here are dense, especially on the landward side, where they are entangled with an encroaching belt of grand, old sea-hibiscus, providing attractive roosting sites for mangrove birds. A plan exists to build a wooden walkway, near the projected visitor centre, across the mangroves to the sea. Unfortunately, this stretch of mangroves, from Sungei Buloh to Kranji Dam, is out of the boundary of the sanctuary. It is, however, worth preserving because this type of habitat has already become a rarity in Singapore, and is of vital importance in maintaining the diversity of mangrove species within the sanctuary.

Singapore ·rity: the ·eater ·inted- ·ipe.

The Sungei Buloh Sanctuary is a success story for the Malayan Nature Society and nature lovers in Singapore. It is significant in that it is the first sizable piece of land to be established as a nature sanctuary in this economically booming nation since the departure of the British colonial government — a sanctuary resulting from the co-operation of a public organisation and the state.

Kranji Bund: Kranji Reservoir lies to the southeast of the Sungei Buloh Bird Sanctuary. The reservoir was created in 1975 by damming the Kranji River at the mouth and is now managed by the Public Utilities Board (PUB). The adjacent farmlands are intact and traditional vegetable farming is still practised. Scattered around the various coves and inlets of the reservoir are some of the most extensive freshwater marshes in Singapore — a relatively rare type of habitat in the Malay peninsula.

Freshwater life flourishes in these wetlands. A remnant population of Estuarine Crocodiles, much-maligned and persecuted by the local people even to this day, still holds out in the less accessible parts of the reservoir. The elusive reptiles hide in the marshes during the day and come out at night to prey on fish in the deeper waters. Monitor Lizards are abundant. Tomans and Aruans — two species of Snakeheads — lurk in the open water of the hyacinth beds.

The most accessible marsh is located near the PUB water-refining plant at the end of Neo Tiew Lane. Park your vehicle in the open space near the plant's gate. A good time to visit the area is the few hours just after dawn and in the evening before darkness sets in. White-breasted Waterhens are abundant along

Neo Tiew Lane and the visitor has plenty of opportunities to watch these dainty-looking birds up close as they trot a short distance ahead before scampering into the grassy edges. From the shrubs and fruit trees of the adjacent farms come the vociferous chirpings of the Common Tailorbirds. The nests of the Baya Weavers — curvaceous, bulbous structures made of reeds — can be seen pendulating from the fronds of the coconut trees near the water-refining plant.

Walk along the bund which runs towards the north. It was created by the PUB in order to form the canal on the left which channels the waters of the reservoir to the refining plant. Water hyacinths proliferate in the canal and require to be cleared regularly. Beyond the canal is a stretch of reeds and grasses, where large flocks of Baya Weavers scramble and hop over one another as they forage for seeds. To the right is the open water of the reservoir where Little Terns fly around, now and then hovering with fluttering wings before diving for fish. During the migrating season, the reservoir teems with White-winged Terns cruising over the water while Brahminy Kites circle overhead, and an Osprey may flap by low over the water.

Move along the bund and, Cinnamon and Yellow Bitterns burst out from the reedy edges of the canal and fly away to land further up. In the migrating season, Great and Black-browed Reed Warblers lurk nervously in the thick clumps of grasses, scolding away in squeaky voices. Richard's Pippits run along the dusty track, and are accompanied, during the migratory season, by Yellow Wagtails in their dowdy winter plumage. Zitting Cistcolas and Scaly-breasted as well as Chestnut Munias scatter from the grasses on the bund when disturbed.

Look out for the White-bellied Sea Eagles and Ospreys that regularly use the crossbars of the towers of the British Broadcasting Corporation on which to tear up and gobble up their catch. During the migrating season, a few Greater

A Brahminy Kite swoops in for the kil

Spotted Eagles can be seen circling high in the sky above this area.

Immediately beyond the BBC towers, to the left and right of the bund, is an impressive stretch of marshland crowded with reeds, simpoh shrubs and ferns and with water hyacinths carpeting the water. Here, water birds abound. Purple, Grey and Little Herons stand conspicuously at the edges of the reedy clumps or by the rocky slopes of the bund, waiting patiently for a fish to spear. Eye-browed and Ruddy-breasted Crakes stalk on the hyacinth beds. Common Moorhens swim in the open pools and gangling Purple Swamphens, with reddish frontal shields, bills and legs, are obvious as they forage in their slow clumsy manner on the denser masses of hyacinths.

A stroll to the end of the bund may be rewarding because of the Grey and Purple Herons which gather regularly here in large numbers. They are not shy and can be closely approached. At dawn, poachers busy with their rods or carrying their catches — some of which are as thick as a man's thigh — may be encountered.

It has to be hoped for that the marshes around the Kranji Bund will be preserved, for they constitute an extensive, easily accessible quiet corner of Singapore with an abundance of life-forms that have adapted to a freshwater wetland habitat. A morning or evening spent here is a refreshing experience. The open skies and the absence of tall buildings impart a sense of space and just for a while, it's difficult to image that this is really Singapore.

Pulau Ubin: Away from the main island there is one other notable area for nature lovers. The island of **Pulau Ubin** lies at the eastern end of the Straits of Johore, just off the northeast coast of Singapore. A 15-minute boat ride from Changi Jetty is all it takes to reach it. Left behind by rapid developments on the main island, Pulau Ubin has become the last stronghold of old Singapore. Despite scars caused by heavy quarrying in some parts of the island, its pastoral charm, with its tapestry of sandy winding roads, fish and prawn ponds, durian, coconut and rubber plantations, *kampung* houses, mangroves and secondary forests, remains intact.

Stroll or cycle along the narrow roads and listen to the resonant, bubbly calls of Straw-headed Bulbuls erupting here and there from the mangrove patches — a sound that has become the hallmark of the island. Common Goldenbacks and Laced Woodpeckers, species becoming scarce on the main island, thrive on Pulau Ubin, foraging for insects in the mangroves as well as in the woods and coconut plantations. Here, Magpie Robins, once a common feature of the suburban and rural landscapes of the main island but now almost extinct there, seem to hold their own as well as Singapore's only population of the Red Junglefowl. Copper-throated Sunbirds and Stork-billed Kingfishers, both scarce on the main island, can easily be seen in the mangroves. Because of its rural charm and its interesting birds Pulau Ubin is an ideal spot for outdoor recreation and nature appreciation in urbanised Singapore.

A Yellow Bittern.

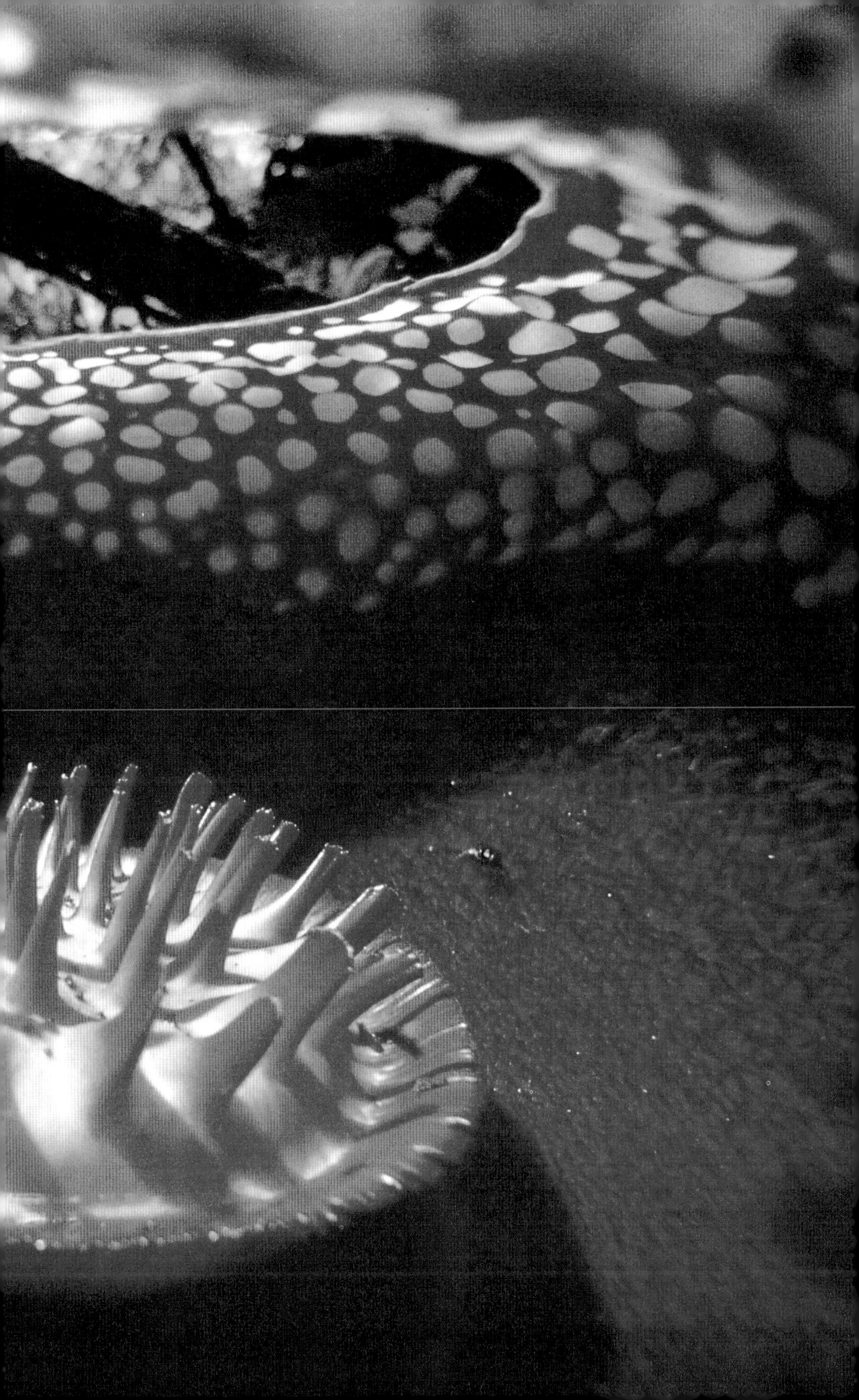

Mount Kinabalu National Park

Occupying the northern corner of Borneo, the Malaysian state of Sabah contains some of the finest and most accessible remaining examples of Southeast Asian rain forest. Lying between 4° and 7° north of the equator, Sabah is also known as "The Land Below the Wind", because it is not directly affected by the typhoons that lash the Philippine islands to the north.

Mount Kinabalu National Park in Sabah is well-known for having the highest mountain between Burma and Irian Jaya within its boundaries. At 4,101 metres Mount Kinabalu is a spectacular landmark that draws visitors from all over the world, and which lays claim to extremely rich plant and animal communities.

The richness of the mountain's plant and animal life was recognised as early as 1851 when Sir Hugh Low, then the Colonial Secretary in Labuan, became the first person to climb the towering peaks and to send specimens of its magnificent plants, many of which now bear his name, back to Europe. Low's barometer failed on the climb, however, and although he made the ascent a second time with his friend Spenser St. John, the highest summit still eluded him. It was left to John Whitehead, the first zoologist to visit the mountain, to reach the true summit, now called Low's Peak, in 1888. Whitehead spent two years on and around Mount Kinabalu and many of its more spectacular animals and birds were named after him. Other explorers and naturalists followed, though by 1951, 100 years after Low's first visit, only 53 others had been recorded. Today the park receives over 170,000 visitors a year, nearly 10 percent of whom climb the mountain.

receding ages: fly's-eye iew inside *afflesia*; at he treeline n Mount inabalu. **eft**, outheast sia's highest eak. **Right**, vell-wrapped valkers.

Geologically, the mountain is very young. Its formation started when a large mass of liquid rock was squeezed upwards by earth movements several million years ago. As it rose it solidified as a granite intrusion into the softer surrounding sandstones and shales and as these weathered down, the hard granite of Kinabalu stood out above. During the Ice Age we know that the Summit Plateau was glaciated as the marks of ice-flows and frost-fractures can still be seen, having created the fantastic sculptures and peaks that are so spectacular. Today the summit is divided into two arms separated by the vast depths of Low's Gully, a chasm that almost splits the mountain in two. It is the western plateau that most people visit, for this is a relatively easy two-day walk. The eastern plateau, with steel ladders to help the non-mountaineers, is now more accessible than in the past but those wishing to ascend it still require a good head for heights.

The most popular activity of Kinabalu visitors is, of course, to climb to the summit, a trip which takes two days, with one night spent at 3,300 metres in the well-equipped modern climbers resthouse at **Laban Rata**. Making the climb is also one of the best ways to see the variety of plants and animals that inhabit the different regions.

The plant life on Mount Kinabalu falls into four different zones. The lowland forest around the **Poring Hot Springs** on the park's eastern boundary is rich in Sabah's main timber trees — the dipterocarps. Wild fruit trees such as relatives of the durian, mangosteen and langsat are not uncommon and the wild Borneo cinnamon is also found here.

The forest canopy is high, little light reaches the ground and what undergrowth there is, is sparse and easy to walk through. The largest flower in the world, the Rafflesia grows here as a leafless parasite on wild forest vines.

Above about 1,200 metres the montane oak-chestnut-conifer forest starts. There are over 40 species of the oak family (*Fagaceae*) recorded from Mount Kinabalu including the trig-oak, (*Trigonobalanus verticellatus*), a living fossil link between the oaks, chestnuts and beaches. It is quite common as a coppicing tree along the trails, with leaves in whorls of three.

Those who are lucky may even see the pink-flowered *Balanophora*, a root-parasite on the forest trees. Rhododendrons and orchids are also common, often as epiphytes hanging over streams. The tree canopy is markedly lower than in the lowland forest letting more light through to allow a tangled shrubby undergrowth to develop with palms and pandans. Along ridge tops where nutrients have been leached out of the soils, pitcher-plants are not uncommon and the mossy forest starts to develop.

Park Headquarters lies at an altitude of 1,500 metres on the southern boundary in this zone and this is where most people start the summit climb. Walking around the labelled collection of living plants cultivated in the Mountain Garden is one of the best introductions to the mountain flora in this zone. Guided nature trail walks are also available at regular times each day.

Around 1,800 metres a thick mossy forest begins to develop, especially along streams and in sheltered gullies. The summit trail mainly follows the ridge-top so it is not until climbers reach

The Ferret-badger is endemic to Kinabalu.

about 2,400 metres where the trail contours round a steep ridge that they will see much mossy forest, where tree-ferns are abundant, as are climbing bamboos, and mosses lie thick on the branches. This forest continues, the species changing with increasing altitude, until about 3,300 metres when the tree-line is reached. However, on the summit trail at about 2,700 metres a zone of ultramafic soil causes a sudden change in the vegetation which becomes much more open. This is one of the most interesting parts of the trail. This is also the zone of the spectacular pitcher-plant *Nepenthes villosa*, another species that is largely restricted to ultramafic soils.

Higher up, around 3,000 metres as the trees become even more stunted, spectacular views of the mountain can be had in clear weather. Haviland's Oak, with rusty young leaves becomes common as a stunted shrub as does the trunkless Haviland's Tree-fern.

On other parts of the mountain the forest merges gradually into high level mossy forest dominated by the box-leaved rhododendron (*Rhododendron buxifolium*). On the summit trail this change occurs as soon as the edge of the ultramafic soil zone is reached at about 3,200 metres. This is near the **Paka Cave** where Low and Whitehead slept. The detour from the main trail that goes past it is often more interesting for botanists and birdwatchers and takes only a little longer. For today's climbers the first overnight stop lies at 3,300 metres at Laban Rata at the edge of the tree line. The tree line is determined, not by snow, but by lack of soil. The hard granite weathers very slowly and there is virtually no soil on the summit slabs except in small sandy depressions where stunted grasses grow and in some sheltered gullies where a few shrubs survive.

Few animals are seen in the summit region of the park but the animal life is of as much interest as the plant life even though it may be more difficult to see.

Nearly all of Borneo's mammals are recorded from the park area. Most of

og, tree rns and ugs are all art of the xperience.

these are from lowland forest in the north eastern parts of the park. At Poring Hot Springs, the visitor is most likely to see animals. Here, a series of hot, mineral rich springs have been channeled into open air bath tubs and a tree canopy walkway has been built nearby. Visitor accomodation is also available. The one main trail is about two-and-a-half hours walk to the beautiful **Langanan Waterfall**, and it is here, on the less frequented parts of the trail, near the waterfall that one is most likely to see wildlife.

Orang-utan and Honey Bear were seen in the past, though it is rare to sight them now. Unfortunately, villages border nearly all the eastern part of the park and hunting is prevalent along the boundary. Red Leaf-monkeys and Barking Deer are still seen fairly regularly and occasionally a Flying Lemur has been sighted near the hot spring pools at dusk. A pile of large tumbled boulders, in which small colonies of bats live, lies a half-hour walk along the Langanan Waterfall trail. Other smaller mammals, which are easier to see near the hot pools, are squirrels and tree-shrews. Prevost's Squirrel with orange and black fur is particularly attractive.

Over 280 species of birds are recorded from the park area, a number which is further enriched by visiting migrants every year. The birds at Poring are similar to those of the lowland forest elsewhere in Sabah but at the park headquarters they change completely. This is probably the best place in Borneo to see montane species and up to 20 of the commonest ones can be encountered by walking along the headquarters complex of roads in the early morning.

Several species are endemic to Borneo, though not to Kinabalu, as the majority have also been recorded on Sabah's next two highest mountains — Trus Madi to the southeast and Tamboyukon to the north, both just over 2,400 metres, as well as the highlands further south in Sarawak and Kalimantan. The Red-breasted Tree Partridge and the Crimson-headed Wood Partridge are both more commonly heard

The Chestnu capped Laughing-thrush.

than seen. The Black-breasted Triller, another uncommon Borneo endemic, occasionally arrives in a mixed flock which hunts around the Park Headquarters for a few days before moving on. Whitehead's Trogon, Whitehead's Broadbill and Whitehead's Spiderhunter are some of the more striking endemic birds seen from time to time. Perhaps one of the most interesting of all the birds is the rare Kinabalu Frogmouth, which has been seen only once, and which, on the basis of its call, may be an undescribed species.

Most insect-eating birds find a rich source of food at the lights along the headquarters complex of roads. Insects are numerous, especially moths and beetles on dark nights after rain. In the early mornings flycatchers, drongos, laughing-thrushes and bulbuls are easily seen while warblers and tailorbirds churr from the bushes.

Birds are not the only hunters around these lights. Mountain Tree shrews and squirrels search here for food and are also often caught rummaging in the rubbish bins by the visitor chalets. Tree-shrews are easy to tell by their long pointed noses and long, rather thin, scruffy tails. The Bornean Mountain Ground Squirrel has a similar long nose but a shorter much bushier tail. The commonest squirrel is Jentinck's with a white eye-ring and a white patch behind the ears. Another distinctive species is Whitehead's Pygmy Squirrel identified by its small size and long ear tufts. Occasionally, the rare Kinabalu Squirrel may be seen with a rich russet underneath and the fortunate ones may catch a glimpse of the beautiful Giant Squirrel, the size of a cat. There are two truly montane mammals in the park which do not occur in the lowlands. These are the Ferret Badger and the Kinabalu Rat.

he rare oon-moth an be found the forest f Kinabalu.

Higher up the mountain there are fewer species and the composition changes again. Large blue earthworms are often seen on the trail above 2,000 metres. They are the food for one of the most unusual animals of all — the Giant Red Kinabalu Leech. Leeches are not common on the mountain, but this remarkable creature is occasionally seen between 2,000 metres and 3,000 metres in very wet weather. It can reach over 30 cm in length and its only source of food are the large earthworms.

Also characteristic of the upper zones are the little Green Mountain Blackeye and the handsome Red-chested Mountain Blackbird, though both of which have also been seen lower down in times of drought. The Pale-faced Bulbul, common at park headquarters, extends all the way up to 3,300 metres. Little warblers often hop about curiously in shrubs near the climbers. However, the Kinabalu Friendly Warbler which was famous for hopping at the feet of early climbers is now not so friendly and has retreated to less disturbed areas of the mountain. Another high level Borneo bird which is seen more often at this altitude than lower down is the slaty-blue White-browed Shortwing. On the summit plateau, animal life, like plant life, becomes even more sparse until, on the highest slopes, nothing moves but the wind or an occasional swift.

KINABALU'S PLANT KINGDOM

Power Flower: Most botanists will agree that *Rafflesia* is one of the greatest wonders in the botanical world. *Rafflesia* is a member of the cosmopolitan parasitic family *Rafflesiaceae* which consists of about eight genera found in the old world tropics.

The genus *Rafflesia* consists of 14 species, all of which are confined to Southeast Asia. The Sumatran and Bornean species, *R. arnoldii* is the most famous of all as in addition to holding the world record for the largest flower (over a metre in diameter) this species started the scientific tale of *Rafflesia*. The story begins with the discovery of *R. arnoldii* in Sumatra on 20 May 1818 by Sir Stamford Raffles and Dr. Joseph Arnold after whom the plant is named.

Rafflesia lives as a parasite, devoid of leaf, stem or root, specifically on *Tetrastigma*, a wild relative of the grape vine. *Rafflesia's* life cycle begins as a tiny thumb-sized bud which emerges from the root or stem of its host plant. The bud takes nine to 18 months to develop into its mature stage which resembles a medium-sized cabbage in appearance and size. The dramatic blooming process usually occurs on a rainy night and normally takes the whole night, with the critical moment of "popping-up" being around midnight. A blooming *Rafflesia*, displaying its breathtaking size and beauty, is an awesome sight seen only by a privileged few.

Unfortunately, the *Rafflesia* bloom does not last long. Its begins to show signs of deterioration two to three days after opening when the flower starts to emit a very faint smell of rotting meat. This attracts its pollinators, carrion flies of the genera *Lucilia* and *Chrysomya* which carry pollen from a male flower to a female. It is therefore essential for two different-sexed *Rafflesia* in the same vicinity to bloom simultaneously in order for pollination to occur. It is a marvel that pollination occurs wtih the rarity of even one bloom, let alone two, from different sexes. If successfully pollinated, the female flower produces thousands of minute seeds, embedded in a fleshy fruit body, which mature after six months. The sweet-acidic fruits, tasting and smelling like rotten coconut, are feasted on by squirrels and other rodents. The minute seeds that stick to these visitors are dispersed to another host liana. The seeds may undergo a dormant phase of up to 18 months or longer, embedded in the root or stem of the host plant, before emerging as a bud and starting the whole cycle again.

Because of its rarity and vulnerability, the *Rafflesia* is threatened with extinction. At least two species have not been observed for 50 years and may already be extinct. Active conservation measures are vigorously undertaken by local authorities, especially in Sabah, where these unique plants stand the best chance of survival. *Rafflesia* locating and monitoring projects and *Rafflesia* conservation campaigns are vigorously conducted.

eft, *afflesia* in loom. **Right**, *lickingeria omata*, an rchid of the owland orest.

Three localities have been declared as *Rafflesia* sanctuaries, in addition to a similar sanctuary in Sumatra and another in Sarawak. *Rafflesia* in bloom can sometimes be seen in Mount Kinabalu National Park.

Wild orchids: Renewed botanical studies of the wild orchids of Borneo, and Sabah in particular, have now increased the estimate of Bornean species from 2,000 to nearly 3,000 species, with at least 1,500 to 2,000 of these growing in Sabah. This estimate represents about 10 percent of the world's total of 25,000 to 30,000 species.

This is largely due to the great variety of habitats which range from coastal swamp and peat swamp forests to majestic lowland dipterocarp forests, intermingled with riverine and podsolic heath forests, the latter being very rich in orchid species. Limestone and serpentine hills and ranges of sandstone mountains rise steeply into the interior, with the mighty granitic massif of Mount Kinabalu overlooking the whole of Sabah.

With so many ecological niches it is not surprising that a very diverse orchid flora, containing members of all the six sub-families of the orchid family, flourishes in Sabah. Mount Kinabalu has about 1,200 species of all these sub-families and is a treasure-trove of evolutionary diversity.

It is no wonder that increasing numbers of orchid enthusiasts now visit this state to marvel at these jewels of the plant world. To make this easier for them the State Government, in addition to being concerned with the conservation of this rich orchid flora, threatened by logging and deforestation and other developments, has set up six orchid centres that are conveniently sited along the most popular tour routes. These centres are concerned with preserving the more endangered, rare and horticultural species, but the collections also form a basis for taxonomic studies and for the enumeration of species.

Visitors to the Kinabalu Park can see many of this mountain's species, at 1,500 metres in a Mountain Garden at

Below left to right, some of Sabah's varied and beautiful orchids.

the park headquarters and also in a landscaped orchid enclosure at Poring.

Of particular interest in these collections are several rare and beautiful slipper orchids including the most aristocratic of them all, *Paphiopedilum rothschildianum*. The rat-tailed *Paraphalaenopsis labukensis* and the scarlet flowered *Renanthera bella* are other species found here. Many of the beautiful necklace orchids of the genus *Coelogyne* are exhibited and are also often seen on the trails or when climbing the mountain. The dainty *Dendrochilum* orchid is represented by more than 30 species. Among the diverse terrestrial orchids, of note are the exquisite helmet orchids with *Corybas pictus* nestling in moss on rocks and banks. This can be seen in the Mountain Garden together with two beautiful jewel orchids, *Macodes lowii* and *Anoectochilus longicalcaratus*.

Another tour route is to follow the unique and scenic train journey to Tenom in the "Land of the Muruts", which winds along the Padas Gorge. The **Tenom Orchid Centre**, 15 km from Tenom, contains more than 450 species in 103 genera including all the known Bornean species of *Vanda*, *Renanthera* and *Phalaenopsis*. Amongst the Bornean specials in this garden is the litter-trapping giant-leaved *Bulbophyllum beccarii*, and the "Elephant Ears" orchid *Phalaenopsis gigantea*. The dainty "moon" orchid *Phalaenopsis amabilis* is locally abundant.

From Tenom the visitor returns to Kota Kinabalu by road through the scenic Tambunan Valley with its terraced rice fields and then over the Crocker Range with wonderful views of hill and montane forest. In the **Sinsuron Mountain Garden**, at 1,700 metres, mountain orchids, pitcher-plants and rhododendrons are being established in their natural habitat — moss forest with oaks and conifers. Trails, picnic shelters, an information room and other facilities enable a pleasant stopover in this cool mountain forest.

Visitors to the Sepilok Orang-Utan Sanctuary near Sandakan on the east

coast of Sabah can visit another lowland orchid garden at the Sepilok Forest Research Centre. Over 200 species are established nearby at several commercial orchid nurseries. A much smaller collection of lowland orchid species is being established at the Danum Valley Field Centre near Lahad Datu, also on the east coast.

An increasing number of specialist orchid groups now visit Sabah. Visits by internationally known orchidologists have also led to increased study of the great diversity of species.

Rhododendrons: The Greeks called them "rose-coloured trees" but many of Kinabalu's rhododendrons are epiphytes or straggly shrubs, very different from the trees of the Himalayan and Chinese species that are so familiar in temperate gardens.

Thirty-five species of rhododendrons are found in Sabah. Of these 25 occur on Kinabalu, with five not being found on any other mountains. All the Borneo rhododendrons are grouped in the Section *Vireya* and the young leaves are densely covered in scales, a characteristic of the *vireya* rhododendrons. These scales may drop off at an early stage, or may be retained, usually on the underside of the mature leaf, to a greater or lesser degree, or at the base of the flower. The function of the scales is unclear but it is possible that they protect young developing leaves from the intense rays of the sun and the ultra-violet light which is very strong on tropical mountains.

The rhododendron species here range from the tiny Heath Rhododendron (*R. ericoides*) with small brilliant scarlet tubular flowers that is found only on Kinabalu's upper reaches, to the magnificent peachy-golden blossoms of Low's Rhododendron (*R. lowii*) that grows in the high level moss forest above 2,400 metres on Kinabalu, almost glowing in the dim mist.

Several species occur around park headquarters but the road to the power station is probably one of the best places to see them. Look out for rhododendrons growing on the open banks or in

Rhododendro[n] lowii **is foun[d] only above 2,500 metre[s]**

the scrubby vegetation at the tops of the road cuttings. Long white flowers indicate *R. suaveolens*. A rare bright pink form of this has also been recorded from this road. The salmon-pink flowers with longish petals belong to *R. crassifolium* which prefers the more open areas as does the tiny, orange-pink flowered *R. borneense*. Large red or orange flowers with yellow centres are those of *R. javanicum*, a species with many different varieties in Borneo. Subspecies *kinabaluense* is one of the loveliest. Then there are the glossy, thimble-sized blossoms of *H. bagobonum*, the scaly-leaved *R. fallacinum* with delicate orange flowers, and the similar *R. polyanthemum* though this has more succulent, scurfy leaves and pinker flowers. *R. stenophyllum* with its dainty bell-shaped blooms and slender long leaves in whorls is a species that is often seen along the lower part of the Summit Trail on Kinabalu.

Heath rhododendron, endemic to Mount Kinabalu.

Pitcher-plants: Tropical pitcher-plants or *Nepenthes* occur mainly in Southeast Asia. The centre of diversity seems to be on the Sunda Shelf and especially Mount Kinabalu, where 10 of the 30 Borneo species occur. Outlying species occur in Sri Lanka, northern India, southern China and Hongkong, New Caledonia, Madagascar and the Seychelles.

The name *Nepenthes* comes from the Greek and means "removing sorrow", a reference to Homer's *Odyssey*, in which Helen drugs wine, to relieve the grief of men. In Malaysia and Indonesia, pitcher-plants are often called "*periok kera*" or Monkeys' Cooking Pots.

The first tropical pitcher-plant of which we have written records is *Nepenthes madagascariensis*, described in 1658 by the then Governor of Madagascar. Rumphius in 1747, published one of the earliest recognisable drawings of a pitcher-plant from Indonesia, which he called "*cantharifera*", meaning "tankard-bearer". It was not until 1753, however, that Linnaeus described *Nepenthes*, basing his description on the Sri Lankan *N. distillatoria*.

Cultivation of *Nepenthes* did not start until 1789 when plants were imported to Europe and grown at Kew. The Victorian craze for stove-plants and hothouses, developing in the 1800s led to large sums of money being paid for new and rare species. The major nursery firms, in fierce competition with each other, even employed their own collectors who were sometimes instructed to lay false trails for rival firms.

The pitchers of pitcher-plants are often mistaken for flowers by non-botanists but though pitcher-plants do have flowers these are generally rather small and inconspicuous. The pitchers themselves develop from tendrils at the tips of the leaves, and it is the size and shape of the pitcher that provides most clues to its identity, as colour can be very variable. *N. rafflesiana*, a handsome species from lowland forest in Malaysia, Borneo, Sumatra and New Guinea, can range from almost white to dark purple with any number of red or green flocks in between.

Pitcher-plants generally grow in areas of poor soil and are among the oddest curiosities of the plant world.

They are insectivorous and supplement their nutrient-poor diets with insects caught in a digestive fluid in the bottom of the pitcher. The pitcher consists of a cup with a lid. In many species, the upper half of the inside of the cup is covered with a waxy slippery surface, while the lower half is dotted with glands that produce the digestive fluid and later absorb the products of digestion. The lid does not open until the digestive fluid inside is fully developed, and it then remains open until the pitcher itself dries up and dies. Possibly the lid serves to keep rainwater out in some species. Perhaps its main function in many species is to produce nectar from glands on the underside. The average life-span of a single pitcher is between six months for the smaller lowland species to 18 months for some of the larger, more spectacular mountain ones.

Most species have two types of pitchers — lower ones, that rest on the ground and are usually squat in shape, and upper funnel-shaped, serial pitchers which offer less wind resistance. These upper pitchers often have a curl in the tendril that coils round a convenient branch and helps to hold the pitcher upright when it is full of liquid.

Nectar glands under the lid and inside the rim of the mouth of the pitcher (the peristome), attract insects. Usually pitcher-plants are thought of only as deceptive traps luring insects to their death. Recent research has shown that this may be an exaggeration. Pitcher-plants are usually found in nutrient-poor habitats which are unsuitable for many other nectar-producing plants. A large majority of insect visitors leave without having been in any danger of falling into the digestive fluid. Most of these have been searching for nectar from the glands under the lid. Only a few that try to reach the nectar glands under the peristome may fall in and drown. Thus, by providing a nectar source in areas where other sources of nectar may be few, the insect community as a whole benefits even if a few individuals are lost.

Below and right, Pitcher Plants come in various shapes and sizes — these are *Nepenthes lowii*, from the high mountains of Borneo.

Once an insect falls in, it is prevented from escaping by the overlapping, downward-pointing glands on the inside of the pitcher and the layer of loose wax scales that cling to its feet, making it impossible for it to get a grip. Insects are not the only victims — centipedes, frogs, and occasionally dead rats have been found.

Insects and other animals also make use of the pitcher in other ways. Mosquitoes, flies and frogs lay eggs inside them and certain spiders construct webs underneath the mouth of the pitcher to catch the insects that fall in. Though they may deprive the plant of its prey in this way, the waste products of these thieves probably provide almost as many nutrients as would the insect prey. More than 150 different species, mostly fly and mosquito larvae, have been recorded living in association with pitcher-plants at some stage in their life-cycle.

Nowhere are pitcher plants more spectacular than on Mount Kinabalu. Nine species grow within the Kinabalu area ranging from the largest in the world to some of the strangest. Pride of place goes to the magnificent Rajah Brooke. Largest in the world, this species can hold two litres (four pints) of water in its crimson cups. *N. rajah* grows in areas of serpentine ultramafic soil, between 1,200 metres and 1,800 metres. None of these areas is close to the park headquarters or to the main trails so most visitors will have to be satisfied with seeing smaller plants in the park headquarters.

Other species growing wild at park headquarters are the graceful little *N. tentaculata* with pitchers up to 13 cm long and the slender, narrow-lidded *N. fusca*, up to 18 cm long, named for the dark blotches on its pitchers.

Higher up the Summit Trail on Mount Kinabalu, between 2,700 metres and 3,300 metres, *N. villosa* with handsome pitchers, nearly always resting on the ground or nestled in moss at the base of trees is the commonest species. The most striking feature of this species is the amazing peristome of raised flanges, tapering to a double row of dagger-sharp points projecting down inside the cup. *N. villosa* is a Kinabalu endemic and one of the highest growing pitcher-plants in the world.

Closely related is *N. edwardsiana*, with longer, slimmer pitchers which are nearly always aerial, but with the same highly developed peristome. It is rarely seen along the trails but seems to prefer more undisturbed areas, and is found at lower altitudes than *N. villosa*.

Low's pitcher-plant (*N. lowii*), named after Sir Hugh Low who first collected the four most spectacular Kinabalu species, is perhaps the most curious of all with its constricted "waist" and white exudate under the lid which is found in no other species. This grows as a forest climber not only on Kinabalu but also on Borneo's other high mountains. It used to be common on Kinabalu's Summit Trail between 1,800 metres and 2,200 metres. Now it is rarely seen because thoughtless climbers have picked pitchers as souvenirs even though this is forbidden under the Park regulations.

Sepilok Sanctuary

Sepilok Forest Reserve, situated at the end of a 30-minute drive on a sealed road from the bustling town of Sandakan, has acquired a reputation as a place to see Orang-utans in their natural habitat, the tropical rain forest. Indeed, this is *the* place to see most conveniently the red ape of Borneo and Sumatra, in the wild. But the Orang-utan rehabilitation centre at Sepilok is the product of a series of chance historical events and the reserve itself, a microcosm of the Borneo rain forest, has much else to offer.

Sepilok covers only 43 square km. The rocks on which the Sepilok forest grows are sandstone with occasional mudstone, tilted and jumbled by the earth's movements, and eroded by 310 cm of rainfall yearly, into all manner of slopes and soil profiles. Except on alluvium along river banks, soil fertility is generally low. Just above sea level in Sepilok, on flat and moderately-sloping land, is lowland dipterocarp forest, rich in plant species, trees, climbing plants, shrubs and epiphytes, and with an uneven but generally tall canopy. Different in many subtle ways is sandstone hill dipterocarp forest, which grows on the steeper sandstone slopes and ridges that occur in Sepilok and Borneo generally. Most of the ridges in Sepilok are below 150 metres, but they are as steep as many of the mountain ranges 10 times that height in the interior. There are fewer and different tree species here than in the lowland forest. The canopy is more even and lower, plants flower very rarely and many produce hard, green fruits. Animal life is scarce. On elevated, gently-sloping, pure sandstone, from which almost all nutrients have long since leached away, is heath forest, low in stature with densely-packed, small trees, pitcher plants and hardly any animal life. Heath forest occurs only in the eastern half of Sepilok, while the southern side of the reserve is fringed by the mangrove swamps of Sandakan Bay.

he Orang-
tan (left)
; a natural
ttention-
rabber
ight).

All large wild animal species are either scarce or absent from Sepilok. Even the Orang-utan is not abundant here. For the largest mammal species, like elephants, rhinos or bears, the reason is clear — Sepilok is too small to support breeding populations. For other, smaller species, there is another reason. A good example is provided by those monkeys known as langurs or Leaf Monkeys, the latter name referring to the fact that about a half of their food consists of leaves. Most of the balance is seeds. A study of the Red Leaf Monkey at Sepilok revealed that this attractive primate lives in groups of about six to eight individuals which occupy a range of about 0.8 square km. A closely-related species, the Banded Leaf Monkey of Peninsular Malaysia, lives in groups of 15 to 25 individuals which occupy a range of barely 0.3 square km. The reason for the substantial difference in density of these monkeys seems to be that Sepilok, like forest generally in Sabah, is rich in dipterocarp trees — which provide almost no food for

mammals — but poor in plants of the legume family, a group generally favoured by primates and many other mammals for food. The Grey Leaf Monkey, which also occurs at Sepilok, is so rare that it is hardly ever seen. This situation is entirely natural and not caused by hunting, which has never been a problem at Sepilok. There are two other kinds of monkeys in Sepilok; the Long-tailed Macaque and the Pig-tailed Macaque. Both are adaptable species which live in groups of 20 or more, and which include a variety of fruits and small animals in their diet. Truly wild macaques as well as semi-tame ones are found in the reserve.

Another primate which may be seen in Sepilok is the Bornean Gibbon, a small, graceful, long-armed ape which invariably lives in family groups of father, mother and one or two youngsters and occupies a territory of about 0.3 square km. The female Gibbon utters loud, high-pitched bubbling cries in the early hours of daylight, a characteristic sound of the Borneo rain forest. At night, two small relatives of the primates, the Slow Loris and the Bornean Tarsier, are active in the Sepilok Forest.

At last count, over 90 species of mammals and over 200 species of birds were known to occur within Sepilok Forest Reserve. The Bornean Bristlehead, a bizarre orange-and-red-headed bird which roams the forest in small groups, whining and cawing, is a notable member of the Sepilok bird list. While dipterocarp trees are a prominent and rich component of the Sepilok flora, perhaps of greatest interest is the Borneo ironwood tree, known locally as *belian*. It is amongst the hardest and most durable of woods in the world and large *belian* trees are many hundreds of years old. The first bridge over the first stream on the trail through Sepilok Forest Reserve to the mangrove is a belian trunk which started life before the first Europeans set foot in Sabah and fell years before the first tourist set foot in Sepilok.

Many of the world's best parks and reserves owe much to foresighted indi-

Park headquarters at Sepilok.

viduals who convinced sceptical governments to preserve wilderness areas for their scenery and wildlife. Paradoxically, Sepilok owes its existence to economic circumstances. In the 1930s, the town of Sandakan was the capital of British North Borneo, an anachronism ruled by a company granted a charter by the British government. The British North Borneo Company, ever anxious to stimulate development of Sandakan's hinterland, encouraged anyone who was willing to clear the forest could settle and cultivate the land. But the government also realised that the town would need guaranteed supplies of timber. An area of 23 square km was established in 1931 as Kabili-Sepilok Forest Reserve. Subsequent additions and excisions of forest land left the 43 square km which today constitutes the area known as Sepilok.

In the 1950s, Barabara Harrisson, wife of the Curator of the Sarawak Museum, began to rescue young Orang-utans being kept locally as pets, and the idea grew of training these animals to fend for themselves so that they might readapt to life in the wild. In 1962, with the backing of the newly-formed World Wildlife Fund, Harrisson visited North Borneo and, on the strength of surveys in the Sandakan and Mount Kinabalu areas, reported that Orang-utans were rare, declining and were endangered with extinction.

naller but st as ractive, Bornean bbon.

In 1963, North Borneo became Sabah, an independent state within Malaysia, and a Game Branch was created in the Forest Department for the conservation of wild animals. A new law prohibited anyone from catching or keeping an Orang-utan. Many young Orang-utans which had been caught during logging or forest clearance and kept in captivity were confiscated — but something had to be done with them. Encouraged by Harrisson, the new head of the Game Branch, Stanley de Silva, chose Sepilok as the place where these Orang-utans would be brought and trained to live a natural life in the forest. The Orang-utan rehabilitation project at Sepilok has continued uninterrupted until the present day, financed entirely by the Sabah government, and well over 200 orphaned Orang-utans have been welcomed. Starting in 1986, the Japanese government provided the services of a wildlife veterinarian and a new clinic and laboratory at Sepilok. In 1988, the Game Branch was upgraded to an independent Wildlife Department.

The rehabilitation process enjoys varying degrees of success. Some Orang-utans succumb to viral infections, malaria or other diseases, but the full-time presence of veterinarians has greatly improved matters. Some of the apes disappear entirely before they are fully mature, while others remain and are partially dependent on food supplied by wildlife staff. And some are successfully rehabilitated to survive unaided in the forest. A few have mated with wild Orang-utans and produced babies.

Orang-utans continue to come into Sepilok, nowadays not from captivity, but directly from areas where forest is being cleared for agriculture. Sepilok serves to give these unfortunate animals life in the forest rather than an uncertain future in a plantation or a cage. Critics have pointed out that rehabilitating Orang-utans and allowing visitors to Sepilok are not strictly compatible. This is true — but Sepilok has come to serve as a link between people and Sabah's wildlife, where the young can gain their first insights into nature and the importance of caring for the natural environment. In the long term, this will perhaps be the most effective way of ensuring that today's protected areas remain protected forever.

A Solitary Life: Not much was known of the ecology or behaviour of Orang-utans in the wild until the late 1960s and early 1970s. Then, spurred by a growing interest amongst scientists worldwide in the life of apes and the evolution of human beings, several enterprising researchers made camps in the rain forests of Sabah, Kalimantan and Sumatra in order to watch these apes for periods of a year or more. These researchers found that wild Orang-utans generally live a solitary existence. They are not territorial, however, and several Orang-utans share any one area of forest. Apart from a youngster's first four or five years, when it rarely ventures far from its mother, the only times that Orang-utans meet are when dining in favourite fruit trees and, even more rarely, when males and females form temporary mating bonds.

Typically, an Orang-utan's day will start just after dawn and consist of a leisurely journey of less than a kilometre through the forest canopy, taking in several food trees, and ending before dusk. The Orang-utan prefers fruit as food, and a wide variety is eaten, but small amounts of young leaves, bark, flowers and insects are also enjoyed. Late in the afternoon, Orang-utans perform an essential task which is unique amongst apes and monkeys in Asia: they make a nest in which to sleep through the night. The nest is actually little more than a springy platform, made by snapping and bending small branches and twigs towards a fork in the crown of a tree. These nests can be

The Flying Lemur is without clos relatives on earth.

anything from four to 40 metres above the ground.

All pioneering studies of wild Orang-utans were done under virtually identical ecological circumstances — in tall, undisturbed lowland dipterocarp forest near either a fairly large river or the coast, much like Sepilok. During the years after these studies, vast areas of this kind of forest were logged for their valuable hardwoods and much has been cleared to make way for plantations of oil palm, cocoa and rubber. Conservationists became increasingly alarmed for the future of the red ape. Some warned that the species was becoming perilously close to extinction. Others, extrapolating from earlier studies, said that great numbers of Orang-utans existed in the still-vast dipterocarp forests of interior Borneo, but predicted that they would die out as the forests were logged. During 1986 to 1988, the World Wildlife Fund (WWF) Malaysia, helped by a donation from its counterpart in the U.S.A. and backed by the Sabah Forestry and Wildlife Departments, attempted to make a realistic assessment of the Orang-utan situation in the state of Sabah.

'estern arsier ırprised by ıe glare of a ırchlight.

For this study, the fact that Orang-utans make distinctive nests in the trees was a real blessing. It was already known that the nests can be spotted not only from the ground but also from a helicopter. The Royal Malaysian Air Force helped out by flying the first survey, where a team plotted distribution of nests over vast areas of forest, an exercise which would have taken months for survey teams on the ground. Later a method was devised for estimating the number of Orang-utans per square km of forest from the number of nests counted per minute from a helicopter rented from a local company. By combining these methods with surveys on the ground and interviews with people living and working in rural areas, it was possible to build up a clear picture of how Orang-utans are faring in Sabah. Many new facts emerged. For example, Orang-utans occur in barely half of Sabah's forests. In almost all areas where they are absent, this is because of natural circumstances, not because of hunting or logging. The abundance of Orang-utans varies greatly from place to place, with coastal swamp forests supporting an average of 10 times as many Orang-utans as in an equivalent area of hill forest in the interior. It was also found that Orang-utans are easily able to tolerate logging for timber extraction, whereas they are wiped out by forest clearance for agriculture. It was estimated that between 10,000 and 20,000 Orang-utans remain in Sabah. This seems to be a large number, but when one considers that the forest areas where Orang-utans are most abundant are the same areas favoured for agriculture, then the long-term picture is not so rosy. The WWF study concluded that Orang-utans are uncommon in most of the existing parks and reserves and that the best way to ensure the species' long-term survival is to establish a sanctuary in the region where they are naturally most abundant — the lower Kinabatangan region in eastern Sabah.

Danum Valley and Other Parks

The **Danum Valley Conservation Area** comprises 438 square km of mostly lowland forest in the upper reaches of the **Segama River** (of which the Danum River is a major tributary). This region has long been recognised as an outstanding area for wildlife.

The **Danum Valley Field Centre** was opened on the Segama River in 1986 and is managed jointly by the Sabah Foundation, the Sabah Forest Department, the Sabah campus of the National University and the Royal Society, London.

Forest to the east of the Field Centre and indeed all around the perimeter of the Conservation Area has either been logged or is scheduled for timber production within the next few years.

Most of the Danum Valley Conservation Area is covered by lowland dipterocarp forest.

The mammal list for Danum Valley includes all 10 species of primate found in Sabah, elephant, banteng, Sumatran Rhinocerous, Clouded Leopard and Honey Bear. It is a particularly good area to look for wild Orang-utans. Other species commonly seen near the Field Centre include Giant Flying Squirrel, Sambar (known locally as *payau*), Barking Deer (*kijang*), Mousedeer, Bearded Pig and occasionally otters.

Among birds, Danum Valley is exceptionally rich in hornbills. At the Field Centre a Buffy Fish Owl is a regular night-time visitor in search of moths attracted to the lights. Other rarities include the Great Argus Pheasant, Black-and-crimson and Blue-headed Pittas, the Bornean Bristlehead and the Bornean and Black-throated Wren-babblers. Over 220 species have been recorded to date.

Substantial areas of Danum Valley remain virtually unexplored. The highest point, a modest 1,093 metres, is **Mount Danum** which is one of a series of outcrops of ultramafic rocks which sweep from the eastern flank of Mount Kinabalu through the hills around Telupid and then east to Mount Silam and the islands of Darvel Bay.

Mount Danum can be reached by a 22-km trail cut in 1987 by an international team of adventurers. This group also built a six-bed cabin on the Danum River at the base of the mountain. Permission is required from the Field Centre manager to use this trail and cabin which is a two-day walk from the Field Centre. Five days are needed for the round trip hike to the top of Mount Danum and a guide is required.

Another dramatic physical feature of the Conservation Area is **Dismal Gorge** — so named by a group of disappointed gold prospectors who in 1889 spent 10 days hauling their boat around the three km gorge. Although only about six km from the Field Centre it remains very difficult to reach and has only once been run by canoe and raft, by a team dropped in upstream by helicopter. Access may become easier in the next few years with the logging of forest nearby on the east bank of the Segama River.

Mount Silam: Fifteen km west of

...ves to the ...ont: left, Buffy Fish-...wl and right, Red Leaf ...lonkey.

Lahad Datu and beside the turn off to Danum Valley, travellers on the Tawau highway will pass the distinctive rounded peak of **Mount Silam**. This is a glorious stretch of coastal scenery. For a short stretch, the road provides the only break in a profile of primary forest which climbs 1,080 metres over eight km from the mangrove-fronted seashore to the top of the mountain. Because of its small size and proximity to the sea, this cloud-capped peak of only modest height boasts lower montane and even upper montane forest on its higher reaches. Better still, a road provides access three-quarters of the way up the mountain to a telecommunications station. Although blocked by a locked gate at the bottom, a key may be borrowed by arrangement with the Telekom Department in Lahad Datu.

Madai Caves: Twenty km beyond Mount Silam and equally visible from the road to Tawau is a limestone outcrop called **Madai**. A five km diversion by rough road takes the visitor to a cave system where cave swiftlet nests are collected. During the breeding months of March to May and September to November, a small village at the base of the caves is inhabited by traditional Idahan collectors.

Maliau Basin: Worthy of mention here, if only to tantalise the visitor, is the magnificent montane vastness of the **Maliau Basin**. Like Danum Valley 90 km to the east this is a Conservation Area within the Sabah Foundation forest concession.

The Basin is near circular in shape and 390 square km in area. It is rendered virtually inaccessible on foot by steep flanking escarpments up to 1,500 metres in height and a steep gorge on the southeast side where it is drained by the **Maliau River**. The highest point is **Mount Lotung** (1,800 metres) on the north rim. Within the Basin are some spectacular waterfalls and an almost unvisited environment, much of it a montane heath forest, rich in rhododendrons, orchids and pitcher plants.

Tawau Hills Park: The **Tawau Hills Park** covers the main volcanic cones of the Tawau district in southeast Sabah which rise to a height of 1,300 metres. These volcanoes are witness to a fiery past but they have long been quiescent and are now overgrown with forest.

Much of the park outside the central hilly portion has been logged, and the park headquarters itself is on the edge of large cocoa and oil-palm estates that extend right up to the boundary. Trails lead along the river and into the forest to some hot springs, another reminder of the area's past upheavals.

There is no overnight accommodation, though there are several picnic shelters for day visitors, for whom the waterfall at park headquarters is the major attraction.

Marine Marvel: An hour's drive east of Tawau is **Semporna**, which is the traditional home town to a large number of Sabah's maritime people — the Suluks and Bajau ("sea Gypsies"), many of whom have relatives across the nearby border with the Philippines. The whole area is studded with islands and reefs of diverse origins. Immediately east of Semporna town and separated by a

Swamp forest still covers much of Borneo.

narrow channel is a flat raised coral reef island, **Pulau Bum-Bum**. Eight km further east are the jagged islands of **Pulau Bodgaya** and **Pulau Bohey Dulang** which are the remains of an extinct volcanic crater. Most of these islands have good coral reefs for snorkeling or diving although security conditions need to be checked with the police.

Fortunately for the coral reef enthusiast **Pulau Sipadan**, 25 km south of Semporna, is safe, accessible and in a class by itself. This tiny speck on the map, a mere 12 hectares (30 acres) in size, is really the tip of an undersea mount that rises vertically from ocean floor depths of 600 metres. On the north side of the island this undersea cliff begins just 30 metres offshore, so that the bows of a 15 metre launch can be grounded while the stern is literally over several hundred metres of water.

Sipadan is Malaysia's only oceanic island and Jacques Cousteau has described it as one of the finest reefs in the world. Visibility in the water is often 70 metres or more and a range of pelagic species, such as sharks, tuna, and barracuda mingle with the myriad reef species of angel fish, damselfish, parrotfish, garoupa, moray eels and so on. Giant clams and other large shells are still common. Sipadan also has a good nesting population of green turtles.

Sipadan has survived in its pristine state because of its small size and remote location. There are now plans to protect it more formally as a National Park. However, for the past few years Borneo Divers, a company based in Kota Kinabalu, has been running diving trips to the island.

Other wildlife areas in Sabah: Until a few decades ago, eastern Sabah probably supported the greatest variety and abundance of large mammals of any part of Borneo. This situation appears to have come about mainly because of relatively fertile soils and numerous natural salt licks in the region, although the absence of hunting peoples except along the large rivers, may also have contributed. Still present in relative abundance, although now declining, are

he Danum alley field tation.

the Asian Elephant, Sumatran Rhinoceros, Banteng , Sambar Deer, two species of Barking Deer and two of mousedeer, Bearded Pigs, Sun Bears, Clouded Leopards, Orang-utans and Proboscis Monkeys.

Sabah is home to almost all of Borneo's elephants. There are an estimated 2,000 of them in the state, mostly in the east, where they form one of the largest remaining concentrations of the species left in the world. Only mature males have ivory and poaching has never been a problem. However, the rhino population has declined drastically during the past century, mainly as a result of excessive hunting for their horns but the Banteng has survived remarkably well, despite the constant pressure of hunting for its meat. As much of eastern Sabah has been judged suitable for permanent agriculture, the greatest threat to the long-term survival of the region's fauna and flora is loss of forest, except where it is protected by law for timber production or wildlife protection.

Tabin Wildlife Reserve was established in 1984 to help conserve rhinos, elephants and Banteng. About 1,225 square km in extent and situated in the **Dent Peninsula**, most of this reserve was selectively logged between 1965 and 1989, when the last logging licence expired. An 86 square km "core area" of magnificent undisturbed dipterocarp forest remains in the middle. Tabin contains at least seven natural salt licks, two of which are mineral-rich "mud volcanoes" covering more than two hectares (five acres).

Kulamba Wildlife Reserve was established at the same time as Tabin for the protection of an extensive coastal freshwater swamp forest ecosystem. Although rather small, Kulamba helps to conserve populations of Storm's Storks, Flying Foxes, Orang-utans and Banteng.

The floodplain of the **Kinabatangan River** is a low-lying wilderness area, much of which is still under natural forest cover and dotted with small lakes, sandstone hills and limestone outcrops. Only some parts of this area are protected as Forest Reserve, but there are proposals for a larger wildlife sanctuary. The area contains the greatest concentrations in Malaysia of Proboscis Monkeys, Orang-utans, elephants, darters and Hairy-nosed Otters. Within this area are the **Gomantong Caves**, two large and numerous small, set in a limestone outcrop. Home to about a million bats and another million swiftlets, these caves have been Borneo's major source of edible birds' nests for at least 200 years.

In contrast to the wild east, the coastal plains of western Sabah have been tamed by centuries of human habitation and rice cultivation. In one area to the north of Kota Kinabalu, is the **Kota Belud Bird Sanctuary**, 120 square km in area and including the **Tempasuk Plain** and **Kerah Swamp**, where the local Bajau and Illanun peoples not only grow rice but rear thousands of buffalo, cattle and horses. The man-made ecosystem of grasslands, marshes, ricefields and ponds which has evolved is ideal for many kinds of birds which are

A Yellow-breasted Flowerpecke

not adapted to forest life. More than 120 species, resident and migrant, including waders, ducks, raptors, terns, wagtails, pipits and warblers have been recorded in this sanctuary.

Named after Malaysia's first prime minister, **Tunku Abdul Rahman Park** comprises five islands and their surrounding seas and coral reefs located off Kota Kinabalu. The islands, except for the largest, **Pulau Gaya**, are covered in secondary vegetation and their main attraction is the clear water and sandy beaches.

Pulau Gaya is the only island still covered with primary forest, but this, typically of island forests, is impoverished, much of it growing on dry sandy soils. There is a good trail system on Gaya and sometimes Pied Hornbills are seen.

The Park Headquarters with visitor chalets and the other facilities is situated on **Manukan Island**, the second largest in the group.

The **Pulau Tiga Park** lies 48 km south of Kota Kinabalu and comprises **Pulau Tiga** itself, a long island with three low hills, and the two smaller islands of **Pulau Kalampunian Damit** (or Snake Island) and **Pulau Kalampunian Besar**, each of which is a few minutes boat-ride to the north. Pulau Tiga was built up from the flows of the three mud volcanoes that form the three low hills. Though two now are dormant, the third bubbles gently, building a three foot cone, but the mud is cold.

A network of trails covers most of the island which is especially good for birdwatching. Pied Hornbills and megapodes are common and easy to see, and the large round piles of sand in which the megapodes lay their eggs are even more common. A flock of frigate birds hovers over the island from September to April before disappearing to unknown breeding grounds. At night they roost on Snake Island. Fruit-eating pigeons gorge on the abundant fruiting trees in season and nightjars can often be heard in the evenings. They lay their eggs on the bare ground in the clearing of Park Headquarters on Pulau Tiga but they are so well-camouflaged that it is almost impossible to see them.

Migrant birds also find Pulau Tiga a good resting place and species such as the Great Crested Terns roost on the sandbars at low tide, while Snake Island, as its name implies, is home to more than 100 black-and-white Striped Sea-snakes (*Laticauda colubrina*).

The **Crocker Range National Park**, covering an area of almost 1,400 square km, is the largest single totally protected area in Sabah and is an important water catchment reserve. Unfortunately, little development has been carried out. There are no visitor facilities, and even the boundaries have not yet been fully surveyed.

The **Rafflesia Forest Reserve** lies between Km 61 and Km 64 of the Kota Kinabalu to Tambman road but flowers are now rarely seen. At the top of the pass, at 1,700 metres a small orchid garden is being developed which will display wild orchid species from the Crocker Range in an enriched forest with trails and labelled plants. This area is a good place to see montane birds.

A Binturong manoeuvres itself into a favoured spot.

Mulu National Park

Mulu is a Must: The East Malaysia state of Sarawak evokes romantic images of white rajahs and head-hunting tribes. The Rajah Brookes, governed for more than 100 years, leaving a legacy that is still evident today. Although head-hunging is no longer practised, Sarawak remains a land of verdant rain forest, riverine village, and some unique wildlife. Travelling into the **Mulu National Park** is something of an adventure. Start by voyaging from Miri, 519 km from the capital **Kuching**, on a commercial express boat up the **Baram River**. For mile after mile, the banks are stacked with hardwood logs. It is two-and-a-half hours to **Marudi**, then another four hours to **Long Terawan**, where visitors switch vessels to a chartered longboat. The Baram River leads to the **Tutoh River**. Not much forest remains along the rivers. It has all been taken out by the logging companies except for a few hardwood trees that tower above the wasteland which has been left behind. Among the bird species observed during the journey, the Greater Coucal, a huge scrubland cuckoo, is the most common one.

After Long Terawan, the scenery changes. Twin 40 horsepower outboards are needed to power the narrow longboat up against the current. An abundance of kingfishers, especially the Stork-billed variety, Straw-headed Bulbuls, Hill Mynas and, occasionally Black-and-red Broadbills are seen along the river banks. Green Imperial Pigeons and various species of hornbills indicate that primary forest is not too far away. Turn into the **Melinau River** and land at the Mulu Park headquarters at **Long Pala**. The river no longer looks like chocolate-milk. There is no logging beyond this point and the waters at the edge of a vast expanse of primary forest are clean. In the background looms the famous limestone cliffs and the summit of Mount Mulu.

Mulu is by far the largest national park in Sarawak and features a geological combination of sandstone and limestone formations. The sandstone mountains consist of compressed sand and silt deposits from eroded rock. The limestone is made of sea-shells and coral debris built up, compacted and uplifted over millions of years. The difference in the two types of sediments is evident at Mulu where approximately five million years ago Mount Mulu (sandstone) and Mount Api (limestone) were about the same height. Because of the massive rainfall (5,000 mm to 6,000 mm per year) Mulu is now 2,376 metres above sea level, while Api stands at 1,750 metres.

The soft and porous limestone has also been eroded into enormous caves. A visit to Mulu National Park usually includes inspection of some of these. The **Deer Cave**, the world's largest cave passage, is only three km from Long Pala. The **Eagle Cave**, next to Deer Cave, is small but packed with drip-stone structures and some fine examples of photo karst, the weird phenomenon where tiny columns of limestone "grow" towards the light. The **Clearwater Cave**, further up Melinau River, is the longest cave system in Southeast Asia. A survey team recently travelled over 50 km into this world of darkness. The **Sarawak Chamber**, further east and nearer to Mount Api, is the largest cave-cavity in the world but is currently closed to the public. It is estimated that only about 30 percent of the caves of Mulu have been surveyed.

On the northern side of Mount Api the **Pinnacles** form a different limestone spectacle. They are sharply pointed pillars that stretch to 45 metres up through the surrounding sub-montane bush vegetation. Visiting the Pinnacles normally takes three days. Travel up the Melinau River and walk through the forest to **Camp 5** on day one: ascend on day two: and back to base on day three. It is a pretty tough walk that, in places, becomes a climb. The elevation changes more than 1,000 meters in two to three hours, but it is worth making the effort as the view over Mulu from the top is magnificent.

The journey to and from Camp 5 is

eft,
ındscape
tease the
nagination
– Mulu's
innacles.

through the alluvial rain forest typical of Mulu. This forest on dense, clay soil holds more kinds of trees per hectare than any other type of forest. The undergrowth is quite open and it is easy to leave the trail.

The trip to Mount Mulu takes four days via **Camp Paku** and **Summit Camp**, but it can be condensed for those who are fit and in a hurry. The view is not that spectacular but includes the montane zone where the forest turns into stunted, alpine growth. Fewer mammals and birds live in this type of forest but the proportion of rare and endemic ones is greater.

The forest in this part of Sarawak is home to the Penan tribe of natives who made the headlines, some years ago, when they protested against the logging operations in "their" forest. In 1987 and again in 1988 they and other indigenous people such as the Iban, Kayan and Kenyah tribes placed barricades across logging roads in the Baram district. A Swiss national, Bruno Manser, staying illegally in Sarawak, joined the Penans and created further publicity for their cause. However, they are up against some powerful forces. One of the major concession holders declared on the occasion that "...logging is good for the forest..." which "...is sustainable and will go on forever and ever". Unfortunately, this is not only a misconception, but a platform of official politics. Meanwhile, logging in Sarawak continues. Even selective logging destroys 50 percent of the tree cover. Primary forest never regenerates completely, and in another decade there will be little such forest left in Sarawak.

Most Penans are now settled in villages and longhouses (there is one such village at the Mulu park boundary) but some still roam the forest in a nomadic way and are allowed into the park to hunt. They can be seen on the trails armed with shotguns or blowpipes in search of large mammals such as Bearded Pig, Sambar Deer and monkeys. They are small, strong-looking people who seldom speak when you are near, not even to each other. Some now work for the government, collecting sand at the river banks, hauling cement for construction projects in the park or operating the chainsaws that have made them alter their lifestyle so drastically.

The fauna of Mulu has been well known, since a 15-month expedition by the Royal Geographical Society and the Sarawak Government in 1977/1978 made an extensive survey of the park in search of wildlife. Traps were set out to inventory small, nocturnal mammals and mistnets used to catch and identify elusive forest birds.

The mammal checklist includes 67 species, mostly bats and rats which can often only be identified by measuring their skull or the pattern of their teeth. Larger mammals were also noted: the Sun Bear, Slow Loris, Western Tarsier, Yellow-throated Marten and several species of civets.

Diurnal animals that the casual visitor is more likely to encounter include (Long-tailed and Pig-tailed) Macaque plus three species of Leaf Monkey. Two of them, the Grey and Red Leaf Monkey, are endemic to Borneo, as is the

View over Mulu National Park.

Bornean Gibbon. Close to Mount Mulu, the gibbons' territorial howling carries a long way in the early mornings. But they are shy animals that will spot you before you see them, and they soon start to retreat quietly back through the canopies, their long arms reaching out for support. Even the squirrels at Mulu, which include some species endemic to Borneo, especially at higher elevations, are of interest. Look out for the charming little plain pigmy squirrel: it only grows to about eight cm long.

The check-list of the birds of Mulu comes close to a the complete list of Bornean avifauna. All the trogons (six species), hornbills (eight), barbets (nine), broadbills (eight) are there plus almost all the bulbuls (20 species) and babblers (32). The list totals a staggering 262 birds, including 20 Bornean endemics, mostly from the montane habitats. Although visitors will many birds at Mulu, the problem is that they do not exactly come out and perch in front of you. Forest birds offer the ultimate challenge for the serious birdwatcher.

The most successful order of rain forest animals are the bats. At Mulu they occur in the millions. Most roost during the day high inside the Deer Cave. At dusk they emerge, some days into the sunlight at 5:30 p.m., other days late in the evening at 7 p.m. They come streaming out in flocks of thousands, and just when one thinks "that's a lot of bats..." the main flock unloads. They are the wrinkled-lipped bats and visitors watch in amazement as they keep pouring out of the cave for 10 or 12 minutes. The Crested Goshawk sits ready at the cave exit and flys out to grab a few but the supreme hunter is the Bat Hawk, a large, falcon-like hawk that pursues the bats with superb flying skills long after they have dispersed throughout the forest. It is an outstanding spectacle that the visitor feels privileged to have witnessed. Although the journey back through the darkness, stumbling along the muddy forest trails with just a flashlight to guide you, is a trifle taxing, it will certainly be the highlight of a trip to Mulu.

t dusk, housands f Wrinkled-ipped Bats tream from he Deer Cave.

NIAH NATIONAL PARK

The **Niah Caves** are in the lee of the 400-metre-high **Subis Mountain**. They are the main attraction in the **Niah National Park**, which was originally formed as a historic monument in 1958 when red haematite rock painting were discovered inside what is now called the **Painted Cave**. The paintings date back 1,200 years and are the only such expressions by early man found on the island of Borneo. Other archaeological finds included a 37,000-year-old human skull dug out from guano deposits. In 1974, the Niah Caves and 78 square km of the surrounding rain forest were gazetted a National Park.

The Park Office is at **Pangkalan Lubang** near **Batu Niah** town, a two-hour drive from Miri.To enter the park the Niah River must be crossed and then it is a four km walk to the caves on a neat plankwalk. Along the way, other trails branch out to the Subis and Tangap rivers. The forest is quite good and in the morning bulbuls, hornbills, trogons and other forest birds can probably be seen. Many common Bornean mammals such as the Lesser Mousedeer and Pig-tailed Macaque also occur. This macaque is somewhat larger and heavier than the more wide-spread Long-tailed Macaque. It prefers to stay close to mature forest and when disturbed jumps down onto the forest floor and runs away — other monkeys always seek safety in the tree canopies. The Bornean Gibbon is also at Niah. It may be heard calling from a distance early in the morning. Look for Flying Foxes which, with their 60 cm wingspan and slow wing-beat, somewhat resemble large birds of prey, but are of course bats, the world's largest bat and totally frugivorous. Like all fruit-eating bats the Flying Fox has a pretty, dog-like face. It is the insectivorous bats that have large ears and prominent nose-leaves used in their echo location that make them look ugly — or interesting; beauty is in the beholder.

The Niah Caves were inhabited by Man some 40,000 years ago.

The caves are carved out of soft limestone eroded by slightly acidic rainwater. The water can dissolve the limestone which settles back into stalactites, stone icicles suspended from cracks in the cave ceiling, or drip down to form stalagmites, pointed, conical formations on the floor of the cave. Where the water flows, rather than drips, frozen stone patterns develop as so-called flowstone. The limestone even takes the shape of pearls when it forms around grains of sand in a pool inside the caves.

The underground world does not exist in total isolation from the rich forest outside. On the contrary, all life inside the caves depends on food-supplies brought in continuously from the forest. The bats and the swiftlets that roost and breed inside the caves all feed on insects outside and deposit huge amounts of guano — a total of one ton per day — during their resting periods. This guano in turn provides energy for a wealth of invertebrate creepy-crawlies that live on the cave floor. Cave cockroaches, scorpions, crickets, and Assassin Bugs abound. The Cave Gecko feeds on the invertebrates and the rare snake-species Cave Racer even feeds on the bats.

There is also human competition for the guano which is a cheap much sought-after natural fertilizer. Sweeping the old, compressed guano off the floor keeps a few dozen people busy. The guano is collected in the **Great Cave** which, consisting of many different corridors and chambers, is the largest of the Niah Caves. A convenient network of boardwalks has been constructed along the main trails which makes it easy to get around. The **Painted Cave**, a few hundred metres further into the forest, does not hold any wildlife.

Individuals differ as to how they respond to this peculiar cave environment. Some people end up spending weeks in cave expeditions penetrating deep into the limestone formations of Borneo. Others breathe a sign of relief when a brief visit is over and they return

A breeding colony of Black-nest Swiftlets.

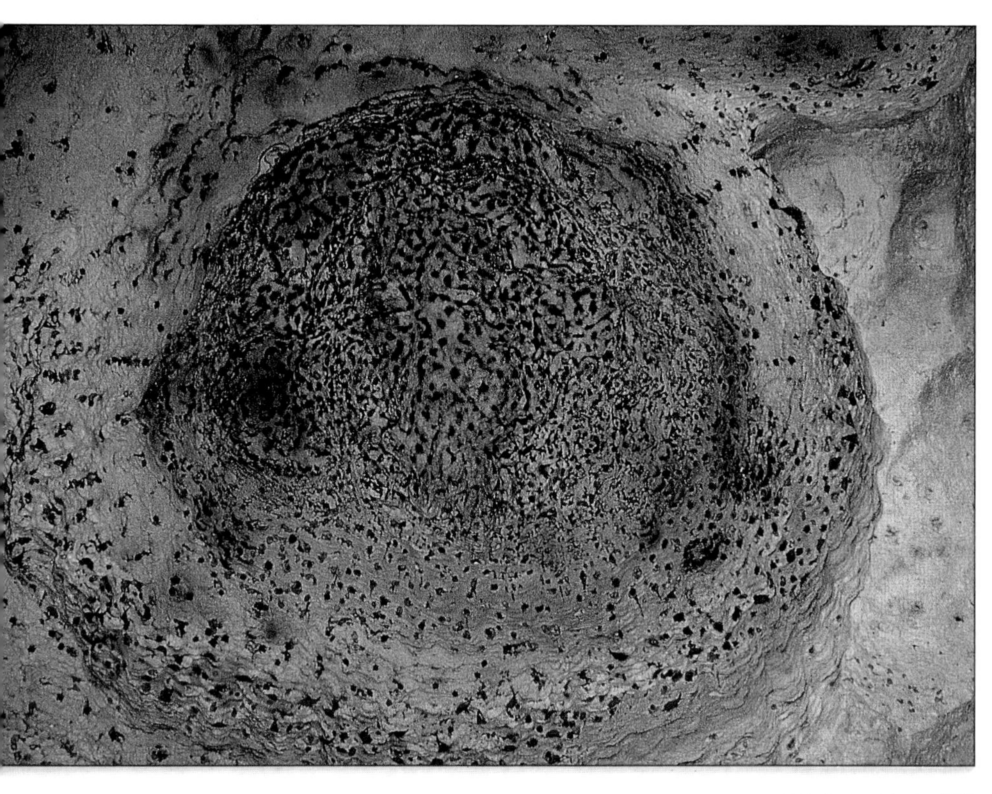

PROBOSCIS MONKEYS

Proboscis Monkeys rank among the world's most unusual animals, and seeing them is a must for any naturalist visiting Borneo.

Proboscis Monkeys certainly moved early naturalists to comment. In 1848, Hugh Low said that "it is a very fine monkey, in size approaching the Orang-utan, but much less disgusting in appearance". Over the next hundred years, descriptions of the Proboscis Monkey included "of singular and ridiculous aspect", "musically gifted", "choleric", "very striking", "highly ludicrous", "vile porty-looking" and "grotesque honker of the Bornean swamps".

The male Proboscis Monkey's most striking features are his huge nose and belly, which have earned him the name *orang belanda* or "Dutchman" in parts of Borneo. The reason why he has such a large nose is almost certainly a result of sexual selection; quite simply, the females like it. He is a big animal, weighing up to 24 kg, and the dark coat on his back and greatly expanded belly make him appear to be wearing a bomber jacket bought in his youth which he has now outgrown. Females only weigh about 10 kg, have smaller, snubby noses and more subtle colouring. In the field, Proboscis Monkeys are easily recognisable as the only monkeys in Borneo with white tails. They spend most of their time in the trees, only occasionally coming to the ground to move between patches of forest. They are highly proficient swimmers, aided by partly-webbed back feet, and can even swim considerable distances underwater.

Proboscis monkeys only occur on Borneo. Even here they are restricted mainly to the swamp and riverine forests of the coastal plain. Rivers are very important to them, and every night, they sleep in the trees adjacent to the river. This can be spectacular because they not only use the riversides for sleep, but also as display grounds. Proboscis Monkeys live in harems, comprising one male and up to eight females and their offspring. Each evening at the riverside, different harems often come together. The males display to attract females, who sometimes move between harems, and also to warn off rival males. They roar loudly, crashing through the trees, often breaking branches loudly and deliberately as they go.

Proboscis Monkeys feed on young leaves and dry, starchy fruits and seeds. They are fussy eaters, and their foods are scarce in the coastal forests. This means that they have to range widely to find enough, so large areas need to be protected if the animals are to survive. Unfortunately, reserves in coastal areas are too small to protect populations fully. In northern Borneo, Proboscis Monkeys are endangered. Sabah, Sarawak and Brunei Darussalam among them contain no more than 3,000 individuals, and that number is still declining. The number remaining in Kalimantan is unknown. Large areas of swamp forest remain, but the status of Proboscis Monkeys there is not known.

Proboscis Monkeys can be seen in several areas. In Sarawak, they are in Bako National Park, and also in the Sarawak Mangroves Forest Reserve near Kampung Salak. Both are close to Kuching. In Sabah, they can be seen in the Kinabatangan area in the east, and Sandakan tour operators take visitors to see the animals near Sukau. In Kalimantan, they are found in the Mount Palung and Tanjung Puting National Parks. In all of these areas except Bako, the animals are seen most easily by boat as they gather by the rivers in the evening. A river cruise alongside groups of Proboscis Monkeys as dusk approaches is a spectacular end to a naturalist's day.

to the sunshine, the fresh air and the colours of the forest outside.

Bats, however, appear to like caves and about 500,000 of them call Niah Caves home. Eight species are represented but most are Naked Bats and Cantor's Roundleaf Bats. All bats, other than a small number of 20,000 Cave Nectar Bats, are insectivorous. Somewhat difficult to see, because they mostly rest in the highest domes out of flashlight reach, are groups of the tiny Bornean Horseshoe bats.

While the bats feed at night the swiftlets work the dayshift. Many swiftlets can be seen at the caves in the daytime, especially during the breeding season which coincides with the rainy period (September to April) when there are plenty of insects about. A swiftlet eats half its own bodyweight per day in insects; at Niah that comes to a collective total of 11 tons per day.

The white-bellied swiftlets, which are incapable of echo-location and nest under the ceiling close to the bright cave entrance, are the most obvious. Further into the caves the black-nest swiftlets take over. They do not use their echolocation system for catching flying insects the way bats do, but it is sophisticated enough to guide them through the caves. Swifts always keep in contact with high-pitched squeaks, but in Niah visitors will clearly hear how they stop calling and instead "switch on" to echolocation when they enter the darkness. Then, a loud and rapid clicking noise is all that is heard.

The swiftlets all "cement" their diminutive nests onto the cave walls with a type of saliva produced from a special gland. The black-nest swiftlet mixes many feathers in its nest, which is therefore not as commercially valuable as the edible-nests that are made of saliva. The former, however, can be cleaned and the nests at Niah are harvested ever year, once at the beginning of the breeding season, around October, and then again, after the season ends, in May.

At Niah, the nest collectors climb up thin wooden poles to the roof of the caves, about 20 metres up. The nests are scraped off the roof under the flickering light of a candle. It takes about 100 nests to make a kilo of saleable birds' nest, which can sell for a few hundred US dollars or more in the shops. Top quality nests have been known to fetch US $8,000 a kilo in Hong Kong!

At dusk all bats leave the caves to forage in the forest, something many visitors to Niah like to watch. Nighttime is always special in the forest, when the wind dies down and the frogs and the insects start calling and the mammals become active.

At Niah the bats seem to sneak out the side exits and quickly disperse around the trees. The swiftlets come in soon after and actually provide the more impressive display. It is almost dark when they seem to fall out of the sky in thousands and swoop through the main entrance. It is one of the wonders of nature that they manage to locate their own nest in the dark, among hundreds of thousands of other nests, without colliding with another bird. It is an outstanding spectacle that leaves the onlooker puzzled and amazed.

eft, adopting n almost umanlike osition. ight, facing p to Cantor's oundleaf orseshoe at.

BAKO NATIONAL PARK

Bako National Park, established in 1957, is Sarawak's oldest and smallest park and a great place for the visitor to experience unique Bornean vegetation and wildlife.

The jumping-off point for Bako is Kuching. A two-hour ferry-ride or 40-minute bus journey takes the traveller to **Kampung Bako** from where it is a 30-minute-long boat trip to the park headquarters. No roads lead into this park. Alight at **Telok Assam**, where the silence is deafening and the visitor is surrounded by nothing but mangroves and thick, coastal forest covering the steep hills. Listen closely for the peculiar honking call of the Proboscis Monkey coming from behind the landing pier.

At Telok Assam are the park headquarters, information centre and accommodation for visitors and staff. The trail-system at Bako is exceptionally good with 16 different routes totalling 30 km of trails cover virtually the entire park. The trails that are kept clear and are well maintained with colour-coded markings and bridges across swampy patches. The geology here is dominated by sandstone formations. The coastline is punctuated with deep bays and narrow sandy beaches exposed at low tide, a type of landscape found in few other places on Borneo.

At sheltered corners of the coast, mud deposits allow pioneer mangrove to expand into the sea. A little further inland, but still in the the tidal zone, the mangrove forest takes over with patches of nipa palms growing among the large lobster mounds. The short distance from Telok Assam to **Telok Delima** includes a variety of habitats.

Continue on the **Lintang Trail** up through open *padang* terrain with scrub-like growth and many rare carnivorous plants including pitcher plants, sundews, bladderworts. The hardwood trees in the dipterocarp section of the forest may not be as wide and tall as in other parts of Sarawak but the

Picturesque Bako National Park.

variety of plants is unmatched anywhere: 25 distinct types of vegetation have been described at Bako.

Telok Delima is the best spot to see Proboscis Monkeys, which are endemic to Borneo. Like all leaf monkeys, Proboscis Monkeys are difficult to keep in captivity. They come close to the sea in the evenings to sleep in the forest just behind the shore. Proboscis Monkeys are unmistakable for their orange fur and large bellies, and are very noisy especially when they climb through the low trees but they can also sit still for long periods and observe the intruder from the canopies. This is one monkey that does not come into camp begging for food. If you are lucky enough, you may get a good look at a few; this chance meeting could well be the highlight of your Bako visit. Two groups move about the Telok Delima area, another one north of park headquarters along the **Telok Paku Trail**. Some individuals transfer between the groups but the total population is not more than 30 to 50 animals.

Bearded Pig nbles by on ark grounds.

Many other animals at Bako, attracted by the open beach vegetation, come into the camp at Telok Assam. The Bearded Pig, the only pig found on Borneo, moves into camp on occasions. A programme to translocate the "domesticated" animals back into the forest failed and has been abandoned because the pigs are harmless and quite timid.

Lesser Mousedeer can be spotted at night with a flashlight as they come to the edges of the forest. The nocturnal Flying Lemur, which is locally called *colugo*, is also found. The entire order of Flying Lemurs consists of only two species and is confined to the Southeast Asian region. To see one of these rare primate related mammals clinging to a tree trunk or gliding across to another tree is a treasured experience.

Around Bako are small caves in the sandstone outcrops where 10 bat species, including rare ones, rest during the day. Two species of otter also live in the park. The large Hairy-nosed Otter is rare, but the smaller Oriental Small-clawed Otter can be seen early in the mornings and in the evenings. Directions from the park rangers are required in order to find the places along the coast where the otters appear.

Currently, 153 bird species are on the Bako checklist, including migrants such as plovers, sandpipers and terns found near the coastline. Many arboreal birds can be seen in the open country around the Telok Assam camp. Early in the morning, White-rumped Shama and Magpie Robin come out to sing and Mangrove Blue Flycatchers are always catching insects around the chalets but there are still enough mosquitoes left in the evenings to make conditions a little uncomfortable.

Other flycatcher species, babblers, minivets, flycatcher-strikes, ioras and warblers compete for insects. Barbets, Asian Fairy-bluebirds and bulbuls emerge from the forest to feed on fruiting trees in the camp. Along the trails are different kingfisher, cuckoo, woodpecker, flowerpecker and sunbird species. Keep an eye open for the small endemic, seed-eating Dusky Munia, which is unique to Borneo.

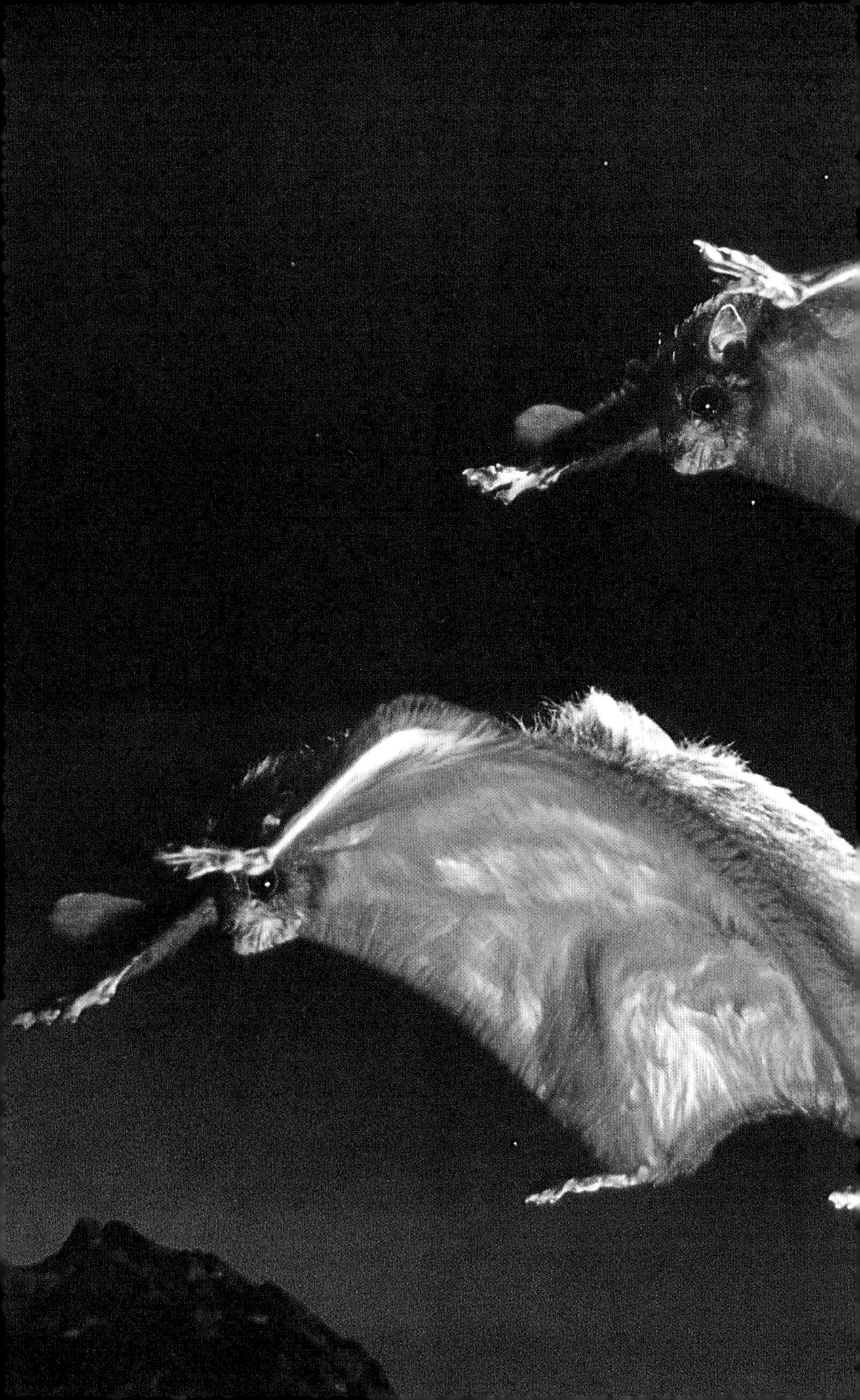

SUMATRA PARKS

Indonesia is by far the largest country of the region with a west-east extension similar to that of the United States of America. The western part of the country, Sumatra, West-Java and Kalimantan, is in a region of tropical wet climate, while areas east of central Java become increasingly dry with a seasonal climate. The easternmost part of the country, New-Guinea and its surrounding islands (an area not covered in this book), again lies in a region of moist tropical climate.

Sumatra is the westernmost of the main islands of the Indonesian archipelago. It is Indonesia's second largest island. Straddling the equator, it stretches 1,700 km between latitudes 6 °N and 6 °S. A wide array of habitats are found, supporting a broad altitudinal range of vegetation types ranging from lowland rain forest to alpine vegetation. Sumatra is home to a number of conservation areas which are of major international importance being among the largest and richest reserves in terms of biological diversity.

Undoubtedly the most famous reserve is Mount Leuser National Park. It and the Kerinci National Park together cover more than 25,000 square km and provide strict protection for the endangered Sumatran Rhinoceros.

On the east coast are two large conservation areas, Berbak and Sembilang, protecting examples of lowland swamp forest. Off the west coast are the Mentawai Islands. As a result of their long separation from the Sumatran mainland, these islands, which include Siberut, are home to a number of unique wildlife species. The latter island is part of West Sumatra province, which has some of the most spectacular scenery in Indonesia.

Mount Leuser National Park: The Mount Leuser National Park was established in 1979 with a total area of almost 10,000 square km, divided into four management areas. The greater part of the area had already been given various degrees of protection since 1934, to protect the largest remaining area of undisturbed tropical forest in the region. Leuser provides a vital refuge for Sumatra's largest mammal species such as the Sumatran Rhino, elephant, and tiger. Like Kerinci-Seblat National Park, it covers part of the Barisan mountain range that runs the length of the island of Sumatra. **Mount Leuser** is Sumatra's second highest mountain, with its summit at 3,445 metres. The park has become world famous for its rehabilitation centres for the highly endangered Orang-utan. Two such centres exist, the first of which was opened in 1971 in **Ketambe**, the second at **Bohorok** in 1973. The centres, managed by the Indonesian park authorities, with funds from the Frankfurt Zoological Society and WWF, seek to reintroduce confiscated Orang-utans, mainly young animals, and have them rehabilitated to live in the wild.

The Leuser area is characterised by heavy folding in a southeast-northwesterly direction, parallel to the main Barisan chain. It has given rise to the three major mountain chains of Serbolangit, West Alas and Leuser Sirupalli which are divided by a series of low lying valleys, sometimes with beautiful gorges, where limestone formations are found. These valleys include the densely populated enclave of the Alas-Renun Rift, which cuts the park into an eastern and western half. Near Kutacane, which is six km wide, the rift is widest. The valley of the Mamas River is about 24 km long and about 10 kilometres at its widest point. Officially, the settlements in the Alas valley are not part of the park. As these valleys are completely surrounded by lowland rain forests, incursions by farmers are a recurrent management problem. Most of the Leuser forest is drained by the Alas River into the Indian Ocean, while the most easterly part of the park runs off to the Straits of Malacca.

Not surprisingly, with the park encompassing such a large array of habitat types and ecosystems ranging from lowland swamp forest to alpine scrub, Leuser contains rich flora and fauna.

Preceding pages: airborne Giant Squirrels; coastline of Meru Betiri National Park. Left, a Sumatran Orang-utan.

In total the park lists an impressive 320 species of birds, 176 species of mammals, 194 species of reptiles and 52 species of amphibians. As such, Leuser harbours some 55 percent of all Sumatran fauna elements. Although holding a smaller population of Sumatran Rhino than the Kerinci National Park, the estimated 200 specimens may still represent up to 20 percent of the entire world population.

The Sumatran Elephant is mainly found along the Silukluk and Manas river valleys as these sites are rich in salt licks. Its main distribution, however, is located outside the park in the lower Alas River, where a total of 200 are known. Still, there is a good chance to see elephants in the wild as they are known to use certain wildlife trails in the park.

Two large species of deer are the Sambar, found almost everywhere, while the Barking Deer prefers lower altitudes. The two smaller deer species Lesser Mousedeer (*Kancil*) and Large Mousedeer (*Napu*) are only common in the lower parts of the park. Large carnivores include the Malayan Sunbear, which has probably a fairly high density in the park, and the Wild Dog. They are common in the Upper Manas area between 1,000 metres and 1,500 metres. Other noteworthy animals are Golden Cat, Leopard Cat and Clouded Leopard. As many as seven species of primates have been recorded.

Wildlife observation in the valleys are most rewarding with easy sightings of otter, Leopard Cat, Hog Badger and many species of monkeys, of which the agile and pretty faced Thomas' Leaf Monkey and the Long-tailed Macaque are most common.

Access to the park is limited. The best area to visit, with easy access from Medan is the **Bohorok**, **Bukit Lawang Orang-Utan rehabilitation centre**. From Medan, the asphalt road leads all the way to Bukit Lawang. Many visitors facilities, including a camping ground and several well-kept bungalows exist. Many trails are available for jungle trekking. Swimming in the clear, cool

Bukit Tinggi, gateway to Kerinci National Park.

Bohorok River is a great treat after a day of hiking.

A good tarmac road runs from Kabanjahe to Kutacane, and gives access to the central section of the complex. A further 25 km to the northeast is the world renowned **Ketambe Research Station**.

The climate of the Leuser Park, like most of mountainous Sumatra, is characterised by a high annual rainfall and absence of a distinct dry season. At lower altitudes, temperatures are high and constant. Cloudiness is common with mist in the mountains during most of the day. Frost has been reported in the mountain valleys and at the summits due to strong radiation during clear, windless nights.

Kerinci-Seblat National Park: Some 200 kilometers to the east of **Padang**, the capital of West Sumatra province, lies the **Kerinci-Seblat National Park**. One of Indonesia's biggest reserves, covering almost 15,000 square km, it embraces parts of no less than four provinces. It incorporates most of the Bukit Barisan mountain range, Sumatra's backbone, and Bengkulu and Padang provinces. As the site has been protected since 1929, most of the reserve's forests have been spared the onslaught of commerce. Safeguarding vital water-catchment areas of Sumatra's two largest rivers: Batang Hari and Musi, its long-term protection is of paramount importance and is of serious government concern. Millions of people, living in the populous provinces of South Sumatra and Jambi are dependent upon a stable water supply for drinking and agricultural purposes. In 1982, in recognition of its environmental importance, the government declared it a National Park.

The spectacular volcanic cone of **Mount Kerinici** at 3,800 metres is one of the park's most striking physical features. Other high peaks include Mount Rasam (2,566 metres) and **Mount Tujuh** (2,604 metres). There are many large rivers, with rapids and scenic waterfalls, sulphur fumeroles and hot springs. The crater lake of Mount Tujuh

Fungi of the genus *Cookeina* occur in all tropical forests.

at 1,996 metres is 11 square km in size and varies in depth between eight metres and 40 metres. Of special botanical interest is **Lake Bentu**, a forested bog. East and West of the central Barisan chain, the mountain landscape slopes to the inland eastern plains of Sumatra and to the coast respectively. The lowest parts of the park are 50 metres above sea level. Part of the park is a flat bottomed valley enclave covering a total of 1,400 square km, which includes **Kerinci Lake**, which covers about 60 square km. Situated at 783 metres, the enclave has a population of some 300,000 inhabitants, posing a serious threat to the hill forests of the volcanic lake.

Kerinci provides a broad spectrum of Indo-Malayan flora and fauna. Of particular interest is the *rafflesia*, the world's largest flower and *amorphophallus*, the world' tallest flower.

Kerinci is the heartland of one of the world's rarest mammals: the Sumatran Rhinoceros, thought not to exceed 500 to 1,000 in number. Due to decades of uncontrolled hunting and habitat destruction, they now exist outside the park only in small isolated populations. Kerinci-Seblat probably contains the largest contiguous world population of rhinos with as many as 500 beasts still roaming the jungle. Protection of this species and its habitat is the primary objective of this national park. Other rare and vulnerable mammals include the elephant, tiger, Clouded Leopard, tapir and six species of monkey. One of the most rewarding monkeys to observe is the Siamang. Its loud and melodious morning calls when it is setting out its territories will make a lasting impression. Both males and females have a large distinctive naked air pouch beneath the throat, which they use to vocalise. It is a highly acrobatic mammal which likes to jump from branch to branch. The Clouded Leopard is a magnificent but little studied big cat because it is largely nocturnal and arboreal. It is much more threatened with extinction than any other Asian felid as it is much less capable of adapting to environ-

In Sumatra, remember to brake for tigers.

ments dominated by man. The park also harbours packs of Asian Wild Dog which range over large areas.

At least 11 Sumatran mammals are restricted to the mountains, including three species endemic to Sumatra: two species of Kerinci Rat and the Sumatran hare, more accurately named the Short-eared Rabbit. Another mountain mammal worth mentioning is the serow, a solitary species of goat, related to the takin and goral, two species of goat in Asia. The serow lives on steep forested slopes browsing herbs.

The park also supports a rich avifauna, including the endemic Schneider's Pitta, rediscovered in 1988 after 70 years. Other rare bird species include Kerinci Scops-owl (known from only one specimen collected), Salvadori's and Argus Pheasant, six species of hornbill and many beautiful coloured trogons and barbets.

Kerinci has great potential for both international and domestic wildlife tourism, being close to both Singapore and Jakarta, but only at **Sungai Penuh**, accessible by air or road, are modest lodging facilities available.

Climate: Due to its geographic position, climatical conditions are extremely variable within the area and range from humid tropical to temperate, depending on altitude. In the higher regions the temperate climate contributes to the presence of cloud forests. On the summits of most volcanoes, the weather can be very rough with temperatures down to 5°C, and thick clouds covering the mountains after 9 a.m. during the dry season. The Kerinci enclave, from where excellent jungle trekking can be enjoyed, has a cool climate.

Berbak Wildlife Reserve: The **Berbak Wildlife Reserve** is located on the east coast of Jambi Province, fringing the Straits of Malacca. Established in 1935, this 1,750 square km area represents prime peat forests and associated fauna, of an extent not found elsewhere in the Indo-Malayan region. Its southern border is formed by the **Benu River**, which also delineates the borderline between Jambi and South Sumatra province. Its

Heavy rain cuts deep canyons into the volcanic soil of western Sumatra.

northern boundary was formerly formed by the Berbak River and Jambung promontory. Due to early incursions by agricultural settlers in the 1950s, the reserve boundaries have twice been modified and 250 square km were degazetted. Berbak forms part of the continuous lowland area found along the eastern coast of Sumatra. Covering some 100,000 square km, it stretches almost one-fifth of the length of this island, from Sumatra's most northern tip in Aceh, down to Lampung province but very little of this huge area is accorded protection.

The Berbak Reserve is unique for its vast area of hitherto undisturbed peat swamp forest and has the largest extent of freshwater swamp forest within any protected area. For nine months of the year, large areas of the reserve are inundated, leaving the higher river levies a refuge for large mammals. Peat swamp forest derives its name from the accumulation of organic material. Leaf litter has formed a layer of peat, which in the case of Berbak can reach up to 12 metres. The peat is acidic and dome-shaped, so that the only input from water is from precipitation. The corresponding increasing infertility towards the peat dome is reflected by a decreased canopy height. Peat swamp forests have few trees exceeding 40 metres in height, and are of smaller stature than freshwater swamp forest trees which can reach 50 metres with emergents up to 70 metres. Both forest types are poorer in wildlife than lowland rain forests. The Air Hitam Laut River flows through the reserve and forms an excellent entry point. Nearer the river's estuary the riverine fringe consists of nipa palm, which can reach heights of 15 metres.

The reserve is inhabited by hunter-gatherers of the Kubu tribe, still maintaining their traditional livelihood. These indigenous people are the original inhabitants of the east Sumatra swamps and have a comprehensive knowledge of forest resources.

The reserve is famous for its tigers which, although seldom seen, are not

Swamp fores of eastern Sumatra.

infrequently heard. It is thought to be the most common large mammal of the reserve and lives off wild boar, deer and monkeys.

Berbak has a rich primate population. Six species occur, of which the Siamang is most readily detected because of its impressive loud cry. Both the Estuarine Crocodile and False Gharial are common reptiles of the rivers, which are also home to large snakes and many species of turtles. Monitor lizards are a common sight along the river banks.

About 240 species of birds belonging to 50 families have been recorded in the reserve. All the Sumatran kingfisher and lowland hornbill species have been seen. The mudflats in the outward fringe of the mangroves are important staging and wintering area for a great many water bird species.

Apart from some patrol tracks in the forest, there are no roads. In fact the reserve can best be appreciated while floating downriver. The best time to visit the area is from June to October, when receding water levels make parts of the forest accessible and allow visitors to appreciate the beauty of a jungle walk. Lying near the equator, Berbak is extremely hot and humid and best hours of the day for wildlife observation are early morning and late afternoons.

Sembilang Protection Forest: Although not a nature reserve, but a Protection Forest, Indonesia's lowest grade of conservation area, the Sembilang swamps are of immense biological richness. The area comprises one of the largest swamp forests in Indonesia and includes the very large delta system of the **Banyuasin River**, many tidal creeks and numerous mangrove rivers. It also features vast areas of inter-tidal mudflats along the coast, some of which are more than two kilometres in width at low tides. An area of 3,875 square km is currently under review as a Wildlife Reserve. The proposed Wildlife Reserve lies in the north corner of South Sumatra province and like the Berbak Wildlife Reserve forms part of a continuous belt of tidally influenced lowland forests. The Musi and Banyuasin

Rattan Palm poised over a forest stream.

rivers carry a heavy silt load and deposit rich silts along the shore. As a result coastal accretion is rapid and may reach in some areas 30 metres per year. On these muds, and under the humid tropical climate, mangrove trees colonize. At Sembilang, quite exceptionally, a mangrove belt of up to 35 km wide fringes the coast. The only other place where such extensive belts are known is the Sundabars in Bangladesh. It has undisturbed transitions from the mangrove belt to the back swamps, consisting of freshwater swamp and peat swamp forests.

Sembilang is of global significance as a centre of mega-diversity. Some 35 globally threatened wildlife species are found here. As a prime wetland it is of chief importance for a number of waterbirds. For some, such as the Milky Stork, Sembilang is the last stronghold in the world. The peninsula between Sembilang and Banyuasin River holds at least one large colony. Up to 20 percent of the world population is believed to be confined to this area. Another stork species, Storm's Stork, is even rarer. Although formerly distributed throughout Southeast Asia, large-scale reclamation of mangrove and swamp forest has brought this little-known species to the brink of extinction. Its global population may number less than 500. Sembilang represents its core area, with regular sightings of groups of five to seven birds.

Sembilang is probably best known for its importance as a staging and wintering area for a great many wader species. The rich mudflats in front of the mangroves, teeming with all kinds of invertebrates such as many species of mudskippers, bivalves and crabs, provide food for an exceptionally large variety of resident and migratory waterbirds. So far 27 species have been recorded and a total of one million birds are likely to pass through the region between September and December.

For species such as the Asian Dowitcher, Sembilang's most common wader, the area constitutes its prime wintering grounds. Some 13,000 dow-

Abandoned *kelong* provides a convenient roost.

itchers are believed to utilize Sembilang during the northern hemisphere winter. Other noteworthy species are Nordmann's Greenshank and Great-billed Heron. It is the only area in Southeast Asia still containing a breeding population of Spot-billed Pelicans, and it has the largest world population of Lesser Adjutants. Indeed, the Sembilang area is a birdwatcher's paradise, with over 300, or 90 percent of all Sumatran lowland species recorded. Birdwatching trips are best done by boat. The mangrove fringe is most impressive with trees reaching heights of 35 metres. During the early morning hours many mangrove birds can be seen. With some luck the endangered White-winged Wood Duck can be seen flying over the mangrove river. Hornbill species include the Helmeted Hornbill, although rarely seen as it is a rather shy bird. Indeed, many of the 300 species are difficult to detect as they prefer the upper storeys of the swamp forest. The area is also rich in birds of prey, of which not less than 19 have been recorded including the magnificent White-bellied Sea-eagle, the largest of birds of prey and Grey-headed Fish-eagle. Both species are often found perching on high mangrove trees and fishing poles that overlook open bodies of water.

The coastal waters also hold important numbers of the Indo-Pacific Humpbacked Dolphin and the endangered Irrawaddy Dolphin, both of which can be seen foraging in the shallow sea and mangrove rivers. Herds of up to 10 individuals of the latter species can regularly be seen following the tide many kilometres upriver.

Although the Sembilang area can hardly be penetrated on foot because of the daily tidal inundations, many large mammals including tiger, Clouded Leopard, Sumatran Elephant, tapir and three species of otter are found. In the mangrove belt, three species of monkey occur: Long-tailed Macaque, Pig-tailed Macaque and Silvered Leaf Monkey. This latter one is mainly arboreal feeding on the shoots of the mangrove trees.

stuarine rocodile isplays an npressive et of teeth.

The macaque species are terrestrial and can easily be observed on the riverbanks and mudflats feeding on crabs, seedlings and leaves which have been washed ashore. The swamp forests are home of some other monkey species including White-handed Gibbon and Siamang. The Estuarine Crocodile is still present, although in greatly reduced numbers because of excessive hunting.

Because of the difficulties of access, much of the area is still virtually undisturbed. So far only researchers have visited the area. Tourist infrastructure is limited. The mangroves are best explored by speedboat. This will give the visitor the chance to see many wildlife species of the riverine forest and mudflats. Visitors should stay for at least the first night in **Sungsang**, where very basic accomodation (but great seafood), is offered. Boats can either be charted at Palembang or at Sungsang. Because of the rough sea during the northern monsoon, boat rides along the coast can be bumpy and not without danger. In return the visitor will appreciate one of the last great wildernesses on earth.

Siberut Wildlife Reserve: This 965-square-km wildlife reserve encompasses the greater part of **Siberut Island**. This island, part of the Mentawai Islands chain, is world famous for its four species of endemic monkeys. In fact, 60 percent of the land mammals found in Siberut are endemic at some level. In total, the Mentawai Islands have 10 endemic mammals and one endemic bird. Siberut Island is separated from the Sumatran mainland by the deep Indian Ocean. The island is beautiful with many secluded bays with extended sandy beaches, especially on the east coast, capes, inlets and one of the best coral reefs in the region, which is found off the central east coast. The sea teems with turtles and cetaceans, and herds of dugongs are regularly seen. Four major forest formations occur. Mangrove forest fringes part of the coast, especially on the east side. Inland freshwater swamp forest and lowland dipterocarp rain forest thrive in the island's central part. Barringtonia forest is confined to the west coast, with tall Barringtonia and Casuarina trees.

Siberut's main attraction are four endemic primates: one gibbon (*Hylobates klossii*), one species of leaf monkey (*Presbytis potensiani*), one species of macaque (*Macaque pagensis*), and *Sismias concolor*. No other island of this size has any endemic higher primates, and no other island in the world has nearly as many endemic primates per unit area. The largest mammal on the island is the Sambar Deer. A large variety of birds is found which include many waterbirds such as frigate birds, egrets, herons and the endangered Woolly-necked Stork. Of the total of 112 species of birds recorded, 12 have evolved distinct subspecies. Siberut hosts one fully endemic bird, the Mentawai Scops-owl. Of the reptiles the amphibians worth mentioning are the Estuarine Crocodile and the three species of turtles which breed on the island — green, hawksbill, and leatherback.

The Siberut people are among the most archaic in Indonesia, still retaining their traditional forest-dwelling lifestyle of hunting-and-gathering. The social life of the Siberut, who are wholly egaliterian, centres around the *uma*, a communal long-house which holds a group of people related through common ancestors. Each *uma* is located within a clan territory which comprises a part of a river catchment. The Siberut have an incredible detailed knowledge of useful forest plants and a strong sense of harmony with their environment. They have a different language and totally different customs to those of the Minangkabaus of mainland West Sumatra. Their major source of food is sago palm which grows in the inland swamps.

Access from the mainland is by regular Perintis ships which ply the Padang-Siberut journey once a week, leaving from Padang's Muara harbour on Monday nights. Access has become more convenient recently after an airstrip has been built and travel agencies now operate Padang to Siberut flights.

Right, portrait of a Fishing Cat.

UJUNG KULON NATIONAL PARK

Ujung Kulon National Park is best known as the last place where the Javan Rhinoceros survives. Formerly widespread, this small rhino is now limited to a population of about 50, all living within the park. The casual visitor is unlikely to see a rhino as their hearing is acute and they are extremely shy. However, three-toed footprints in the soft mud — and piles of rhino dung — are often seen.

Peucang Island: Although the park is connected to Java by land, the best way to get there is by sea. It takes six hours from the port of **Labuan**. During the journey flocks of terns overtake the ship, Frigate Birds glide overhead, and silver Flying Fish break the surface of the water and skim along it for tens of metres. Most people stay at **Peucang Island**, where even a day spent relaxing at the guest house can prove profitable in terms of seeing wildlife. Rusa stags with their hinds wander out of the forest in the afternoon and, after dark, startle the unaccustomed with their loud whistling snorts. Long-tailed Macaques and two-metre long Monitor Lizards surround the guest houses begging for food — or simply stealing it. Southern Pied Hornbills flap past with their wheezy wingbeat advertising their approach and White-bellied Sea-eagles perch in the trees by the shore.

A gentle walk across the island offers excellent examples of strangler figs in all stages of development, from the first few tendrils reaching for the ground, to a mighty network of thick branches clasping the host tree.

The beach near the guest house is a classic curve of fine white sand shaded by pandanus trees. The sea is warm, shallow and calm, while a few outcrops of coral provide something of interest for snorkellers.

The mainland is a few minutes boat ride from Peucang. A path leads inland first through the mangrove swamp where land-crabs scuttle away into their enormous burrows at the sound of feet. Nipa palm fronds wave above the pencil-like roots of mangrove trees which stick up out of the mud at low tide. Not far from the shore is the grazing ground of **Ciujungkulon**, with an old (but still usable) watchtower at one side. In the early morning and late afternoon herds of banteng come to graze here.

The other main grazing ground is at **Cihandeleum** on the western side of the peninsula. Here are banteng, pigs, and peafowl. At both places Magpie Robins hop from the low branches of the trees to the ground and back, White-breasted Wood-swallows with their distinctive triangular shape glide out from the nest-hole in a dead tree, and a Banded Broadbill, with its extraordinary yellow and maroon plumage, blue eyes and turquoise bill can be spotted skulking along a branch in thickly-foliaged trees. The endemic Javan Kingfisher, one of the largest kingfishers and attractive with its purple and turquoise colouring, also occurs here.

There is another guesthouse at **Handeleum**, also on an island close to the

eft, an xtremely are Javan hino. Right, lalaysian ight Heron.

mainland. The landing-place normally used is near the mouth of the **Cigenter River**, a short walk from the Cihandeleum grazing ground. Landing on the beach can be quite an exciting experience as the rangers take people ashore from the larger boat in a tiny dug-out canoe, powered only by paddles.

Once through the surf, however, the canoe proves a perfect craft for exploring the Cigenter River. For most of the year this is silent and slow-moving, and the gentle dipping of the rangers' paddles hardly disturbs the peace.

Cibunar and the South Coast: From **Ciujungkulon**, in the north, a pleasant day's walk is to head south across the peninsula to the beach at **Cibunar**. The walk is about three hours in either direction and involves a small amount of scrambling into and out of riverbeds and a short, easy climb over **Mount Payung**. Forest birds, most often making their presence known by their calls, are present in large numbers. An Asian Fairy-bluebird may give its sweet, tuneful song and quiet, patient waiting may allow a glimpse of it through the trees, its bright blue crown and back contrasting with its black wings and underside.

Where the path leads on to the southern coast, the sea shows a completely different aspect from its gentle lapping along the sheltered northern shore. Blocks of black volcanic rock stick out into the sea eroding into pocks and knobbles as the waves crash down on them. Other than tiny Christmas Island, several hundred kilometres away, the closest land-mass to the south is Antarctica, so nothing breaks the force of the waves until here. Starting at **Taman Jaya** and following the coast path all the way round the peninsular is an alternative way of getting into the park. It takes two days to reach Peucang on foot, even taking the longer route via the dramatic west coast rather than the shorter one across Mount Payung.

Krakatau: The park includes what is left of the volcano **Krakatau**. On 27 August 1883 it erupted, blowing most of itself into the atmosphere and forming a gaping, red-hot hole in the sea-

Much of Ujung Kulon National Park is untouched by humans.

bed. The water rushed back into this and a second massive explosion occurred, resulting in a tidal wave which swept over the neighbouring coastlines and killed 36,000 people. The wave was even recorded in the Atlantic all the way to the shores of Britain, and the tremendous noise of the eruption was heard in Brisbane, 4,000 km, away. Indonesia is dotted with such volcanoes: if such an eruption and tidal wave occurred today near Java, ten times as many people would be killed.

Three vestiges of the original island surround the new peak of **Anak Kratatau** (Child of Krakatau) which appeared from the sea in 1928 and which has since grown to a height of 150 metres. It has been the sight of considerable scientific research, as successive waves of biologists and zoologists have visited the island to study the successive groups of plants and animals which colonised. it. A substantial stand of trees occupies the beach and grasses and bushes are beginning to stabilise the steep slopes of dry black lava. Insects have been carried in by the wind and by birds, and rats and snakes are also present. A number of bird species breed on the island. Peregrine Falcons find the crater walls a congenial place to nest, and the harsh screech of the Collared Kingfisher indicates its presence in the coastal area.

Anak Krakatau can be something of a disappointment after the long and often uncomfortable sea-journey needed to reach it. It is not one of Java's most spectacular volcanoes and is rather difficult to climb. Even the initial scramble up the first slope taxes most people, while the final climb up loose, sliding volcanic gravel and boulders is frankly unpleasant. To look down into the crater at the summit can make the effort worth it, yet the experience can be further marred by dramatic, swirling clouds of throat-catching sulphurous fumes.

Nevertheless, visiting the site of such a historic eruption is an inspiring experience , and can easily be accomplished as a day-trip from the coastal settlements of Carita, Anyer or Labuan.

eopards
unt under
over of
arkness.

MOUNT GEDE-PANGRANGO NATIONAL PARK

This mountain reserve features twin peaks of **Mount Gede** and **Mount Pangrango**. It has the best developed trails of any national park in Indonesia. It is just two hours south of Jakarta along the main Bogor-Bandung road, and is by far the most popular park with local residents. Mount Gede-Pangrango is an excellent place to visit because of its accessibility and its history, and because its variety of resources offer a satisfying visit to those with anything from half a day to a week to spare. As an added attraction it has a comfortable, Dutch-built guest-house.

Adjacent to the park are the historic **Cibodas Botanic Gardens**, established by the Dutch in 1830 as a high-altitude subsidiary to the gardens at Bogor.

Mount Gede-Pangrango was first protected by the Dutch in the mid-19th century as a hunting preserve. At that time West Java was almost entirely covered by primary rain forest. Little now remains but the sub-montane, montane, and sub-alpine moss forests of the park are the best surviving examples in Java. Tigers and rhinos used to roam here, but the biggest animals found today are wild pigs, deer, leopards and the endemic Javan Gibbon.

A surfaced path into the park leads first of all to the **Cibeureum Waterfall**, reached in an hour from the botanic gardens. The forest is in excellent condition here; the waterfall itself is a lovely place, with the streams falling out of the forest down a sheer cliff covered in water-loving plants kept constantly damp by the spray. Sitting quietly here should allow several birds to be spotted: a flock of bright scarlet and black Sunda Minivets, or a pair of Lesser Forktails hunting for insects under the fall of water. Mount Gede-Pangrango is home to 260 species of bird, including 20 of the 23 Javan en-

Hiker about to attempt a minor crossing.

demics. Keen bird-watchers head for this park as the best place to spot these endemics. They might see a family of Chestnut-bellied Partridges crossing the path in search of food on the forest floor, while the three endemic barbets — the Brown-throated, Javan, and Blue-crowned — will be more easily heard than seen as they give their repetitive calls from the tops of the trees. Java's smallest bird, the Pygmy Tit, can be identified by its tiny size, and the Red-fronted Laughing Thrush gives its presence away by its noisy chuckling.

Higher Ground: There are three recognised camping grounds on the mountain, the lowest of which is **Kandang Batu**. This is reached about two hours after Cibeureum, along a trail which becomes steeper and rougher with increasing height. Near Kandang Batu the path crosses some hot springs where water heated underground bubbles to the surface. At **Kandang Badak**, a kilometre or so beyond Kandang Batu, is the second camping ground. Campers sleeping at these sites may be woken by the exhilarating, whooping calls of the Javan gibbon, and during the day may see families of these or other primates.

The forest at these altitudes is excellent for viewing various large, attractively-marked doves and pigeons, including the Pink-necked Fruit Dove and the Wedge-tailed Green Pigeon.

At Kandang Badak the path divides, one path leading up Pangrango, at 3,029 metres slightly the higher of the two peaks, and the other leading up Gede, 2,958 metres. As the trail climbs higher it gives an excellent illustration of how vegetation changes with altitude. Kandang Badak is at 2,400 metres above sea-level, and below it the path passes through sub-montane to montane forest, while above it are the stunted, moss-draped trees of sub-alpine cloud forest.

Pangrango is now extinct, though Gede is still active — a fact advertised by the choking sulphurous fumes gusting from the crater walls, home to the rare Peregrine Falcon. Just below the summit of Mount Gede on the opposite side from the main path is an *alun-alun*, or open grassy area, known as **Suryakencana** and used as the park's third camping site. Growing all over the *alun-alun* is the Javan edelweiss, for which the area is famous: an unmistakable grey-green plant with white flowers. Camping at Suryakencana is a bracing experience for which warm gear is essential — night temperatures can fall below freezing, and hail-storms can occur. But to watch the dawn from here, as the sun gradually lightens the inhabited plains hundreds of metres below, makes it worth the discomfort.

From Suryakencana it is possible to walk down the other side of the mountain, to the south, leaving the park at **Mount Putri**, on a road which leads to the town of **Sukabumi**, a shorter but much steeper path.

Because of the large numbers of visitors to the park, the mountain trails are closed for part of the year to allow the vegetation to recuperate. Exact dates depend on the weather, but the park is generally closed between late December and March, which are the wettest months.

he scene n the upper nontane orest of Mount Gede.

BROMO-TENGGER NATIONAL PARK

For spectacular scenery, few places can match the primordial volcanic landscapes of **Bromo-Tengger National Park** in East Java — an overnight excursion from Surabaya or Bali. A lodge has been built on the lip of a massive caldera, literally suspended above the clouds. The cluster of volcanoes is topped by the still-active **Mount Semeru**, at 3,676 metres, the highest mountain in Java. It is honoured by Hindus as Maha Meru, the greatest of all gods. The mountain complex further includes the Bromo and Tengger area with its many legends and its Tenggerese people, one of the few surviving mountain tribes from the Majapahit-Hindu period. **Mount Bromo** is the scene of an annual pilgrimage when thousands come to the crater edge on "Kasada" day on the 14th or 15th of the 12th month of the Javanese calendar. The climate is very pleasant and is a welcome relief from the hot and moist lowlands. The landscape of several volcanoes, in particular Mount Bromo, surrounded by the Tengger "sand sea", forms a most attractive scenic area in the middle of Java. Most visitors start off for Bromo in the very early morning and leave by foot or on horseback from Ngadisari and Cemoralawang. They descend into the Tengger caldera and cross the sand sea for eight km before climbing to Mount Bromo's crater edge to catch the breathtaking sunrise. As the sun rises over the "sea", there is the feeling of being at the "world's first morning". The park includes a mountain complex of 510 square km which is very important as a water catchment area for the highly productive and fertile areas downstream which supports one of the densest populations of Indonesia. The vegetation of the park contains little primary forest. The lower slopes are mainly covered in commercial forest plantation and generally disturbed forest to about 1,500 metres. Part of the area is covered by attractive pine-

Catching early mornin rays at Bromo's crater edge.

like woodlands of *Casuarina junghuhniana*, named after the famous naturalist Junghuhn, who helped to develop medicine to fight malaria. The upper part of Semeru, the sand sea and Mount Bromo and Mount Batok are mainly bare.

The sand sea, enclosed by 200-metre-to-600 metre steep walls, is the caldera floor of a former Tengger volcano and is eight km to 10 km in diameter. Bromo and Batok are deeply fissured and consist of younger volcanic cones now topping the sand sea. Bromo is still active and its smoke bellows skywards.

Visitors who explore the park and climb Mount Semeru pass through the land of the Tenggerese people who grow vegetables for the local markets, cultivating very steep hillsides at the Ngadas enclave and flatter parts at the high plateau of **Ranu Pani** at 2,200 metres which, in colonial times, was one large dairy farm. From Ranu Pani or Rani Lake a path leads to Mount Semeru and then through a scenically attractive area including **Ranu Kumbolo**, a substantial lake enclosed by wooded hills. The region only includes very small areas of lowland forest, such as on the south foothills of Semeru near the Kumbolo lake (960 metres), and does not have a long list of flora and fauna. However, it is of considerable scientific importance as nine plant species, such as *Melastoma zoolingeri*, endemic to the Tengger area, have not been found anywhere else in Java. Other plants mentioned by van Steenis in his mountain flora of Java include *Begonia laciniata*, *Carpesium cernuum*, *Epilobium cinereum*, *Geranium homeanum*, *Hoplismenus undulatifolius*, *Rumex brownii*, *Stellaria vestita* and *Tylophora adnata*. Wildlife to be seen in this area includes leopard, Barking Deer, Rusa Deer and pigs, although numbers are rather low.

Mount Semeru erupts about every eight minutes with longer intervals after bigger eruptions which may occur after about every fourth eruption. Gazing into the crater is possible for one or two minutes only.

The park features a succession of volcanic cones.

Baluran National Park

Just opposite Bali, the famous tourist island of Indonesia, stands **Baluran National Park**, at the northeastern tip of Java. Baluran is a small, easily accessible area and one of the prettiest parks in Indonesia. The extinct Baluran volcano with its exploded top is surrounded by a large savanna and woodland country not unlike an East African plain with tall acacia trees over the short and long grass areas. A unique feature of the park is its variety of habitats over short distances. Within less than 10 km one can see splendid coral reefs, sandy, stony and muddy beaches with mangrove, sheltered bays and more exposed coasts, tall beach forest, grass savanna and woodland, thorn forest, evergreen moist forest, and monsoon forest on the mountain slopes and at the bottom of the crater. Herds of deer roam over the savanna and woodland areas. Wild pigs, Barking Deer, Green Peafowl, Green and Red Junglefowl are abundant and easily seen. The area's most conspicuous large mammals are the water buffalo and the banteng.

Hunting pressure and disturbances have kept animal populations somewhat below optimal densities even since the park became protected in 1936. The area is the driest of the region with only two wet months per year. Competition between man and animal for water resources has been severe and man has disturbed and depleted a number of vital water sources for the wild animals. Only a dozen open water holes are found along the coast and one at the foothills of Mount Baluran in the centre of the park at Talpat and year-long running water found deep in the crater at the River Kacip. This condition provides excellent opportunities for wildlife viewing as animals have to cross the open areas in search of water and many of them must visit the water sources every one or two days.

The tiger, extinct in Bali and for long very rare in Java, is found in Baluran one of its best hunting areas and refuges. Last seen in Baluran in the early 1970s it may still survive at **Meru Betiri Park** in the south coast, where tracks were last recorded in 1988, but numbers are not likely to exceed a few individuals. Baluran is still connected with the Meru Betiri Park through the forests of Maelang, Ijen Merapi and Rauung.

A visit to the watchtower at the small hill of **Bekol** at dawn when the world comes gradually to life is a great experience. The 20-metre high hill gives a view over a wide savanna area with Mount Baluran at one side and the coastal plain at the other. On top of the hill, next to the tower, is a water basin where many animals come to drink. Squirrels arrive along fixed pathways through the trees. The most frequently seen birds are flocks of Peaceful Doves, Javan Turtle-doves and bulbuls. The fantails have their own way of getting water by drinking a mouthful while swiftly diving below the surface. Later in the morning, Green Peafowl arrive after having descended from the horizontal branches of the acacia trees in the savanna edge and savanna woodland. In the savanna flocks of deer and single Barking Deers and Wild Pigs can be spotted. On some days, groups of water buffalo and banteng appear and flocks of Fruit Pigeons and Black-winged Mynas pass overhead.

The Black-winged Myna, a relative of the Bali Starling has a symbiotic relationship with Rusa Deer picking insects from its back. Only a few kilometers away, on the other side of the Bali strait in a similar type of dry savanna-woodland, a few dozen Bali Starlings survive. Strangely, this beautiful white bird has never managed to cross the narrow Bali Strait or to settle in Baluran.

Two species of primates occur, the Black Leaf Monkey and the Long-tailed Macaque.

Leopards are still common in Baluran. Several can be heard calling in the mating season. Their major habitats are the foothills of the mountain, the crater, the thorn forest, the southeast part of the park and the river areas in the west.

Water Colours: Baluran has a coastline of some 40 km with an amazing variety

eft, big
anteng bull
t Baluran.

of coastal habitats. Mud flats, mangrove, coral reefs, coral sand, black volcanic sand, beaches with large volcanic boulders, white sand beaches, small bays and lagoons, are all found in subtle pattern along the coast. Fringing reefs are found over half of the coastline. Reef areas at **Gatal** and **Bamah-Kelor** are good snorkelling areas and within easy access. The reef area at **Labuan Merak** is famous for its garden eels. The mangrove is only a fringe in most places, but substantial at **Bilik** and south of Bamah-Kelor. The mangrove is dominated by *Rhizophora apiculata* and *mucronata*. Locally pure stands of small leaved *Pemphis acidula* occupy the high coral sandbars behind the beach. A small area of bare saltflats occur north of Pandegan at Mesigit west of Bilik.

The original beach forest has shrunk as the result of fire and cutting, but is well developed at Bamah-Kelor and several other places. Locations with a high ground water table occur at a few places behind the sand bars of the coastline and give an intricate pattern of mangrove, beachforest and freshwater swamp. The beach and freshwater swamp areas support a tall luxurious evergreen forest with undergrowth of palms, notably *gebang* palm (*Corypha utan*) at the drier open places and *Caryota mitis* palms at moist locations.

This forest is rich in birds. Hornbills, orioles and kingfishers are commonly seen and heard. Green Imperial Pigeons, Fruit Pigeons, and Pink-necked Pigeons come to rest at night. Almost all mammal species will visit this area which has a number of permanent water holes at regular distances all along the coast.

The savanna is largely confined to black cotton soil, which is very muddy and sticky in the wet season and very dry with deep cracks in the dry season. Very few species can stand the swelling and shrinking forces of this extreme type of soil. The tree species of the savanna are mainly restricted to Pilang (*Acacia leucophoea*), Klampis (*Acacia tomentosa*), Mimbo (*Azidarachta in-*

Baluran provides some basic accommodati for visitors.

dica) and Kesambi (*Schleichera oleosa*). Typical birds in the savanna are Spangled Drongos, Shrikes, Turtle Doves, starlings, tailorbirds, coucals and both Green and Red Jungle Fowl. The Red Jungle Fowl prefers the denser parts of the vegetation, though there is considerable overlap in habitat. The Spangled Drongo can often be seen fiercely defending its territory, even against large birds of prey and crows. Singing Bushlarks flutter in the air and with their sweet trilling endless songs set the atmosphere for this habitat. At dusk, the eyes of the Savanna Nightjar light up the road to the coast.

Monsoon Forest: The monsoon forest consists of a thorny, hardly penetrable forest in the flat areas, and shows an increasing number of evergreen trees in the moister locations at the higher part of the volcano and in the crater. The thorn forest has many lianas and is full of prickly thorn bushes and low trees with generally little undergrowth. These forests are extensive in the southeast part of the park. They support high densities of small wildlife, in particular birds, due to their many fruit trees. Also common are small carnivores and frugivores such as the Javan Mongoose and civet cats. Thorn forest and monsoon forest are a very rare habitat. Outside reserves, most of the monsoon forest has gone — converted to agricultural land or monospecies forest land. Even inside the park many areas have been influenced by fire and converted to savanna.

Baluran can be visited all year round. However, in December to March, the black cotton soils of the savanna are very sticky and hang in big lumps on your shoes, making walking difficult. Baluran offers superb snorkelling and scuba diving opportunities starting from **Bamah**, and within reach by a road up to the coast, or from Gatal entering from the northwest side of the park. Visits to the coastal area, starting from Bamah are best made at low tide when an enormous variety of coastal habitats can be explored on foot north to Labuan Merak.

ry savanna rrain ominates e scenery.

BALI BARAT NATIONAL PARK

The **Bali Barat National Park** features one of the world's rarest and most endangered birds. It is the last stronghold of Bali's only endemic vertebrate animal species, the Bali Starling. Due to poaching, this splendid bird is currently vanishing in the wild from "hundreds of birds" in 1925 to about 125-180 birds in 1984 and a low 23-30 in October 1989. In sharp contrast, the world's captive population — they are sold on the black market in Jakarta for US$250 — is estimated at a healthy 700 birds.

In 1947, some 200 square km in the extreme northwestern tip of Bali was announced as a Nature Park, basically for the sake of the scenery. In 1974, four islets were added to the conservation area, by then a Nature Reserve. Eventually, in 1985, some 500 square km of hitherto Protected Forest on the east-west mountain ridge and about 60 square km of marine area were added and the whole promoted to National Park.

Although the park covers about 700 square km, the most accessible area is the 150-square-km **Prapat Agung** peninsula in the northwest. There are a number of patrol trails suitable for nature watching. The park's list of birds features 160 different species. The tranquil scenery is in strong contrast to the island's vibrant culture.

Roaming Bali Barat is a rewarding, yet not simple undertaking. Sturdy footwear to negotiate the razor-sharp coral rocks hidden under tussocks of tall grass, a bottle of water and a hat will be useful. Others may consider a snorkelling mask, fins and a T-shirt to marvel at the park's underwater world around **Pulau Menjangan** (Deer Island).

A good start for an introductory one-day excursion is the guardpost in **Sumber Klampok** village, 15 minutes by public minibus from Gilimanuk on the road to Singaraja. Directly opposite this post an old dirt road runs straight north for about two kilometres to the park's

Guest-house at Bali Barat has a native flavour.

Research Centre and the Bali Starling Project's Pre-release Training Centre.

Invariably, the first bird seen is the Yellow-vented Bulbul. The bushes bustle with this species. Also seen from this road are the Ashy Tailorbird, the Olive-backed Sunbird and the Brown-throated Sunbird. In the planted *Sawo-kecik Manilkara kauki* forest to the left can be heard the Black-naped Oriole which on Bali has a completely black abdomen. This popular cagebird could be called "the blackbird of Indonesia". Also worth looking at are Pied Fantails tumbling through the undergrowth.

Birds Galore: After one kilometre the open plain, **Tigal Bunder**, connecting Bali's mainland to the peninsula, is reached. Here the skies are spotted with White-bellied Swiftlets and Edible-nest Swiftlets. The Pacific Swallow, the Deft-crested Tree-swift and the White-breasted Wood-swallow, with its peculiar triangular wings, may also be seen. During the northern hemisphere's winter, the Pacific Swallow is often accompanied by the similar, but much heavier, Red-rumped Swallow or the slightly larger but brighter Barn Swallow.

Beyond the regrowing palm savanna on the left is a small patch of the original forest which covered this plain before Dutch settlers largely cleared it for a coconut plantation early this century. Thanks to its promotion to national park status, uncontrolled cattle grazing stopped here in 1985. Perched on the low shrubs and young palms are many Long-tailed Shrikes: the local sub-species has a gray hind-crown, nape and hind-neck. More often heard than seen is the Striated Warber.

Abundant virtually everywhere in Indonesia is the noisy Collared Kingfisher and from June through September the very similar, but silent Sacred Kingfisher, somewhat smaller and with its white underparts tinged buffy-brown. The deep blue Javan Kingfisher, endemic to Java and Bali, the sparkling blue diminutive Small Kingfisher, endemic to Java, Bali, Lombok and Kangean Island, and the Rapid Rufous-backed Kingfisher are regular sightings on this seasonally flooded plain. However, the Stork-billed Kingfisher with its heavy dagger-shaped red bill is a rare visitor.

Both the acrobatic Black Drongo with its red eyes and the spangled drongo with its milky-white eyes are inhabitants of the forest edge around this plain. In the more dense monsoon forest, a little further north, the metallic green Racket-tailed Treepie looks at first sight like a drongo, but is actually a member of the family of the crows.

Quietly perched in solitary trees are Dollarbirds. At sunset, the Savanna Nightjar begins its squealing *chir-chir* call, while a little later the Collared Scops-owl may come out with its mellow, disyllabic *boo-ek* calls.

After nine in the morning, when the air has sufficiently warmed up, the melancholic high-pitched *kweeee-kweeee* of the Crested Serpent-eagle high overhead, begs attention. Apparently not affected by the widespread environmental degradation it is one of the commonest raptors in West Indonesia. In late October and early Novem-

ypical park egetation — avanna and angrove.

ber, a massive west-east migration of many birds of prey comes over this land.

Having reached the crossroads in front of the Research Centre the visitor might wish to pay a brief visit to the Pre-release Training Centre which is 300 metres to the right. Young Bali Starlings, bred in captivity at the Surabaya Zoo in East Java, are trained to adapt to the wild.

Five km west of the Research Centre is Gilimanuk's ferry port. On this stretch, there is a good chance of seeing a number of Rusa Deer. Together with the common macaque in the canopy and the Wild Pig, the rusa was the main prey of the extinct Bali tiger, that had its distribution centre here.

Sitting on the road are Javan Turtle-doves, whose more common counterpart, the Spotted Dove, prefers rural areas. Further on, it may be possible to spot the Banded Pitta, a secretive ground-dwelling forest bird.

Nearer the coast is a spot where, as late as 1983, one could marvel at a roost of 50 Bali Starlings. Nowadays, the fortunate may see an odd pair here, probably at the end of the dry season. Chances are better around the **Teluk Kelor** guardpost on the north coast, where the starling's favourite habitat is found — a savanna with a matrix of tall-grass and groups of *Pilang* (*Acacia leucophloea*) trees with their undergrowth and patches of monsoon forest in the slightly more moist valleys. The typical flat crown and white bark of this acacia, above the parched golden yellow grass, gives the landscape a distinctly African appearance. The park guide will point out the trail from the dirt road north across the peninsula to the Teluk Kelor area.

Waiting at the north coast is a band of Silvered Leaf Monkeys. Alternatively, the Teluk Kelor area can be reached more comfortably by motorboat from **Labuan Lalang** village, offering the possibility for a refreshing dive over the corals around **Menjangan Island** on the return trip. In addition, this 165 hectare-island has 40 to 50 barking Deer and is the only place on Bali where

Rare in the wild but thriving in captivity — the Bali Starling.

one can meet the Mangrove White-eye, a species apparently confined to the mangroves of lesser islands.

Before turning to the left or south into the Bay of Gilimanuk, a large brown-and-white sea-bird — a Brown Booby — may be seen sailing low over the waves. On the sandbank in the entrance of the bay many terns, mostly Great Crested and Common, loaf at low tide.

Some wader species can be spotted on the mudflats, and about six Lesser Adjutants have taken up residence here. A walk along the fringes of the mangroves is likely to flush out one or two Great Thick-knees, which, breeding in the park, reaches the eastern limit of its world distribution in this bay.

Across the bay is the **Nusantara II Hotel**, owned by the colourful "Mr. Moustache", alias Pak Made, the former head of the wardens who, together with his staff, received the *Golden Spoonbill* Award in 1986 for helping to protect the Bali Starling when the bird's plight had not yet attracted the world's attention.

Visitors may consider paying a visit to **Makam Jayaprana**, at **Teluk Terima** village, there is a small temple in the hills. An additional incentive for climbing the 450-step staircase is the magnificent view over Teluk Terima Bay and Menjangan Island.

Dolphin watching North Bali: The popular sea resort of **Lovina Beach**, 11 km west of Singaraja on the north coast of Bali, offers a new attraction to travellers: dolphin watching. The rich calm sea not only teems with all kinds of fish and crustacean species but also features large schools of dolphins. Most abundantly seen is the Common Dolphin. Native fishermen formerly captured these dolphins from small wooden dugout canoes. Now they have abandoned dolphin hunting and, with the same boats, take tourists to the feeding areas of the dolphins, some three to five kilometres offshore. Dolphins can be seen almost every day, with the Common Dolphin, a truly aerial acrobat, approaching to within 20 metres of the small canoes.

colony of
ruit Bats.

KOMODO NATIONAL PARK

Lying in the island-strewn strait between Sumbawa and Flores, some 500 km east of Bali, is **Komodo National Park**. The park's main island of **Komodo** is home to the world's largest reptile, the Komodo Dragon (*Varanus Komodoensis*). This giant Monitor Lizard (called *ora* by the natives) is one of the world's oldest species, closely related to the prehistoric dinosaurs of 100 million years ago, although it was discovered only in 1912. Komodo, together with the park's other islands of **Rinca** and **Padar**, are very different from the ever-wet tropical forests of western Indonesia. These islands, part of the Lesser Sundas, lie in the Australian rainshadow and rain falls only between November and March. The lack of water means that the rugged hills of the national park are covered mainly with coarse grass, with very few trees. Herbivores such as deer, pigs, feral water-buffaloes and horses are well adapted to this environment and all are preyed on by the Komodo Dragon.

The dragon can grow to lengths of over three metres and weigh 150 kgs. They are by far the largest predators in the islands. They have a keen sense of smell and are effective scavengers of dead animals, but they also take live prey. They lurk in the grass waiting for an animal to pass by, then knock it down with their muscular tails before sinking their teeth into it. Even if the animal escapes, the dragon's saliva is so poisonous that the wound becomes infected and the animal will almost certainly die. The carcass is swallowed piece by piece, fur, bones and all. The dragons' efficient digestive system breaks everything down, leaving only a finely-processed whitish scat. Dogs are sometimes abandoned on the island and compete with the lizards as scavengers.

The dragons have no fear of people, and there are rare reports of them attacking both local inhabitants and visitors. Not far from the park's guest-houses, a

Arriving at Komodo.

viewing site has been established, at which a goat is killed on Sundays and fed to the dragons. Until 1989, a feed took place whenever a group of tourists came along, but as the number of visitors increased the dragons grew lethargic, greeting each fresh offering of goat with indifference. So a more natural feeding pattern with weekly feeds only has been adopted.

The majority of visitors to Komodo come only to see the dragons and stay for as short a time as possible. A longer stay, however, provides a more rewarding experience. The national park is in Wallacea, the overlap between Asian and Australian bio-geographical zones. The overlap is particularly striking amongst the avifauna: Sulphur-crested Cockatoos and Noisy Friar Birds — also found in Australia — mingle with birds of Asian origin such as Jungle Fowl and Monarch Flycatchers. Near the guest-house compound is an artificial-looking mound built by Brush Turkeys, which pile rotting vegetation over their eggs to incubate them.

Little grows on the dry coast of this barren island. There is a small native community, living in the only village on the island, **Kampung Komodo**. Behind the village, a path leads up steep, hot slopes and over plains dotted with lontar palms to the top of the island. Deep red orchids sprout from the trunks of the lontar palms while spiders with a 25 cm span cast their webs across dry streambeds. The walk takes most of a day, but the magnificent view from the top of the island is worth the effort: the brown slopes of Komodo contrast with the dark blue of the open sea, the turquoise shallows, and the pure white beaches. A further scramble through low trees leads to the summit of **Mount Ara**, the highest point on the island at 720 metres. Boats can be hired to a **Lasa Island**, across from the village. The snorkeling and swimming here are superb. And, of course, there is always the chance of spotting a Komodo dragon or two, away from the artificial feeding-site where most tourists see them.

The focus of visitor attention.

DUMOGA-BONE NATIONAL PARK

The forest-clad mountains, beautiful bays, beaches and wildlife of the spider-shaped island of Sulawesi, will inspire travellers. Exploring this island with its primitive species such as Babirusa, Anoa, Cuscus, Tarsier and the Brown Sulawesi Civet — presents the opportunity to see some unique wildlife. Because of its long geological history and isolation from the mainland and other islands, Sulawesi harbours many endemic species. Some 30 percent (or 75 species) of the island's resident birds are endemic as are 62 percent of all mammal species or, excluding bats, 98 percent. Through its unique zoo-geographical position Sulawesi includes both marsupial (or pouch) animals, and placental animals. Sulawesi is also where famed naturalist Alfred Russel Wallace was inspired for most of his ideas on evolution and distribution of species during his famous journeys in the transitional zone of the Indo-Australian region nowadays called Wallacea.

The **Dumoga-Bone National Park** extends over 110 km in the middle of the north arm of Sulawesi and covers 3,000 square km of tropical rain forest at altitudes of 50 metres to 1,970 metres. The park harbours most of the island's wildlife and is classified as being the richest in species of all Sulawesi parks. The 250-km journey from Manado can be made in four to five hours. The journey to the west leads through some wide landscapes with splendid sunsets over the sea at **Air Anjing**, about three hours from Manado.

The park was only initiated in 1977, first to protect the water catchment forest around the Dumoga valley in order to guarantee waterflows for the wet-rice cultivation and later to include what is now the whole Dumoga-Bone National Park with both the headwaters of the Dumoga and the Bone rivers.

The park is home to three of the seven species of Black Macaques in Sulawesi. Babirusa, which locally means "pig-deer", is an ancient animal resembling a pig with curiously curled tusks, and is mainly restricted to lowland riverine areas. It was, and to some extent still is, much hunted by the Minahasan tribe east of the park, who also eat monkeys, snakes, rats, bats and dogs which are offered as food in road restaurants. The Babirusa is more common at the Gorontalo side of the park.

Very few people have ever seen the Tarsier although it is quite common and easily spotted at the forest edge. This small creature, no bigger than a Scops-owl, is only active at night and is called by the Dutch "*spook diertje*". Following its high-pitched sounds when it comes out of its hide at dusk in family groups will lead to the bushes where the large eyes of this funny small creature will reflect in a flashlight. They eat insects, small birds and favour snakes. They can move their head as much as 180 degrees and can jump backwards, turning around in the air before reaching another branch.

Snakes are common including strangler snakes such as pythons, which, in

Left, at the western fringe of the marsupial's distribution — a Bear Cuscus. **Right**, a Harlequin Bat.

Sulawesi, attain a length of 10 metres. Only five species out of some 60 snake species are poisonous.

Watch the Birdie: The Maleo bird, a megapode or "giant foot", is a rare island endemic and common in the Dumoga-Bone park. This peculiar bird digs its nesting holes in hot volcanic soil and near hot water sources. They bury their eggs at depths varying from 30 cm to over one metre in search of the right temperature and moisture levels. When the chicks hatch after some two to three months, and work themselves out on the ground in one to two days, they are already able to fly and will take off into the forest without any parental care. Several nesting places are found in the park where one can hear and see pairs of these alert birds digging their nest holes or surveying the area from a tree branch for enemies such as Monitor Lizards, pigs or humans interested in digging up the eggs. The bird itself is not much bigger than a chicken though the egg is five times as large as a chicken egg.

The Dumoga valley landscape is a patchwork of fertile rice fields with many Cattle Egrets, Javan Pond Heron and occasionally groups of white-necked storks. The area around park headquarters, including the Kosinggolan flood plain, secondary and primary hill forest and ridge tops, is good birding country. A few thousand Cattle Egrets, a smaller number of Pond Herons, egrets and Purple Herons come for the night to rest at the reeds and bushes in the flood plain. Groups of Whistling Tree Ducks and Racket-tailed Parakeets often visit the valley and rice fields at night.

The Kosinggolan swamplands are full of cracking noises and often heard are the Purple Swamphen, White-browed Crake and White-breasted Waterhen. At the drier places one finds the Isabelline Waterhen, the Buff-banded Rail, common along roadsides, and visitors often hear the Barred Rail common in secondary growth, its presence with a cacophony of screeches.

Woolly-necked Storks may be seen circling high above the valley or visiting the rice fields. The forest edge is rich in fruit doves, parrots, lorikeets and drongos. A typical and noisy bird is the Sulawesi Forest Coucal, common in the secondary growth, and the Black-billed Koel, easily recognized as being very vocal and sounding too excited. Colourful birds include the Green-billed Malkoha, the Ornate Lorikeet, sunbirds, flowerpeckers and hornbills. Eight species of kingfishers occur, some of them found only deep in the primary forest or restricted to high altitudes, such as the Bar-headed Wood Kingfisher. Blue and purple colours are typical in many bird and insect species of Sulawesi and excel in birds like the Purple-bearded Bee-eater, a rare bird of the deep primary forest. Birds of prey are still common in Sulawesi and include many endemic species. Often heard is the Sulawesi Scops-owl. Four endemic hawks occur, the Spot-tailed Goshawk being the most common. Commonly seen in the valley are the Black Kite, Brahminy Kite, Black-shouldered Kite, White-bellied Sea-eagle and Sulawesi Serpent-eagle.

Recommended journeys include a visit to the **Tambun Maleo** nesting ground at the valley edge, the **Tumpa/Toraut** research forest area; the area around park headquarters and the Kosinggolan flood plain and a forest walk to the **Tumokang Maleo** nesting ground. A one-day visit is recommended to the separate small **Mount Ambang Reserve**, featuring **Mooat Lake** (1,000 metres) and a climb of about three hours to the active sulphur crater. It is only18 km west of **Kotamobagu**. Visit in the early morning before the clouds roll in.

A visit to **Gorontalo** may include a forest walk to the **Lombongo waterfall** which is three km from the forest edge, and trekking along the Bone River from Tulabolo to Hungayono with precious limestone outcrops, hot water sources and Maleo nesting grounds. The path along the Bone River leads through lowland rain forest and swamp areas to the **Pinogu** enclave, an agricultural settlement of a few thousand hectares about 30 km inside the forest. It can also be gained by car by travelling along a more southern route.

Right, Sulawesi Tarsier invites an unwilling grasshopper for a meal.

BUNAKEN MARINE PARK

The best coral reefs of Indonesia, with its more than 13,000 islands, are found in the eastern part, in particular in Sulawesi and Irian Jaya. A magnificent underwater world is found in **Manado Bay** in North Sulawesi. It comprises the five small islands of Bunaken, Manado Tua, Siladen, Montehage and Naen. It offers a great variety of coral habitats, good for many days of diving or snorkelling. Bunaken and **Manado Tua** islands have the highest tourist value, being the the most readily accessible and good for snorkelling as well as for diving. Manado Tua features a crater lake in the centre of its extinct volcano. The coral reefs of **Bunaken Marine Park** are as rich as those of the Australian Barrier Reef and other highly valued reefs of the world. In some parts the visitor may see up to 50 genera of coral within half an hour. Barrier reefs, fringing reefs, lagoonal reefs, mangrove and seagrass areas are all together. The most outstanding feature is the rich invertebrate fauna of the vertical fore reef. A series of wide crevices, also called the grooves, split the outer wall of the sheer vertical drop-off from one metre to 100 metres at the edge of the island shelf. The walls have many small caves and overhangs and are covered by a bewildering variety of colourful sponges, tunicates, criniods, hydrozoans, alcyonarians, antipathians and anemones. **Naen Island** has the greatest variety of reef types and is surrounded by a well developed barrier reef which encloses a lagoon. Much of the shore of Bunaken, Montehage and Naen have fringes of mangrove. About 10 square km of mangrove covers the northern half of Montehage.

Visibility in the water is good up to at least 10 metres, and as the currents are not strong at all, swimmers will be able to concentrate fully on viewing the marine life. Most seasons are good to visit the park, except perhaps when it rains during a few days in the months of

Diving the clear waters at Bunaken.

October to February. The Bunaken area is one of the few areas in Indonesia with experienced local divers and some of them have an extensive knowledge of the coral life.

Probably in contrast to the deep rain forest, which for some people is a moist green cathedral full of noises but without many animals to be seen, coral reefs will seldom disappoint. There is much to be seen including colourful sponges, anemones, flat black coral, massive schools of pyramid fishes, seahorses, clams, blue spotted rays and tuna fish, or turtles visiting the seagrass areas. Three species of sea turtle can be found in the area, namely green, hawksbill and leatherback. Many snails can be found such as cyprea, conus, trochus and turbo species. Lambis, chiragra and ovula ovum are common at certain sites.

Also included in the Bunaken marine Park is the **Arakan reef**, some three hours by boat to the west of Manado, and **Teluk Tanawangko**, up to the bay of Amurang-Tumpaan. This coastal area with extensive shallow coral is not yet much visited. Seagrass areas may still harbour a small population of dugongs (sea cows), a distant relative of the elephant, and perhaps the inspiration behind the many tales of mermaid sightings.

Restaurants and hotels are found in Manado and west of Manado at the Malalayang coast. The development of the Bunaken Marine Park, officially established in 1986 with a total area of 750 square km (Bunaken) and 138 square km (Arakan reef) owes much to Loky Herlambang, who started diving activities more than 20 years ago and trained many divers at the Nusantara Diving Club. He convinced the local government that marine tourism together with protection of coral reefs held a good future for Manado Bay. The village of **Molas**, just north of Manado, and the village of **Malalayang**, west of Manado, form the main locations with diving centres for departures to visit the Marine Park. Boat rides to Bunaken take half an hour from Molas and over an hour from Malalayang.

arine esident — a hale Shark.

PHILIPPINES PARKS

High population density, easy accessibility from all parts of the country for timber transport by sea to its principal customer Japan, and the lack of enforcement of laws to protect its remaining forests have led to the destruction of much of the natural vegetation of the Philippines. However, increasing local and international interest in gazetted national parks together with the interest that is aroused, as anywhere in the world, from critically endangered species, such as the Philippine Eagle, may spell hope for the future of the country's natural resources.

Mount Apo National Park

At 2,954 metres, **Mount Apo** in eastern **Mindanao** is the highest and one of the richest national parks in the Philippines. Created in 1936 to protect the volcanically-formed mountain and the surrounding rain forest, the park is easily accessible from **Davao City** to the east and has four trails to the summit.

The vegetation of the lower reaches of Mount Apo has been considerably disturbed by logging and agricultural activities, the latter supporting bananas, coconut and hemp. Remnants of original dipterocarp forest (*Shorea spp.*, *Parashorea spp.*, and *Dipterocarpus spp.*) assume a canopy of 50 metres or more in height. Between 1,000 metres and 1,500 metres, the closed canopy of the forest ranges from 10 metres to 20 metres high with emergent trees of up 40 metres. The dominant tree genera include *Shorea*, *Quercus*, *Cinnamomium*, *Tristania*, *Eugenia*, *Astrocalyx* and *Ficus* — the strangling fig or *balete*. Some of the older Ficus are hollow, having killed the host tree. These are left undisturbed by the park occupants as they believe them to be inhabited by spirits. Tree ferns (*Cyathea* and *Saurauia*) are found around forest edges along with some palm trees. The conifer almacega

Preceding pages: in the high mountain ranges; attractive man-made landscape. Mount Apo (below left and below) is known as *Sendawa*, or Mountain of Sulfur.

(*Agathis dammara*), much prized for its wood, exists only in scattered remnants.

Up to 2,700 metres the trees become smaller with gnarled and twisted stems covered with mosses, lichens, liverworts, ferns, pitcher plants (*Nepenthes*), and epiphytic orchids. Dominant tree species include *Syzgium, Dacrycarpus, Agathis* and the oaks *Podocarpus and Lithocarpus.* Often referred to as mossy forest, this part of Mount Apo is a network of moss-covered roots with the forest floor a metre or so below.

The peak area consists of massive boulders interspersed with small bushes of *Vaccinium, Rhododendron* and *Ilex,* and low grasses and sedges on the flatter areas. Steam and sulphur gases issue from numerous vents below the peak.

Mount Apo National Park which is one of two ASEAN National Heritage Parks in the Philippines (the other is Mount Iglit-Baco), supports many endemic plants and animals, some endemic to Mindanao alone. The world-renowned orchid *Vanda sanderiana,* popularly known in the Philippines as *waling-waling,* is believed to have originated from the highlands of Mount Apo. Dawsonia, the world's largest moss, growing up to 25 cm tall, is found on the eastern slope as far as the neighbouring **Mount Talomo**. The pink *Udang-udang* flower (*Impatiens sultani*) dots the forest floor along the trails to the timberline.

To date, some 84 species of birds have been recorded in the Park. Of these, 61 species are endemic to the Philippines and 14 endemic to Mindanao. The most spectacular inhabitant is the Philippine Eagle (*Pithecophaga jeffreryi*). There is one species of megapode, 10 species of dove, five of parrots and three of hornbills. The mid-mountain level is rich in birds of prey, cuckoos, owls, woodpeckers, warblers, flycatchers, orioles, nuthatches, tits, sunbirds, flowerpeckers and white-eyes.

The largest mammals in the park are Sambar Deer, wild pig (*Sus verrucosus*), and Long-tailed Macaque. Other mammals include Flying Lemurs

he hilippines as some of e region's ost scenic ettings.

(*Cynocephalus volans*), a species differing from the one occurring in the rest of Southeast Asia, Civet Cats, flying squirrels, fruit bats and treeshrews. The endangered Philippine Tarsier may be found in intact lowland forest and the insectivore *Podogymnura truei* occurs in the mossy forest.

A number of tribes occupy the western portions of the park. To the west around **Ilomavis**, are the Manobos and in the south and east are the Bagobos. Both are farming communities but still hunt in the forest. The nomadic Aetas (also known as Negritos) live in the north of the park and are dependent upon the forest for their livelihood.

In recent times, there has been a significant influx of lowlanders along the logging roads. They clear and settle logged-over areas, often forcing the original inhabitants to higher elevations. A recent estimate put the total number inhabitants within the park at about 20,000.

A dormant volcano, Mount Apo has many hot springs, waterfalls and mountain lakes, most notable of which are the **Tudaya Falls** and Agco and Venado lakes. With its clay-loam and fertile volcanic soil, the park is characterised by valleys and plateaus of moderate to steep slope.

Although there is no pronounced dry and wet season within the park, May to October are recorded as the wettest months.

Early March to April is the optimum time to climb Mount Apo as the rainfall is slight. Allow four days for the ascent and arrange for guides and provisions ahead or obtain them in **Kidapawen**, North Cotabato — some 110 km from Davao City. At Kidapawen it is advisable to register at the Municipal Building before proceeding by jeepney to Ilomavis. From here it is a three-hour trip through farmland to the first camp at **Lake Agco** (1,200 metres). The lake is fed by hot springs and climbers may spend the night in a cabin by the shore.

The second day requires a climb from Agco to the milky-white **Marbel River** for lunch and on to **Lake Venado** (or

The Crested Serpent-eagle is found throughout Southeast Asia.

Maluno). Along the way, the climb frequently crosses the Marbel River, passes hot springs and some attractive waterfalls. The shores of Venado — a shallow lake at 2,400 metres — offer the camper a place for the night.

There now remains a three-hour hike to the summit through rocky terrain, moss and dwarf trees. The rim of the dormant volcano offers impressive panoramas and a cool reprieve.

Several other routes such as the northeast trail through Baracatan and the southwest through Makilala are feasible for the more experienced climber. Those planning to visit the park should contact the Ministry of Tourism or Amity Tours for information.

High Flier: The Philippine Eagle is one of the world's largest eagles with a wing-span of nearly two metres. It is over one metre in height and weighs up to seven kg. A mated pair is thought to need a feeding range of 60 to 100 square km.

The eagle often soars above forest canopy or the clearings of lowland forest and feeds upon a wide variety of forest animals including cobras and bats. It builds a huge nest upon large epiphytes growing on the tallest trees and produces one egg per nesting cycle (September to January) and rears offspring over two years. Incubation lasts 60 days and is undertaken largely by the female. In the early stages after hatching, the female does most of the feeding, but as the chick grows into an eaglet, the male takes turns in feeding. The eaglet remains in the nest for about six months during which time it is entirely dependent upon its parent for food.

The plight of the Philippine Eagle is well known. Now confined to Leyte, Samar and to portions of Luzon and Mindanao, it has been reduced to between 100 and 300 in number by loss of forest habitat and by collection and hunting. A conservation campaign initiated in 1970 succeeded in slowing its decline. However, the continuing destruction of the remaining forest and trophy hunting constitute a serious threat and Mount Apo National Park — the only park known to be inhabited by

aced with xtinction: e agnificent hilippine agle.

the eagle — is barely large enough to support two or three pairs. At the Philippine Eagle Research and Nature Center in Toril, Davao City, the visitor may see captive eagles which the centre hopes to breed and perhaps prevent extinction of the species.

Iglit-Baco National Park

Iglit-Baco National Park, created in 1970, lies near the centre of the island of **Mindoro** and falls between two climatic zones. The western half of the park lies in a zone with a pronounced rainy season in June to November and a dry season in November to April. The eastern half is in a zone with rainfall relatively evenly distributed throughout the year.

A closely-spaced group of 10 low mountains, with slopes to the northeast and southwest, traverses the middle of the park. The central portion of the park is dominated by **Mount Baco** some 2,488 metres above sea level. **Mount Mangibok** is the second highest mountain in the park at 1,432 metres. The eastern portion of the park is drained by the Bongabon River and the south by the Bugsenga and Lumintao rivers.

Due to the imposition of two climatic patterns, the western portion of the park is grassland — the main species being *Themeda triandra* (*punaw*), *Imperata cylindrica* (*cogon*) and *Saccharum spontaneum* (wild sugarcane) — and the eastern portion mainly composed of evergreen forest and dry season deciduous shrubs and trees of the families *Leguminosae*, *Euphorbiaceae*, *Dipterocarpaceaee*, *Combretaeae* and *Casuarinaceae*. Cattle ranching, upland farming, fuelwood gathering and human settlements have contributed to the rapid deforestration and denudation of the park.

The park supports the habitat of the endemic Tamarau (*Anoa mindorensis*) which reaches about one metre in height and has short thick-set horns, slightly curved at the tips. Though its general appearance resembles a diminutive water buffalo, it is stockily built and its

A prickly porcupine.

courage and ferocity more than compensate for its small stature. It frequents lowlands near swamps and marshlands but it is found most commonly in open grasslands and forests climbing to elevations of 2,000 metres or more. The Tamarau roams in herds of 10, often with more than one bull. However, monogamy prevails during the mating season from about April to July.

The avifauna includes swifts and swiftlets, kingfishers, hornbills, pigeons, doves and parrots among them Philippine endemics such as Tarictic Hornbill, the Yellow-breasted Fruit Dove, and the Guiaiabero, a parrot. Among the mammals are deer, wild pigs, bats and monkeys. Monitor and Flying Lizards can be easily seen.

Iglit-Baco National Park and its surroundings are the home of the tribal Batangan, the Hinunoo, the Mangyan and the Bangan. The communities practice shifting cultivation of hill rice and hunt animals on the mountains.

The park, which is readily accessible by a 55 km road from San Jose to the south, is excellent for hiking, offering spectacular scenery with its mountains covered in grassland, deep ravines and canyons. The journey takes from four to six hours and brings the visitor to **Bariopoypoy**. A trek from Bariopoypoy to the summit of Mount Iglit takes four to five hours from where it is a further hour to the habitat of the Tamarau.

Despite the often hot, glaring sun, it is advisable to visit the park during the dry season. During the wet season, most of the tributaries and rivers overflow and trails become impossibly slippery and often impassable.

St. Paul's National Park

The island of **Palawan** has landscapes of outstanding natural beauty, the karst limestone landscapes around **El Nido**, in the north and in **St. Paul's National Park** being particularly dramatic. The park includes an eight-km underground river — largest in the world — lush forest vegetation and

he Tokay, large Gecko, a frequent sitor on opical ights.

various endemic birds and mammals. Major attractions are the Palawan Peacock Pheasant, Philippine Cockatoo, Palawan Racket-tail, Hill Myna, megapode and Bear Cat. Other mammals found in the park (and on Borneo — betraying connections between the land masses in the mid- to late- Pleistocene) include the pangolin, a porcupine, treeshrews, otters, badgers, mongooses and the Palm Civet.

Above the underground river are gigantic mountains of limestone formations. Two-thirds of the park is covered with forest having a canopy of 35 metres.

The park is subject to two pronounced seasons. It is dry from November until late April and wet the rest of the year. The best time to visit is from February to early May.

Most people reach St. Paul's via Baheli from Puerto Princesa. The jeepney ride to Baheli takes one-and-a-half hours and from there the *banca* ride is up to three hours each way. The *banca* moves down the Baheli River, through degraded mangrove and nipa swamp, into the vast Ulugan Bay. At the head of the bay, the *banca* proceeds north along spectacular cliffs and forest to a cove in St. Paul's Bay where trees shade a white sandy beach. From the mouth of the river — a blue lagoon — a smaller *banca* conveys the visitor upriver for some three km, passing through huge domed areas and formations of stalactites and stalagmites. Explorers have penetrated some eight km — sometimes wading through guano deposited by the numerous bats — without locating the source of the river. St. Paul's is also accessible via a two-and-a-half hour the *banca* ride from Port Barton and then via a jeepney ride to Cayasan Park station. However, about 14 km of rough road needs to be negotiated before reaching the underground river.

Mount Isarog National Park

Mount Isarog (1,966 metres), an extinct volcano, is the highest peak in the southern part of **Luzon**. Much of the

Mindoro's coral beache are everybody's drear

park is steep and is drained by 18 rivers and nine creeks. The park possesses spectacular waterfalls, natural pools, a series of crystalline caverns and numerous gorges and canyons. It also contains the famous **Malabsay Falls** which, at 40 metres in height, plunge to a depth of 15 metres.

The vegetation changes from mixed lowland and montane forest at 900 metres to lower montane and upper montane forest (at 1,125 metres and 1,350 metres respectively) and mixed montane mossy forest and lower mossy forest at 1,550 metres and 1,750 metres, respectively.

The park supports three species of bats: the Hairy-winged Bat, the Orange Tube-nosed Bat and Narrow-winged Brown Bat. Other wildlife includes monkeys, wild pigs, owls and hawks and various snakes and lizards. Among the birds are species that are not only endemic to the Philippines, but even to Luzon, such as the Green and the Luzon Mountain Racket-tail parrots or Merril's and Mooche's Fruit Dove.

About four hours of hiking and climbing are needed to reach the striking Malabsay Falls, which can sometimes be seen on the right face of the volcano as the visitor approaches the park from the town of Carolina.

Quezon National Park

Quezon National Park, proclaimed in 1940, embraces rolling and rugged terrain with waterfalls, caves and impressive limestone formations (called coral nagiting or animal pen) near the 420-metre peak of **Mount Pinagbanderahan**. Despite disturbance by hunters and poachers the park still preserves outstanding examples of the flora and fauna of Philippine lowland dipterocarp forest.

Wild pig, monkeys and very occasional Sambar Deer can be seen. Butterflies are plentiful and easily observed feeding on the roadside bushes and flowers. The elusive Luzon Bleeding Heart, Red-breasted Pitta, White-browed Shama and Ashy Thrush may

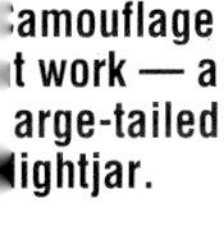

amouflage t work — a arge-tailed ightjar.

be found along the trail to Mount Pinagbanderahan. Careful examination of the trees may reveal the Philippine Trogon, Spotted Wood Kingfisher, Philippine Forest Kingfisher or a roving flock of Rufous Coucals, often chattering like chickens.

Quezon National park experiences heavy rainfall from November to January. Charcoal making, quarrying and fuelwood harvesting are degrading large portions of the Park.

Mount Canlaon National Park

Mount Canlaon National Park, proclaimed in 1934, is situated in the northern part of the island of **Negros**. It is an irregularly hilly and mountainous area broken by deep gullies with impressive cliffs but its outstanding feature is the active volcano whose rim is 2,465 metres above sea level. Hot springs, caves containing stalagmites and stalactites, and cold, crystal-clear rivers and creeks are also special features of the park. **Makawili Peak**, part of the old crater's rim, is the highest point in the park and often obscured by clouds.

The vegetation is divided into transition lowland dipterocarp and lower montane forest at elevations of 750 metres to 1,350 metres, typical montane forest at 1,350 metres to 1,950 metres, and mossy type forest at 1,950 metres to 2,250 metres.

Noteworthy flora includes *sangumai* (*Dendrobium anosmym*) and mariposa (*Phalaenopsis amabilis*) orchids, pitcher plants, ribbon ferns (*Opheoglossum pendulum*) and staghorn ferns (*Platycerium stouli*). Deer, wild pigs, monkeys, giant bats and many species of birds constitute the varied fauna.

The Park experiences a dry season from December to April.

Mount Pulog National Park

Mount Pulog was proclaimed a National Park in 1987. It is 2,930 metres above sea level at its highest point—the highest in Luzon — and is distinctly characterised by deep ravines, gorges and steep terrain.

The summit area of Mount Pulog is open tussock grassland with dwarf *Arundinaria* bamboo and bilaw (*Miscanthus sinensis*) associated with the grasses *Calamagrostis, Anthoxanthum, Microlaena, Deschampsia, Isache* and *Monostachya*. Species of *Ranunculus, Gentiana, Hypericum* and *Euphrasia* suggest a sub-alpine habitat. Mossy forest extends beyond a formation of Benguet pine (*Pinus insularis*) between 360 metres to 660 metres. The park is also home to the unique Cloud Rat (*Crateromys schadenbergi*), a large, long-haired rodent which occupies the mossy forest.

Portions of the park have been disturbed by upland farming. The western side of the park houses the mummy cave and burial grounds of native people residing at the foot of Mount Pulog. The mountain is sometimes used as a reference point for shipping and is considered sacred ground by local inhabitants. There is a pronounced rainy season from June to November.

Left, keeping the night watch, a Collared Scops-owl. Right, Monitor on the move.

TRAVEL TIPS

THAILAND

MALAYSIA

SINGAPORE

INDONESIA

PHILIPPINES

GLOSSARY

BIBLIOGRAPHY

ART/PHOTO CREDITS

INDEX

MAPS

Getting There

BY AIR

Bangkok is a gateway between east and west and a transportation hub for Southeast Asia. Served by more than 40 regularly-scheduled airlines, Thailand has four international airports at Chiang Mai, Phuket, Hat Yai and Bangkok plus domestic airports in 22 towns within Thailand. For example, it is possible to fly Thai International directly from Phuket or Hat Yai to Kuala Lumpur, Penang, Singapore, Bandar Seri Bengawan. Dragonair offers direct service from Hong Kong and Air Lanka flies from Vienna and Sydney. In addition, there is an airport at U-tapao, 30 minutes from Pattaya, that serves charter flights from Singapore.

Thailand's flag carrier, Thai Airways International or THAI serves 44 cities on four continents and enjoys an excellent reputation for in-flight service. Its domestic arm, also called THAI, operates a network of daily flights to 22 of Thailand's major towns aboard a fleet of sleek 737s and Airbuses.

BY SEA

The days when travelers sailed up the Chao Phya River to view the golden spires of Bangkok are long gone. Luxury liners now call at Pattaya and Phuket but have ceased serving Bangkok. Check with a travel agent or shipping company to find those which depart from your city.

BY RAIL

Trains operated by the State Railways of Thailand are clean, cheap and reliable albeit a bit slow. There are only two railroad entry points into Thailand, both from Malaysia on the southern Thai border. The trip north to Bangkok serves as a scenic introduction to Thailand.

The Malay Mail leaves Kuala Lumpur every day at 7.30 a.m. and 8.15 a.m. and 3 p.m., 8.30 p.m., and 10 p.m., arriving seven to nine hours later at Butterworth, the port opposite Malaysia's Penang island at 1.35 p.m., 5.50 p.m., and 9.10 p.m., and at 5.30 a.m. and 6.40 a.m. respectively. A daily

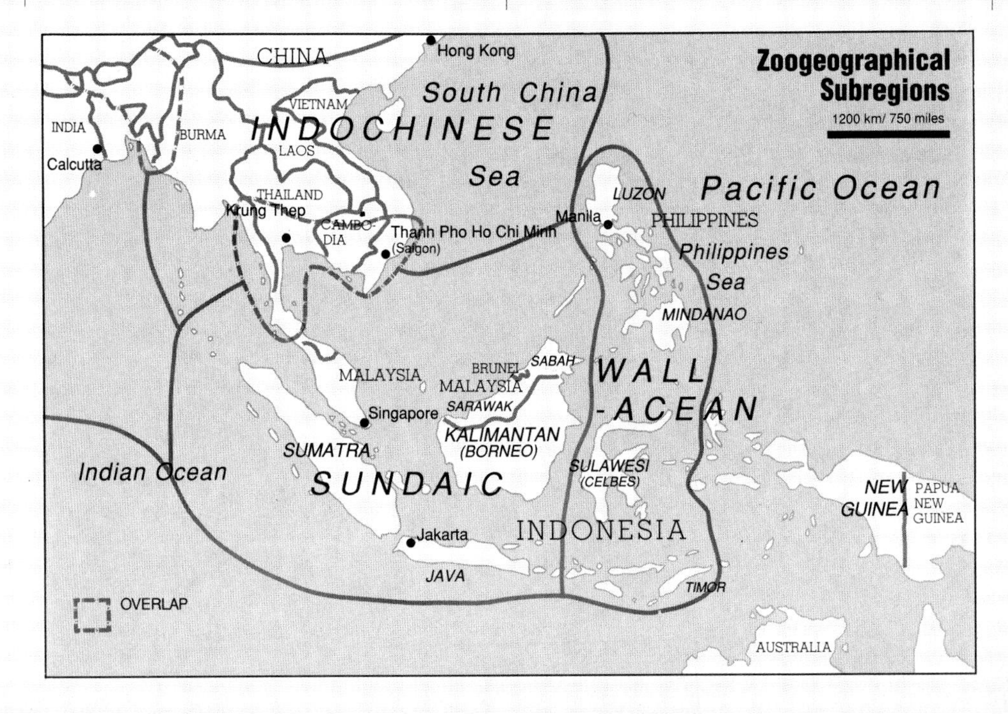

train leaves Butterworth at 1.40 p.m., crossing the border into Thailand and arriving in Bangkok at 8.35 a.m. There are second-class cars with upper and lower sleeping berths at night. There are also air-conditioned first-class sleepers and dining cars serving Thai food.

First-class air-conditioned Butterworth-Bangkok tickets are 1,227 baht (US$48). Second-class air-conditioned sleepers are 681 baht (US$26.70) for an upper berth and 731 baht (US$28.70) for a lower one. Second-class sleepers cooled by fans are 581 baht (US$22.80) for an upper berth and 631 baht (US$24.75) for a lower. Trains leave Bangkok's Hualampong Station daily at 5.15 p.m. for the return trip to Malaysia.

A second, somewhat less convenient but more entertaining train travels from Kuala Lumpur up Malaysia's east coast to the northeastern town of Kota Bharu. Take a taxi across the border to catch the SRT train from the southern Thai town of Sungai Kolok. Trains leave Sugai Kolok at 10.05 a.m. and 10.55 a.m., arriving in Bangkok at 6.35 a.m. and 7.05 a.m. the following day. A berth in an air-conditioned first-class sleeper is 1,108 baht (US$43.50). Second-class air-conditioned sleepers cost 618 baht (US$24.25) for an upper and 678 baht (US$26.60) for a lower berth. Second-class sleepers cooled by fans cost 518 baht (US$20.30) for an upper and 578 baht (US$22.70) for a lower berth.

BY ROAD

Although Thailand shares borders with four countries, only the border point with Malaysia is open to road traffic. Drivers will find most Thai roads modern and well maintained by comparison with those of its neighbours. The Malaysian border closes at 6 p.m. so plan an itinerary accordingly.

TRAVEL ESSENTIALS

PASSPORTS & VISAS

Foreign nationals holding valid passports from the following countries will at the point of entry be granted free, a Transit Visa valid for up to 15 days: Algeria, Argentina, Australia, Austria, Bahrain, Belgium, Brazil, Brunei, Burma, Canada, Denmark, Egypt, Fiji, Finland, France, Greece, Iceland, Indonesia, Ireland, Israel, Italy, Japan, Jordan, Kenya, Kuwait, Luxembourg, Malaysia, Mexico, Netherlands, New Zealand, Nigeria, North Yemen, Norway, Oman, Papua New Guinea, Philippines, Portugal, Qatar, Saudi Arabia, Senegal, Singapore, South Korea, Sudan, Spain, Sweden, Switzerland, Tunisia, Turkey, UAE, UK, USA, Vanuatu, Western Samoa, Germany and Yugoslavia.

Royal Thai Embassies or Consulates around the world can issue tourist visas valid for 60 days which cost 300 baht (US$12). The same embassies can issue transit visas valid for a 30-day stay for 200 baht (US$8).

VISA EXTENSIONS

Visas can be extended by applying at the Immigration Division on Soi Suan Plu in Bangkok, Phuket Road in Phuket and the Airport Road in Chiang Mai (8.30 a.m. to 4 p.m., Monday to Friday) before the visa's expiration date.

Fifteen-day transit visas can be renewed for an additional 7 days for a fee of 500 baht (US$20). Thirty-day transit visas can be extended for an additional 30 days for a fee of 200 baht (US$8). Sixty-day tourist visas can be extended for an additional 30 days for a fee of 300 baht (US$12). For all three, you must apply in person with three passport-sized photographs.

To remain in Thailand beyond 90 days,

you must travel to Penang, Rangoon, Kuala Lumpur, Singapore or Hong Kong and apply for a new visa; plan to spend two days there waiting for the visa process to be completed.

RE-ENTRY VISA

To leave Thailand and return before the expiration of your visa, you must apply for a re-entry permit prior to departure at immigration offices in Bangkok, Chiang Mai, Pattaya, Phuket and Hat Yai. The fee is 500 baht (US$20).

MONEY MATTERS

Currency: The baht, the principal Thai monetary unit, is divided into 100 units called satangs. Banknote denominations include the 500 (purple), 100 (red), 50 (blue), 20 (green) and 10 (brown) baht notes.

The coinage is confusing with a variety of sizes and types for each denomination. There are 10-baht coins (a brass centre surrounded by a ring of silver), 5 baht coins (silver pieces with copper rims in two sizes), three varieties of 1 baht coin (silver; only the small one will fit public telephone coin slots); 2 baht coin (rare, silver, nearly the same size as the small 1-baht and small 5-baht coin); and two sizes of 50 and 25 satang (both brass-coloured).

Exchange Rate: The very stable Thai currency is tied to a basket of international currencies heavily weighted in favor of the US dollar. The rate at the time of going to press was approximately 26 baht to the US dollar. For daily rates, check the *Bangkok Post* or *The Nation Review* newspapers. There are no black market rates.

Cash and traveller's cheques can be changed in hundreds of bank branches throughout the country; rates are more favourable for traveller's cheques than for cash. Banking hours are 8.30 a.m. to 3.30 p.m., Monday-Friday, but nearly every bank maintains sidewalk money changing kiosks in major towns around the country. These are open daily from 8.30 a.m. to 8 p.m. These kiosks can be found in Bangkok as well as Chiang Mai, Pattaya and Phuket. Hotels generally give poor rates in comparison with banks whose rates are set by the Bank of Thailand.

Banks: Thailand has a modern banking system offering most major services. Several foreign banks maintain local representative offices (see below). Money can be imported in cash or traveller's cheques and converted into baht. It is also possible to arrange telex bank drafts from your hometown bank. There is no minimum requirement on the amount of money that must be converted. Money can be reconverted into foreign currency at bank counters at airports in Bangkok, Chiang Mai, Phuket and Hat Yai.

Banks in Bangkok include Thai institutions and branches of foreign banks. Most are equipped to handle telegraph and telex money transfers and a wide range of money services. Banking hours are 8.30 a.m. to 3.30 p.m. Up-country, the services are much more restricted. If you have overseas business to conduct with a bank, do it in Bangkok.

Credit cards: American Express, Diners Club, Mastercard and Visa are widely accepted throughout Bangkok. Many stores levy a surcharge on their use, the highest (3-5 percent) on American Express cards. Credit cards can be used to draw emergency cash at most banks. Each has a local representative office.But be warned: Credit card fraud is a major problem in Thailand. Don't leave them in safe deposit boxes. When making a purchase, demand the carbons, rip them up and throw them away elsewhere.

HEALTH

Visitors entering Thailand are no longer required to show certificates of vaccination for smallpox or cholera. Persons arriving from Africa must show certificates indicating vaccination against yellow fever.

Concerned about the spread of AIDS, the government passed a new regulation in late 1987 barring the entry of persons with the disease. Viewed by many as unenforceable, it has not, to date, been implemented.

Hygiene: However little attention they may pay to keeping the environment clean, Thais place a high value on personal hygiene and understand the dangers of germs and infections. Restaurants catering to foreigners are generally careful with food and drink preparation.

While Bangkok **water** is pure when it leaves the city's modern filtration plant, the pipes that carry it into homes are old. Thus, visitors are advised to drink only bottled water or soft drinks. Both beverages are produced under strict supervision as is the ice used in large hotels and restaurants. Most streetside restaurants are clean; a quick glance should reveal which are and which are not.

With its thriving nightlife and transient population, Bangkok is a magnet for **venereal diseases** and anyone who does not take precautions is asking for trouble. With the spread of AIDS worldwide, there is even more reason to be careful.

Malaria and **dengue fever** persist in the rural areas. When in the hills – especially during the monsoon season – apply mosquito repellent on exposed skin when the sun begins to set.

WHAT TO WEAR/BRING

Clothes should be light and loose; natural blends that breathe are preferable to synthetics. Open shoes (sandals during the height of the rainy season when some Bangkok streets get flooded) and sleeveless dresses for women, short-sleeved shirts for men, are appropriate. Suits are worn for business and in many large hotels but, in general, Thailand lacks the formal dress code of Hong Kong or Tokyo. Casual but neat and clean clothes are suitable for most occasions.

One exception is the clothing code for Buddhist temples and Muslim mosques. Shorts are taboo for women and for men wanting to enter some of the important temples. Those wearing sleeveless dresses may also be barred from luxury hotels and temples. Improperly dressed and unkempt visitors will be turned away from large temples like the Temple of the Emerald Buddha. Dress properly in deference to the religion and to Thai sensitivities. Sunglasses and hats are useful items to protect eyes and sensitive skin from tropical glare.

Bangkok, Chiang Mai and Phuket are modern cities with most of the modern amenities found in similar large cities in Europe or North America. Chapstick and moisturizers are needed in the north during the cool season.

ON ARRIVAL

Thailand's principal airport, Don Muang, is located 22 km (14 miles) north of Bangkok. In September 1987 it opened a modern new terminal building which doubled handling capacity and now speeds visitors through immigration formalities, baggage collection and customs.

On arrival, disembark at an airbridge or at a distant bay and be transported to the terminal by air-conditioned bus. It is a long, level walk to the immigration counters where visas are issued or checked. After completing formalities, descend an escalator to the baggage carousels on the floor below. On either end of this hall are telephones for calls into the city.

Baggage carts along the inner wall are free. After collecting your luggage, proceed through the green (nothing to declare) or red (items to declare) customs lines. Pass through frosted glass doors into the main arrival hall. Group tours move to the right; individual travellers turn to the left. Hotel chauffeurs will be lined up along the outer wall holding placards bearing either the guest's or the group's name.

In this hall are also currency exchange counters paying bank rates, transportation desks for limousines into the city (see the section "From the Airport") and a desk maintained by the Thai Hotels Association where it is possible to book a hotel room in the city.

If transferring to a domestic flight, go to the domestic terminal which is a separate building 500 metres (550 yards) south of the airport. A free shuttle bus leaves every 10-15 minutes from the northern end of the international terminal.

CUSTOMS

The Thai government prohibits the importation of drugs, pornography, dangerous chemicals, firearms and ammunition. Foreigners are allowed to bring with them a maximum equivalent to US$10,000 per person. All amounts higher than this figure must be declared at the point of entry.

Foreign visitors can import without tax, one camera with five rolls of film, 200 cigarettes and one bottle of spirits.

PORTER SERVICES

There are no porters as such but luggage carts are available free for both arriving and departing passengers. Upon request, the airport can provide wheelchairs and other assistance for disabled persons.

RESERVATIONS

Hotel reservations can be made in the airport arrival lounge once you have passed through customs. With the rapid rise in the number of arrivals over the past few years, it is recommended that you book a room in advance, especially during the high season (November-April), and most particularly during the Christmas-New Year holidays and Chinese New Year which falls between late January and early February.

ON DEPARTURE

On departure from Bangkok's international terminal, travellers must pay an airport tax of 200 baht (US$8) at the check-in counter. Those on international flights from Phuket are charged 150 baht (US$6). Domestic airport departure tax is 20 baht (US$0.80).

GETTING ACQUAINTED

GOVERNMENT & ECONOMY

Thailand is a constitutional monarchy headed by His Majesty, King Bhumibol. His power has been reduced considerably from the period before the 1932 Revolution. However, he can, by the force of his personality, influence the direction of important decisions merely by a word or two.

Although he no longer rules as an Absolute Monarch of previous centuries, he is still regarded as one of the three pillars of the society – monarchy, religion and the nation. This concept is represented in the five-banded national flag: the outer red bands symbolizing the nation; the inner white bands the purity of the Buddhist religion; and the thick blue band at the center representing the monarchy.

The decades he has spent in the countryside working with farmers to improve their lands and yields has influenced others to follow his example in serving the people. His mother, the Princess Mother, Her Majesty, Queen Sirikit, and other members of the royal family have also been active in promoting the interests of Thais in the lower economic strata. Thus, the photographs of Their Majesties which hang in nearly every home, shop and office, have been placed there, not out of blind devotion, but out of genuine respect for the Royal Family.

The structure of the government is defined by the 1932 Constitution and the ten ordinances that have followed it (the last, in 1978). Despite the many revisions, the Constitution has remained true to the spirit of the original aim of placing power in the hands of the people, although the exercising of it has favoured certain groups over others.

Modelled loosely on the British system, the Thai government comprises three branches: legislative, executive and judiciary, each acting independently of the others in a system of checks and balances. The legislative branch is composed of a Senate and a House of Representatives. The Senate consists 267 leading members of the society including business people, educators and a heavy preponderance of high-ranking military officers. They must be over 35 years of age and must not be members of political parties. Members serve for life and new ones are selected by the Prime Minister and approved by the King. The House of Representatives comprises 357 members elected by popular vote from each of the 73 provinces of Thailand.

The executive branch is represented by a Prime Minister who may or may not be an elected Member of Parliament. He is selected by a single party or coalition of parties and rules through a cabinet of ministers, the exact number dependent on his own needs. They, in turn, implement their programs through the very powerful Civil

Service. The present Prime Minister, Chatichai Choonhavan, is an ex-artillery general, who was selected by a coalition of parties in August 1988.

The Judiciary comprises a Supreme Court, an appellate court, and a pyramid of provincial and lower courts. It acts independently to interpret points of law and counsels the other two branches on the appropriateness of actions.

Nearly 70 percent of Thailand's 55 million people are farmers who till alluvial land so rich that Thailand is a world leader in the export of tapioca (No. 1), rice (No. 2), rubber (No. 2), canned pineapple (No. 3) and is a top-ranked exporter of sugar, maize and tin.

GEOGRAPHY & POPULATION

Lying between 7°N and 21°N latitude, Thailand has a total land area of 514,000 sq km (198,000 sq miles), nearly the size of France. The country is said to resemble an elephant's head with its trunk forming the southern peninsula. Bangkok, its capital, is sited at its geographic centre, approximately at the elephant's mouth. The country is bordered by Malaysia on the south, Burma on the west, Laos across the Mekong River to the northeast and Cambodia to the east.

The north is marked by low hills and contains the country's tallest peak, Doi Inthanon, standing 2,590 metres (8,490 feet) tall. A range of hills divides Thailand from Burma and forms the western boundary of the broad alluvial Central Plains which is the country's principal rice-growing area. To the east, the Plains rise to the Korat Plateau which covers most of the Northeast. The spine of the southern peninsula is the same range of hills that separate Thailand from Burma, sloping down to the Andaman Sea on the west and the Gulf of Thailand on the east. Thailand has a total of 2,600 km (1,600 miles) of coastline.

Bangkok is situated at 14°N latitude. Like Hungary's Buda and Pest, Bangkok is a city divided into halves by a river, the Chao Phya which separates Bangkok and Thonburi. The city covers a total area of 1,565.2 sq km (602.2 sq miles) of delta land of which no natural area is more than 2 metres (7 feet) above any other. Its population totals 5,832,843 although a semi-permanent migrant population has in recent years swelled that number to an unofficial 8 million. Bangkok functions as the epicentre of the country's political, business and religious life, a primate city some 35 times larger than Thailand's second and third largest cities of Chiang Mai and Khon Kaen.

TIME ZONES

Thailand Standard Time is 7 hours ahead of Greenwich Mean Time. Hence when it is +7 hours GMT in Bangkok, it is:

Hong Kong	+ 1 hour GMT
Tokyo	+ 2
Sydney	+ 3
Honolulu	- 17
Los Angeles	- 14
New York	- 12
London	- 7
Paris	- 6
Bonn	- 6
New Delhi	- 1.5

CLIMATE

There are three seasons in Thailand – hot, rainy and cool. But to the tourist winging in from anywhere north or south of the 30th parallel, Thailand has only one temperature: hot. To make things worse, the temperature drops only a few degrees during the night and is accompanied 24 hours by humidity above 70 percent. Only air-conditioning makes Bangkok and other major towns tolerable during the hot season. The countryside is somewhat cooler, but, surprisingly, the northern regions can be hotter in March and April than Bangkok.

Adding together the yearly daytime highs and the nighttime lows for major world cities, the World Meteorological Organisation has declared Bangkok to be the world's hottest city. When the monsoon rains fall, the country swelters.

The following is a guide to the degree of heat you can expect:

• Hot season (March to mid-June): 27-35°C (80-95°F)

• Rainy season (June to October): 24-32°C (75-90°F)

• Cool season (November to February): 18-32°C (65-90°F) but with less humidity.

CULTURE & CUSTOMS

Thais are remarkably tolerant and forgiving of foreigners' foibles but there are a few things which will rouse them to anger. As they regard the Royal Family with a reverence paralleled in few other countries, they react strongly if they consider any member of royalty has been insulted. Ill-considered remarks or refusing to stand for the Royal Anthem before the start of a movie will earn some very hard stares and perhaps worse.

A similar degree of respect is accorded the second pillar of society, Buddhism. Disrespect towards Buddha images, temples or monks is not taken lightly. Monks observe vows of chastity which prohibit their being touched by women, even their mother. When in the vicinity of a monk, a woman should try to stay clear to avoid accidentally brushing against him. When visiting a temple, it is acceptable for both sexes to wear long pants but not shorts. Unkempt persons are frequently turned away from major temples.

From the Hindu religion has come the belief that the head is the fount of wisdom and the feet are unclean. For this reason, it is insulting to touch another person on the head, point one's feet at him or step over him. Kicking in anger is worse than spitting at him.

When wishing to pass someone who is seated on the floor, bow slightly while walking and point an arm down to indicate the path to be taken. It is also believed that spirits dwell in the raised doorsills of temples and traditional Thai houses and that when one steps on them, the spirits become angry and curse the building with bad luck.

Greetings: The Thai greeting and farewell is "*Sawasdee*", spoken while raising the hands in a prayer-like gesture, the fingertips touching the nose, and bowing the head slightly. It is an easy greeting to master and one which will win smiles.

Dress & Hygiene: Thais believe in personal cleanliness. Even the poorest among them bathe daily and dress cleanly and neatly. They frown on those who do not share this concern for hygiene.

Behaviour in Public: Many years ago, Thai couples showed no intimacy in public. That has changed due to Western influence on the young, but intimacy still does not extend beyond holding hands. As in many traditional societies, displaying open affection in public is a sign of bad manners.

Names: Thais are addressed by their first rather than their last names. The name is preceded by the word "*Khun*", a term which honours him or her. Thus Silpachai Krishnamra would be addressed as Khun Silpachai when someone speaks to him.

Titles: You will find some Thais referred to in newspapers with the letters M.C., M.R. or M.L. preceding their names. These are royal titles normally translated as "prince" or "princess". The five-tier system reserves the highest two titles for the immediate Royal Family. After that comes the nobility, remnants of the noble houses of old. The highest of these three ranks is Mom Chao (M.C.), followed by Mom Rachawong (M.R.) and Mom Luang (M.L.).

The title is not hereditary, thanks to a unique system which guarantees that Thailand will never become top-heavy with princes and princesses. Each succeeding generation is born into the next rank down. Thus, the son or daughter of a Mom Chao is a Mom Rachawong. Soon, Thailand will be a nation of Nai, Mr., and Nang, Miss or Mrs. – a truly democratic realm.

Concepts: A few Thai concepts will give not only an indication of how Thais think but will smoothen a visitor's social interaction with them. Thais strive to maintain equanimity in their lives and go to great lengths to avoid confrontation. The concept is called "*kriangjai*" and suggests an unwillingness to burden someone older or superior with one's problems. It means not giving someone bad news until too late for fear it may upset the recipient.

"*Jai yen*" or "cool heart", an attitude of remaining calm in stressful situations, is a trait admired by Thais. Getting angry or exhibiting a "*jai ron*" (hot heart) is a sign of immaturity and lack of self-control. Reacting to adversity or disappointment with a shrug of the shoulders and saying "mai pen rai" (never mind) is the accepted response to most situations.

"*Sanuk*" means "fun" or "enjoyment" and is the yardstick by which life's activities are measured. If it is not "*sanuk*", it is probably not worth doing.

Thais converse readily with any stranger who shows the least sign of willingness. They may be shy about their language ability but will struggle nonetheless; speak a few words of Thai and they respond even more eagerly. Be prepared, however, for questions considered rude in Western societies such as: How old are you; how much money do you earn; how much does that watch (camera, etc) cost? Thais regard these questions as part of ordinary conversation and will not understand a reluctance to answer them. If, however, one is reticent about divulging personal information, an answer delivered with a smile will suffice; for e.g., to the above question of "How old are you?" with "How old do you think?" (out of politeness, they will usually guess a lower age and one can agree with them).

WEIGHTS & MEASURES

Thailand uses the metric system of metres, grams and litres.

ELECTRICITY

Electrical outlets are rated at 220 volts, 50 cycles and accept either flat-pronged or round-pronged plugs.

BUSINESS HOURS

Government offices are open from 8.30 a.m. to 4.30 p.m., Monday through Friday. Business hours are from 8 or 8.30 a.m. to 5.30 p.m. Monday through Friday. Some businesses are open for half a day on Saturday from 8.30 a.m. to noon.

Department stores are open from 10 a.m. to 9 p.m. seven days a week. Ordinary shops open at 8.30 or 9 a.m. and close between 6 and 8 p.m., depending on location and type of business. Supermarkets on Patpong 2 are open 24 hours. Some pharmacies remain open all night.

Small, open-air coffee shops and restaurants open at 7 a.m. and close at 10 p.m., although some stay open past midnight. Large restaurants generally close at 10 p.m. Most coffeeshops close at midnight though some stay open 24 hours. For a late-night Thai meal, the markets at Bangrak and the vendors on Soi 38, Sukhumvit Road, stay open until the wee hours.

PUBLIC HOLIDAYS

The following dates are observed as official public holidays:

January 1 – **New Year's Day**
Full moon day of February – **Makha Puja**
April 6 – **Chakri Day**
April 13 – **Songkran**
May 1 – **Labour Day**
May 5 – **Coronation Day**
Full moon day of May – **Visakha Puja**
Full moon day of July – **Asalaha Puja**
August 12 – **H.M. the Queen's Birthday**
October 23 – **Chulalongkorn Day**
December 5 – **H.M. the King's Birthday**
December 10 – **Constitution Day**
December 31 – **New Year's Eve**

Chinese New Year in February is not officially recognised as a holiday but many shops are closed for several days.

COMMUNICATIONS

MEDIA

Newspapers and magazines: *The Bangkok Post* and the *Nation Review* are two of the best and most comprehensive daily newspapers in Asia. The *Asian Wall Street Journal* and the *International Herald Tribune* are available at most bookstalls after 4 p.m.

Newsstands in major hotel gift shops carry air-freighted and therefore expensive (at least 50 baht per copy) editions of British, French, German and Italian newspapers. Newsvendors on Sukhumvit Soi 3 offer copies of Arabic newspapers. *Time*, *Newsweek* and other news magazines are readily available.

RADIO

AM radio is devoted entirely to Thai-language programs. FM radio includes several English-language stations playing the latest pop hits. Radio Thailand offers 4½ hours of English-language broadcasts each day on the 97 MHz frequency. Of value to visitors is a daily English-language program of travel tips broadcasted at 6.40 a.m., with news at 7 a.m., 8 a.m., 12.10 p.m., 12.30 p.m., 6.35 p.m. and 7 p.m. In addition, a French-language program of general interest topics and news is broadcast from 11.30 a.m. to noon every day.

TELEVISION

Thailand has five television stations which broadcast from 5 p.m. to 6.30 p.m. and 8 p.m. to around midnight. News, dramas, some foreign sitcoms translated into Thai, quiz shows and talk shows are the primary fare. About the only programs which appeal to foreigners are broadcasts of Thai boxing and other sports at 10.30 p.m. week nights and on Saturday and Sunday afternoons.

Bangkok has five television channels, three of which offer American, Japanese and European programs dubbed in Thai. At 8 p.m., all channels broadcast news beamed by satellite from Europe. The English-language soundtrack is supposed to be broadcasted simultaneously and on FM radio, but for inexplicable reasons, seldom does. The FM frequencies can be found in the daily newspapers.

Large Bangkok hotels have in-house channels that include Cable Network News (CNN) beamed by satellite from the United States.

POSTAL SERVICES

Thailand has a comprehensive and reliable postal service. Major towns offer regular air mail service as well as express courier service that speeds a package to nearly every point on the globe.

In addition to the postal service, Bangkok offers a number of international courier agencies including Federal Express, Purolater, DHL and Skypak.

Main post offices in Bangkok, Phuket and Chiang Mai have special facilities where stamp collectors can browse and buy from a wide selection of beautiful Thai stamps.

In Bangkok, the Central Post Office on New Road between Suriwongse and Siphya roads opens at 7.30 a.m. and closes at 4.30 p.m. Monday through Friday; and from 9 a.m. to noon on Saturday, Sunday and holidays. At the right side of the lobby is a packing service with boxes in various sizes. A department on the right end of the building is open 24 hours and sells stamps and sends telegrams.

Branch post offices are located throughout the country. Hours vary but they generally close at 4 p.m. Some stay open as late as 6 p.m. Post office kiosks along some of the city's busier streets sell stamps, aerograms (8.50 baht each) and ship small parcels. Hotel reception desks will also send letters for no extra charge.

TELEPHONE & FAX

Bangkok has a sophisticated communications system with most overseas calls handled by operators. Most hotels have telephones, telegrams, mail, telex and FAX facilities. Hotel operators can place international telephone calls within minutes; luxury Bangkok hotel rooms have IDD phones. If the hotel lacks such facilities, place your long distance calls at the General Post Office annex on the ground floor of the Nava Building on Soi Braisanee, just north of the GPO itself. It is open 24 hours.

Main post offices in nearly every city offer telegram and telex services to all parts of the world.

FAX services are now available throughout the country. FAX operations are generally run by small shops which also offer long-distance telephone service. They ask a nominal service charge in addition to the normal long-distance rates. Look for their signs along main streets of provincial towns.

EMERGENCIES

SECURITY & CRIME

When in Thailand, **avoid** the following:

• Touts posing as Boy Scouts soliciting donations on Bangkok's sidewalks. The real Boy Scout foundation obtains its funds from other sources.

• Touts on Patpong offering upstairs live sex shows. Once inside, one is handed an exorbitant bill and threatened with mayhem if he protests. Pay, take the receipt. and go immediately to the Tourist Police to gain restitution.

• Persons offering free or very cheap boat rides into the canals. Once you are well into the canal, you are given the choice of paying a high fee or being stranded.

• Persons offering to take you to a gem factory for a "special deal". The gems are usually flawed and there is no way to get your money back.

• Persons on buses or trains offering sweets, fruits or soft drinks. The items are often drugged and the passenger is robbed while unconscious. This is unfortunate because Thais are generous people and it is normal for them to offer food to strangers. Use your discretion.

If you do run into trouble in Bangkok, the police emergency number is 191. There are also Tourist Police assigned specially to assist travellers. They are located at the Tourist Assistance Centre at the Tourism Authority of Thailand headquarters at No. 4 Rajdamnern Nok Avenue and can be reached by telephoning 195 or 282-8129. They also maintain a booth on the Lumpini Park corner of the Rama 4 and Silom roads intersection. Most members of the force speak English.

Their offices can also be found in Chiang Mai and Phuket. In Chiang Mai, they are located below the Tourist Authority of Thailand office at 105/1 Chiang Mai-Lamphun Rd., Tel: 232-508, 222-977. In Phuket, find them at the TAT office at 73-75 Phuket Rd., Tel: 212-213, 211-036.

MEDICAL SERVICES

Accidents & Illnesses: First-class hotels in Bangkok, Chiang Mai and Phuket have doctors on call for medical emergencies. The hospitals in these three cities are the equivalent of those in any major Western city. Intensive Care Units are fully equipped and staffed by doctors to handle emergencies quickly and competently. Nursing care is generally superb because there is a higher staff to patient ratio. Many doctors have been trained in Western hospitals and even those who have not speak good English.

Most small towns have clinics which treat minor ailments and accidents. In the unlikely event that you suffer a criminal attack in Bangkok, you must go to a Police Hospital, normally the one at the Rajprasong intersection. Up-country, most hospitals will treat you.

HOSPITALS

BANGKOK

Samitivej Hospital, 133 Soi 49, Sukhumvit Rd., Tel: 392-0011, 392-0061.

Bangkok Adventist Hospital, 430 Phitsanuloke Rd., Tel: 281-1422, 282-1100.

Bangkok Christian Hospital, 124 Silom Rd., Tel: 233-6981, 233-6907.

CHIANGMAI

Lanna Hospital, off the Superhighway on the northeast side of town, Tel: 211-037.

McCormick Hospital on Kaew Nawarat Rd., also on the northeastern edge of town, Tel: 241-311/2.

Suan Dok Hospital on the corner of Boonruangrit and Suthep roads across from Wat Suan Dok, Tel: 221-122.

The city's biggest hospital is part of Chiang Mai University on Suthep Road.

PHUKET

In Phuket town: **Mission Hospital** on the northern end of Phuket town on Thepkrasatri Rd. (Highway 402 to the airport), Tel: 212-396.

In Patong: **Andaman Hospital**, half a km south of Patong on the road to Karon, Tel: 723-0108. Ambulance available.

CLINICS

There are many polyclinics with specialists in several fields. In Bangkok, the **British Dispensary** at 109 Sukhumvit Rd. (between Sois 5 and 7), Tel: 252-8056, has two British doctors on its staff. In Chiang Mai, you can go to **Chang Puak Polyclinic** at 52/2 Chang Puak Rd., Tel: 210-213. Similar clinics can be found in major towns.

Dental Clinics:
Bangkok: **Dental Polyclinic**, 211-3 New Petchburi Rd., Tel: 314-5070.

Chiang Mai: **Dr. Thavorn's Clinic**, 156 Chang Moi Rd., Tel: 236-443.

Phuket: **Dental Care Clinic**, 62/5 Rasda Center, Rasda Rd. (in Phuket town), Tel: 215-025.

Chiropractor:
Bangkok: 51 Soi Vichan, Silom Rd., Tel: 234-2649.

Snake Bites: There is little chance of being bitten by a poisonous snake in Bangkok or its environs but should it occur, most clinics have anti-venom serum on hand. If they cannot acquire any, travel to the **Saowapha Institute** (Snake Farm) on Rama IV Road. They maintain sera against the bites of six types of cobras and vipers. Up-country clinics maintain a constant supply of anti-venom serum supplied by the Saowapha Institute.

GETTING AROUND

MAPS

Bangkok: Bookshops carry a variety of maps of Thailand. Among the best are those printed by APA and several local publishers. The most accurate map of Bangkok is published by Siam Book Store with no title other than *Especially for Tourists*. It details the routes for ordinary and air-conditioned buses and sells for 60 baht.

There are also a number of specialised maps. The Association of Siamese Architects publishes a series of four colourful hand-drawn *Cultural Maps* with details and information on Bangkok, the Grand Palace, the Canals of Thonburi and Ayutthaya.

D.K. Books issues an English-language book of canal maps by Geo.-Ch. Veran under the title *50 Trips into the Canals of Thailand*. Detailed maps outline journeys using commuter boats which operate along Bangkok's river and canals.

Nancy Chandler's *Market Map* has colourful cartoon maps of all of Bangkok's major markets. This well-traveled lady has filled the margins with useful information designed to make a market trip fun. For historians, the Jim Thompson House sells old maps of Siam; they make handsome home decor items when framed.

Chiang Mai: The best maps are those included as part of a free monthly guide called *Welcome to Chiang Mai*. P&P 89 Promotions' *Tourist Map of Chiang Mai, Rose of the North* has a very good map of the North but a rather muddled map of Chiang Mai. Many interesting items have been left out and the bus map is very difficult to figure out. 40 baht.

Like her Bangkok *Market Map*, Nancy Chandler's Chiang Mai map is filled with useful tips as well as precise directions on how to find the best shopping bargains in the city.

Phuket: The best, *Phuket Planner*, is printed by Compass Publishing Co. and costs 35 baht. It includes detailed maps of the island, town and Phi Phi island.

FROM THE AIRPORT

Transport from Bangkok International Airport: The journey along the expressway into Bangkok can take from 30 to 45 minutes depending on traffic conditions. Many hotels have limousines to pick up guests but otherwise the traveller must find his own transportation. Three services offered just outside the departure hall are listed in descending order of price:

Thai International operates a transportation service to city hotels or private residences on a round-the-clock basis. Air-conditioned limousines are the most convenient transportation and cost just 300 baht (US$12); luxury cars like Volvos are 400 baht (US$16). The extra expense of a limousine might be worthwhile if arriving at the airport during rush hour (which in Bangkok can be any hour of the day but is most acute in the morning and evening). The limousine rental desk is just inside the door in the middle of the hall.

THAI air-conditioned mini-buses cost 100 baht (US$4) per passenger and leave the airport every 30 minutes. Because they have to fight traffic to deliver their passengers to widely-scattered hotels around the city, they can take considerable time to get to your hotel.

THAI also operates a shuttle service between the airport and Pattaya. Buses leave Don Muang Airport at 9 a.m., noon and 7 p.m. They leave major Pattaya hotels at 6.30 a.m., 1 p.m. and 6.30 p.m. Telephone 423-140 or 423-141 in Pattaya to arrange a pick-up from your hotel. The 3-hour journey costs 180 baht.

A second service, **Bhairab**, operates limousines to city hotels and homes for about the same price. Its desk is on the northern end of the terminal, just inside the door.

Ordinary **taxis** are air-conditioned and are the cheapest alternative. Buy a ticket at the desk and on arriving at your destination, pay the driver the amount stated on it. A normal journey costs 200 baht.

Avoid the operators of private or "black plate" taxis who may accost you. Not only are their prices outrageous, there have been incidents in which tourists have been harmed or robbed. For economy-minded travelers, five air-conditioned buses – Nos. 4, 10, 13, and 29 – stop in front of the airport. The trip into town costs 15 baht (US$0.60). The last buses leave at about 8 p.m. (except for No. 4 which stops running at 7 p.m.).

For the very economy minded, a train leaves from the station across the highway every hour until 8:13 p.m. for Hualampong Station in the center of Bangkok. The 40-minute journey costs 5 baht (US$0.20).

Transfers between Bangkok's International and Domestic Airport: A free shuttle bus runs to the Domestic Terminal 500 metres (550 yards) south of the International Terminal at 10-15 minute intervals.

Transfer from Bangkok Hotel to Don Muang Airport: For the return trip to the airport, THAI limousines leave from three downtown offices at 584 Silom Rd., (Tel: 235-4365/6), Montien Hotel (Tel: 233-7060) and Asia Hotel (Tel: 215-0780). Most major hotels have air-conditioned limousines. Taxis make the trip for around 150-200 baht (US$6-8) depending on traffic conditions.

Transport from Chiang Mai Airport: Chiang Mai's airport is a 10-minute drive from the city centre. There is a frequent bus service but it runs a circuitous route and visitors with a lot of luggage might prefer the following modes of transportation:

Major hotels have limousines to ferry guests with reservations (one can make a reservation at airport) to their premises. They charge about 50 baht per person.

Thai International operates a mini-van between the airport and its town office on Prapoklao Road. The cost each way is 20 baht per passenger. You must then find your own way from the office to the hotel

Major hotels maintain mini-vans to ferry guests with reservations (it is possible to make a reservation at the airport) to and from the airport. They charge up to 500 baht (US$20) per car.

Thai International's mini-van runs hourly between the airport and its Phuket town office on Ranong Road. The cost each way is 50 baht (US$2) per passenger. You must then find your own way from the office to the beach.

THAI Ground Services also offers air-conditioned limousines to each beach. The price is computed per vehicle and each vehicle holds four persons. Prices per vehicle (no extra charge for luggage) range from 250 to 450 baht (US$10-18).

DOMESTIC SERVICES

By Air: Thai Airways International (THAI) operates a domestic network serving 22 towns, offering seven daily service to some of them like Chiang Mai and Phuket.

BY RAIL

The State Railways of Thailand operates three principal routes from Hualampong Railway Station. The northern route passes through Ayutthaya, Phitsanuloke, Lampang and terminates at Chiang Mai. The northeastern route passes through Ayutthaya, Saraburi, Nakhon Ratchasima, Khon Kaen, Udon Thani and terminates at Nong Khai. The southern route crosses the Rama 6 bridge and calls at Nakhon Pathom, Petchburi, Hua Hin and Chumphon. It branches at Hat Yai, one branch running southwest through Betong and on down the western coast of Malaysia to Singapore. The southeastern branch goes via Pattani and Yala to the Thai border opposite the Malaysian town of Kota Bharu.

In addition, there is a line from Makkasan to Aranyaprathet on the Cambodian border. Another leaves Bangkok Noi Railway Station for Kanchanaburi and other destinations beyond along the old Death Railway. There is also a short route leaving Wongwian Yai station in Thonburi that travels west along the rim of the Gulf of Thailand to Samut Sakhon and then to Samut Songkram.

Express and Rapid services on the main lines offer first-class air-conditioned or second-class fan-cooled cars with sleeping cabins or berths and dining cars. There are also special air-conditioned express day coaches which travel to key towns along the main lines.

Train Stations in Bangkok:

Hualampong: Rama 4 Rd., Tel: 233-0341.
Bangkok Noi: Tel: 411-3102.

BY ROAD

Air-conditioned bus service is available to most destinations in Thailand. VIP coaches which allow extra leg room are the best for overnight journeys to Phuket and Chiang Mai. Air-conditioned coaches also leave half-hourly from the Eastern bus terminal on Sukhumvit Road (opposite Soi 63) for Pattaya and other points beyond. For the very adventurous, there are fan-cooled buses filled with passengers and chickens and tons of luggage that are used by poorer Thais for their journeys across the country.

To reach a small town from a large one, there are smaller buses or baht buses – pick-up trucks with a passenger compartment on the backs.

PUBLIC TRANSPORT

BANGKOK

Limousines: Most major hotels operate air-conditioned limousine services. Although the prices are about twice those of ordinary taxis, they offer the convenience of English-speaking drivers, and door-to-door service.

Taxis: Bangkok taxis are a reliable, air-conditioned means of traveling around town but the drivers' command of English is often less than perfect. It is often difficult to get him to understand where you want to go although that usually does not prevent his setting a price.

Do not step into a taxi without having first agreed on a price. It can fluctuate depending on the hour of the day and the amount of traffic, rain and one-way streets he must negotiate. A sample fare for 9 a.m. between the Dusit Thani Hotel and the Grand Palace or between the Oriental Hotel and the Jim Thompson House would be 80 baht. The basic fare for all journeys is 30 baht. There is no extra charge for baggage stowage or extra passengers and there is no tipping.

Tuks-Tuks: If the English fluency of taxi drivers is limited, that of *tuk-tuk* (also called *samlor)* drivers is even less. *Tuk-tuks* are the bright blue and yellow three-wheeled taxis whose name comes from the noise their two-cycle engines make. They are fun for short trips but should be avoided for long journeys. Pollution, noise, heat and *samlor* drivers' inability to resist racing can make for an unpleasant trip. For very short trips, the fare starts at 10 baht.

Buses: There are three types of Bangkok buses: air-conditioned, ordinary and mini-buses. They operate every two or three minutes along more than 100 routes and, though crowded, are a good means of travel, especially since they can move along special lanes against traffic on one-way streets.

Air-conditioned: There are more than a dozen air-conditioned bus routes through the city. Fares for the big blue and white buses start at 5 baht for short distances.

Ordinary buses: There are two types: blue and white buses (2 baht) and red and white (3 baht). They travel identical routes and differ only in their prices.

Mini-buses: These ply the same routes as the blue and white buses but are smaller and have less headroom for tall visitors. The fare is 2 baht but rises to 3.50 baht after 10 p.m. Conductors prowl through the aisles collecting fares and issuing tickets. Unfortunately, destinations are only noted in Thai so that a bus map is required. To add to the confusion, the front windows of some buses (like No. 25) carry a blue sign or red sign noting the route number, each colour defining a different route. Most routes cease operation around midnight though some (Nos. 2 and 4) run all night.

Up-country Bus Stations in Bangkok:

Eastern: opp. Soi 63, Sukhumvit Rd., Tel: 391-2504, 392-2520.

Northern and Northeastern: Moh Chit, Phaholyothin Rd., Tel: 271-0101/2.

Southern: Nakhon Chaisri Highway, Tel: 411-0511, 411-0112.

Boats: White express boats with red trim run regular routes at 20-30 minutes' intervals up and down the Chao Phya River, going all the way to Nonthaburi 10 km (6 miles) north of the city. They begin operation at 6 a.m. and cease at 6.30 p.m. Fares are 5 baht for short distances.

Ferries cross the river at dozens of points and cost 50 satang per journey. They begin operation at 6 a.m. and stop at midnight.

CHIANG MAI

Chiang Mai has two types of buses, the yellow and the red minibuses which carry passengers along five routes through the town. Fares are 2 baht each. Consult city maps for routes.

Tuk-tuks, the motorised three-wheel taxis, charge according to distance, starting at 10 baht. You must bargain for the price before you get in.

Samlors, the pedal trishaws, charge 5 baht for short distances.

PHUKET

Picturesque wooden buses ply regular routes from the market to the beaches. They depart every 30 minutes between 8 a.m. and 6 p.m. between Phuket town market and all beaches except Rawai and Nai Harn. Buses to Rawai and Nai Harn leave from the traffic circle on Bangkok Road. They prowl the beach roads in search of passengers. Flag one down. Fares range from 10 to 20 baht.

Tuk-tuks operate between Patong and Karon/Kata and between Kata and Nai Harn for up to 40 baht, depending on distance. Intracity *tuk-tuks* charge 10 baht per person regardless of distance for trips within Phuket town or Patong.

Motorcycle taxis leave from the market on Phuket town's Ranong Road. Drivers in maroon vests convey passengers anywhere in the downtown area for 5 baht per ride. It is a convenient way to get around.

PUBLIC TRANSPORT

Thailand has a good road system with over 50,000 km (31,000 miles) of paved highways and more are being built every year. Road signs are in Thai and English and one should have no difficulty following a map. An international driver's license is required.

Driving on a narrow but busy road can be a terrifying experience with right of way determined by size. It is not unusual for a bus to overtake a truck despite the fact that the oncoming lane is filled with vehicles. It is little wonder that when collisions occur, several dozen lives are lost. Add to that, many of the long-distance drivers consume pep pills by the bucketful and have the throttle to the floor because they are getting paid for beating schedules. One is strongly advised to avoid driving at night for this reason. When dusk comes, pull in at a hotel and get an early start the next morning.

CAR RENTAL

Avis, Hertz and local agencies offer late model cars with and without drivers and with insurance coverage for Bangkok and up-country trips. Prices for a chauffeured Mercedes Benz (Bangkok only) average 4,000 baht per day (2,500 baht for self-drive) and 2,200 baht for a chauffeured Toyota (1,800 baht for self-drive) plus gasoline costs (8.90 baht per liter for Premium at the time of going to press). A deposit of 2,000 baht or more is required.

Up-country, agencies can be found in major towns like Chiang Mai and Phuket. These also rent four-wheel drive jeeps and mini-vans. When renting a jeep, read the fine print carefully and be aware that you are liable for all damages to the vehicle. Ask for First Class insurance which covers both you and the other vehicle involved in a collision.

BANGKOK

Unfortunately, there are no car rental desks at the Bangkok airport. Contacted by telephone, agencies will deliver the car to your hotel and you can fill out the rental forms there.

Avis, 2/12 Wireless Rd., Tel: 255-5300/1. Open daily 8 a.m. to 9 p.m. Its desk at the Dusit Thani Hotel is open 24 hours.

Hertz, 987 Ploenchit Rd., Tel: 253-6251. Open from 8 a.m. to 5 p.m.

Grand Car Rent, 144/3-4 Silom Rd., Tel: 234-9956. Open from 8 a.m. to 9 p.m.

CHIANG MAI

Avis, Head office: 14/14 Huay Kaew Rd., Tel: 222-013, 221-316. Open 8 a.m. to 6 p.m. Branches: Dusit Inn, Tel: 251-033, 251-034; Chiang Inn, Tel: 235-655. Open 8 a.m. to 5 p.m.

Hertz, 12/3 Loi Kroa Rd., Tel: 235-496, 249-473; Novotel Suriwongse, Tel: 236-789, 236-673; Chiang Mai Plaza, Tel: 252-Aod Car Rent; 49 Chang Klang Rd. (opposite the Night Bazaar), Tel: 249-197.

PHUKET

Avis, Airport, Tel: 311-358; Le Meridien Phuket, Relax Bay, Tel: 321-480; Holiday Inn Phuket, Patong, Tel: 321-020; Phuket Cabana Hotel, Patong, Tel: 321-138; Dusit Laguna Phuket, Bang Thao, Tel: 311-174; Phuket Arcadia Hotel, Karon, Tel: 381-038; Phuket Island Resort, Rawai, Tel: 215-950; Club Med, Kata, Tel: 214-830.

Hertz, Airport, Tel: 311-162; Pearl Hotel, Montri Rd. (Phuket town), Tel: 211-044; Pearl Village Resort, Nai Yang, Tel: 311-378; Patong Merlin Hotel, Patong, Tel: 321-070; Thara Patong Hotel, Patong, Tel: 321-520.

MOTORCYCLE RENTAL

Motorcycles can be rented in Pattaya, Chiang Mai and Phuket for economical rates. Remember that when you rent a motorcycle, you must surrender your passport for the duration of the rental period; so change money first.

Motorcycles range in size from small 90 cc models like Honda Dreams and similar brands to giant 750 cc behemoths. The majority are 125 cc trailbikes. Rental outlets can be found along beach roads and main roads in each town. Prices run between 180 and 250 baht per day.

ON FOOT/HITCHHIKING

There are not many places one can walk in Thailand other than national parks like Khao Yai and Phu Kradung. It is possible to hitch a ride on ten-wheel trucks but it is not advisable simply because the drivers often drive in a manner designed to frighten the life out of you; if you are female, you are only asking for trouble. It is strongly recommended that you not travel at night on trucks as drivers are often tanked up on amphetamines and cause some of Thailand's most horrendous accidents.

WHERE TO STAY

HOTELS

Once a scattering of tawdry wooden inns designed more for short-term assignations than overnight accommodation, Thailand's hotels have grown to world standard in terms of facilities and service. Bangkok hotels frequently appear in the World's Best lists, with one of them, The Oriental, consistently topping the ranks.

Charges are normally assessed by the room, not the number of people occupying it, i.e. the charge for a single and a double is the same. Most hotels above 400 baht per night have shopping facilities, tour counters and coffeeshops, and are air-conditioned throughout. Nearly every Pattaya and Phuket hotel listed has a pool and a majority have "snooker clubs" where one can play billiards and pool.

GUEST HOUSES

Until the past decade, hotels were divided into three categories: first-class hotels, backpacker inns and the cheap Chinese and ramshackle hotels found mostly upcountry. The gap between the upper end and lower end of the scale was quite wide but has in the past few years been filled quite well by two new concepts: inns and guesthouses. Inns are found primarily in Bangkok. Small, they offer nothing more than basic amenities and are generally priced between 400 and 800 baht per night, although prices have been rising as room shortages elsewhere in the city increase.

FOOD DIGEST

WHAT TO EAT

The dramatic rise in the number of Thai restaurants opened around the world in the past decade says something about the universality and uniqueness of one of Asia's supreme cuisines. It is no surprise that when gourmets arrive on these shores fresh from Thai dining experiences at home, they fall into a feeding frenzy that lasts their entire stay.

One would expect that a dish would taste better on its own home ground, but what diners soon discover is the diversity of tastes a single dish can have. Varying chefs, varying types of ingredients and a society that, if it knows nothing else, knows food, ensure fresh taste experiences each meal, each as individual as the chef.

The base for most Thai dishes is coconut milk. Ginger, garlic, lemon grass and fiery chillies give Thai dishes a piquancy that can set tender palates aflame. While many of the chilies are mild, their potency is in obverse proportion to their size; the smallest, the *prik khii no* or "rat dropping chillies", are guaranteed to dissolve your sinuses and cloud your vision with tears. For those averse to spicy food, chefs can bland the curries or serve one of the dozens of non-spicy curries.

DINING TIPS

Dining is a communal affair. Meals are best enjoyed with friends in a convivial atmosphere, normally lubricated with beer or whiskey. Thais dawdle over their meals, talking and making an entire evening of the affair. Since dishes are placed in the middle of the table and shared by all, it makes sense to bring a few friends so one can order more dishes and sample more tastes.

Thai food is eaten with rice. Traditionally, the curries were secondary to the meal, a means of pepping up one's tastebuds so one would eat more rice. Even today, rural Thais eat enormous quantities of rice with nothing more than bits of dried salted fish to flavour them. Chillies are also a means of spicing up Thai rice, even though it is, on its own, one of the most flavorous rice in Asia. Try a few spoonfuls of plain rice before you get into the meal and discover just how delicious it is.

Affluent city Thais and foreigners have turned the order of importance around so that rice is secondary to an enjoyment of a curry. The rice serves to absorb the curry or to clear the palate much in the way French eat bread when drinking wine.

In dining, one heaps rice on his plate and then ladles a spoonful or two of a curry onto it. It is considered polite to take one curry at a time, consuming it before ladling another curry onto the rice. Thais eat with the spoon in their right hand and fork in their left, the fork being used to shovel the food onto the spoon for transport to the mouth. Chopsticks are used only for Chinese noodle dishes.

There seems to be some confusion among those who have sampled their first Thai meals abroad, about the proper condiments to add to the food. Much to their surprise, they discover that "peanut sauce", an "indispensable" additive to every dish in Western restaurants, is really of Malayan and Indonesian origin and is used in Thailand only for *satay*. Similarly, instead of salt, Thais rely on *nam plaa* or fish sauce for their salt intake, splashing a bit of it on the rice and mixing it in.

When ordering the dishes described in the following paragraphs, remember that the initial "K" letter is pronounced as a "G" and "Kh" as an aspirated "K".

SPICY DISHES

"*Gaeng*" means curry. The group includes the spiciest of Thai dishes and forms the core of Thai cooking. Among the green curries is *Gaeng Khiew Wan Gai*, a gravy filled with chunks of chicken and tiny pea-sized eggplants. A relative, *Gaeng Khiew Wan Nua*, has beef bits floating in it. *Gaeng Luang*, a category of yellow curries, includes *Gaeng Karee* which is also made with chicken or beef.

Gaeng Phet is a red curry with beef. A close relative is *Penang Nua*, a so-called "dry" curry with beef in a tasty paste.

Among the fiery favourites is *Thom Yam Kung*, a lemony broth teeming with shrimp. It is served in a metal tureen that is wrapped around a mini-furnace heated by charcoal so that it remains piping hot throughout the meal. *Po Tak* ("The Fisherman's Net Bursts") is a cousin of *Thom Yam Kung* – the broth also has squid, mussels, crab and fish.

Gaeng Som is cooked in a sour soup but falls into the category of hot-sour soups. It is filled with bits of fish or shrimp.

Yam is generally translated as salad and is as much meat as vegetable. It is also one of the hottest dishes.

MILD CURRIES

Among the non-spicy dishes is a favorite among foreigners: *Thom Kha Kai*, a thick coconut milk curry of chicken chunks with lemon grass.

Plaamuk Thawd Krathiem Prik Thai is squid fried with garlic and black pepper. When ordering, ask that the garlic (*krathiem*) be fried crispy (*krawp krawp*). The dish is also prepared with fish.

Gaeng Joot is a non-spicy curry, a clear broth filled with glass noodles, minced pork and mushrooms.

Nua Phat Namman Hoi is beef in oyster sauce with a few chopped shallots to add variety.

Muu Phat Priew Wan, sweet and sour pork, is probably of Portuguese origin and arrived in Thailand via Chinese émigrés. It is also possible to order it with Red Snapper (*Plaa Krapong*), beef (*nua*) and shrimp (*kung*).

Homok Talay is a seafood casserole of

fish and shellfish chunks in a coconut mousse steamed in a banana leaf cup.

Gaeng Musselman is an unspiced curry. It consists of pieces of beef or chicken, mixed with potatoes and onions in a brown gravy and resembles a Western stew.

Of Chinese origin but having secured a place in Thai cuisine is *Plaa Jaramet Nung Kiem Bue*, steamed pomfret with Chinese plum and bits of ginger.

Pu Phat Pong Karee is pieces of unshelled, steamed crab slathered in a curry sauce laden with shallots.

Hoi Malang Pu Op Moh Din is a thick, savory coconut milk gravy filled with mouth-watering mussels.

NOODLES & OTHERS

Most noontime dishes are derived from Chinese cuisine. Noodle dishes, a Chinese invention, have been adopted by the Thais. Those served at streetside, open-front shops come in two varieties: wet and dry. When ordering either, specify the wetness by adding the word "*nam*" or "*haeng*" to the dish's name. Thus, a wet *Kuay Tiew* would be *Kuay Tiew Nam.*

Kuay Tiew is a lunchtime favourite, a soup of noodles with balls of fish or bits of beef. *Baa Mii* is egg noodles with bits of meat and vegetable.

The rice-based lunchtime dishes are also Chinese and include *Khao Mun Kai*, boiled rice topped with slices of chicken and bits of ginger; *Khao Moo Daeng*, the same dish with pork slices, and *Khao Kha Moo*, stewed pork with greens on rice.

Then, there are the variants using noodles. *Kuay Tiew Rawt Naa* is broad white noodles boiled and served in a dish with morning glory. *Phat Thai* is noodles fried in a wok with *tofu* and bits of vegetables. *Mii Krawp* is crisp-fried noodles coated in honey and served with bits of vegetables.

Served late at night and early in the morning are two soup-like dishes filled with boiled rice. The rice in *Khao Tom* is watery and is augmented by bits of minced pork and shallots. A close relative is *Jok* in which the rice has been cooked until the liquid becomes viscous like a porridge. Into this mix is tossed ginger, coriander and bits of meat.

REGIONAL DISHES

Each of Thailand's other three regions has its own cuisine. Northern and northeastern cuisine are related to Lao dishes which are eaten with glutinous rice. Southern food is flavoured with the tastes of Malaysia and Muslim cooking.

Northern Cuisine: Northern specialties are generally eaten with *Khao Niew* or "sticky rice" which is kneaded into a ball and dipped into various sauces and curries.

Sai Oua (also called *Naam*) is an oily, spicy pork sausage that epitomises northern cooking. The sausage is roasted over a fire fueled by coconut husks which impart an aroma to the meat. Generally prepared hygienically, it is best to buy it only at better restaurants. Beware the *prik khii nuu* chillies that lurk inside waiting to explode on the tongues of the unwary.

Khao Soy originated in Burma. This egg noodle dish is filled with chunks of beef or chicken and is lightly curried in a gravy of coconut cream and sprinkled with crispy noodles.

Nam Prik Ong combines minced pork with chillies, tomatoes, garlic and shrimp paste. It is served with crisp cucumber slices, parboiled cabbage leaves, and pork rind (the latter another northern snack).

Larb is a minced pork, chicken, beef or fish dish normally associated with northeastern cuisine. While Northeasterners traditionally eat it raw, Northerners cook it thoroughly. It is served with long beans, mint leaves and other vegetables which contrast with its mellow flavour.

Gaeng Hang Lay, another dish of Burmese origin, is one of the spiciest of northern dishes and should be approached with caution. Pork and tamarind flesh give this curry a sweet and sour flavour. The curry is especially suited to dipping with a ball of sticky rice.

Mieng, or fermented tea leaves, is also Burmese and is eaten as an hors d'oeuvre.

Northeastern Cuisine: Northeastern food is simple and spicy. Like northern food, it is eaten with sticky rice which *I-sarn* (northeastern) diners claim weighs heavily on the brain and tends to make one sleepy.

Kai Yang or northeastern roasted chicken has a flavour found in no other chicken.

Basted with herbs and honey, it is roasted over an open fire and chopped into small pieces. Two dips are served with it; one hot and the other sweet.

Larb (described above under "Northern Cuisine").

Nua Yang is beef dried like a jerky. One can chew for hours on a piece and still extract flavour from it.

Som Tam is the dish most associated with the northeast. It is a spicy salad made from raw shredded papaya, dried shrimp, lemon juice and chillies.

Southern Cuisine: *Khao Yam* is rice with *kapi* (a paste made of fermented shrimp).

Phat Phet Sataw. *Sataw* looks like a lima bean but has a slightly bitter but pleasant flavor. This dish is cooked with shrimp or pork with a sprinkling of chillies.

Khao Mok Gai, a Muslim dish, lays roasted chicken on a bed of saffron rice and mixes it with ginger which has been fried lightly to make it crisp.

Khanom Chin is found throughout Thailand but the South claims to have created it. Tiny bits of minced beef are stewed in a red sauce and then served atop rice noodles. It is generally sold in markets in the early morning.

Nam Prik Kung Siap or dried prawn on a stick, is grilled and served with chillies, *kapi* (fermented shrimp paste) and lime.

Gaeng Dtai Plaa was created by bachelor fishermen who wanted a dish that would last them for days. Fish kidneys, chillies and vegetables are blended in a curry sauce and stewed for up to seven days.

Homok Khai Plaa. Fish roe are stirred into a coconut mousse, wrapped in leaves and steamed.

DESSERTS

The traditional Thai dessert and the perfect counter to the heat of the meal and the heat of the night is a simple plate of fruit. A typical plate consists of papaya, pineapple and watermelon, peeled and cut into bite-sized chunks. It may also include banana, tangerine and seasonal fruits like jackfruit, rambutan and mangosteen. For a taste treat, try durian – the rich, spiky fruit whose smell and texture offends but whose taste transcends all its bad aspects.

Coconut milk, tapioca, vermicelli, lotus seeds, water chestnut and fruits are the prime ingredients of luscious sweets sold by sidewalk vendors. Try "ice cream kathit", made of coconut milk.

DRINKING NOTES

Many restaurants catering to western tastes whip up a delicious shake made of pureed fruit, crushed ice and a light syrup. Chilled young coconuts are delicious; drink the juice, then scrape out and eat the tender young flesh. Soft drinks like Coca-Cola are found everywhere. Try Vitamilk, a health drink made from soy bean milk. For a refreshing cooler, order a bottle of soda, a glass of ice and a sliced lime. Squeeze the lime into the glass, add the soda and instantly, your thirst is slaked.

Sip the very strong Thai coffee flavoured with chicory. The odd orange Thai tea is sticky sweet but delicious. On a hot day, the Chinese prefer to drink hot tea, believing that ice is bad for the stomach. Try all three over ice anyway.

Of the many Thai cane whiskeys, Mekhong is the most popular. It is drunk on the rocks, with soda and lime or with a bit of honey added to it. Most foreign liquors are available.

PARKS & RESERVES

DOI SUTHEP–PUI NATIONAL PARK

Location: ca. 15 km west of Chiang Mai. Head west on Huai Kaew Road, past the arboretum and the zoo and follow the road signs to Doi Suthep.

Size: 262.5 sq km

Access: Regular bus service (pick-up) from the Chiang Puah gate in Chiang Mai town to the temple (Wat Phadhat), Doi Suthep and Doi Pui.

Bus fare: 35 baht to the temple; 50 baht to Phuphing.

Taxi fare: About 100-200 baht, and 400 baht if the taxi waits all day or returns in the evening.

Registration: Not necessary unless the visitor wishes to stay at the headquarters. Most visitors make day-trips from Chiang Mai. Contact: National Park Division, Royal Forest Dept., Phahok Yothin Road, Bangkok 1090. Tel.579-4842

Accommodation: Comfortable hotels in Chiangmai, basic accommodation at the headquarters.

Food: Food available in the many restaurants near the temple.

Transport in park: Taxis may be hired from the temple to the Honong hilltribe village or to Phuphing or to the summit of the mountain.

Trails: Well defined trail from HQ to the Kon Tha Than waterfall. Another trail branches off the road at km 16.

Further advice: Many visitors hire motorcycles from Chiang Mai. This is an ideal form of transportation to reach the summit of Doi Pui.

HUAI KHA KHAENG WILDLIFE SANCTUARY

Location: ca. 80 km west of Uthai Thani, Western Thailand. Head north from Bangkok on Highway 32 (ca. 200 km). Turn left for the town of Uthai Thani, a further 16 km to the west. From Uthai Thani, follow the signs to the small market town of Lan Sak, another ca. 80 km. From Lan Sak, it is approximately 20 km to the turning to Huai Kha Khaeng.

From the entrance to the sanctuary headquarters is a further 14 km along a track.

Wildlife Sanctuaries are not usually open to ordinary tourists. Visitors should have some particular purpose in mind (e.g. birdwatching, wildlife photography, research).

Visitors should apply in person and write to the Director, Royal Forest Department, explaining the reasons for their proposed visit and mentioning the likely dates and duration of their stay. Open all year.

Size: 2,575 sq km

Access: Private car or hired taxi from Lan Sak market.

Taxi fare: ca. 100 baht to headquarters.

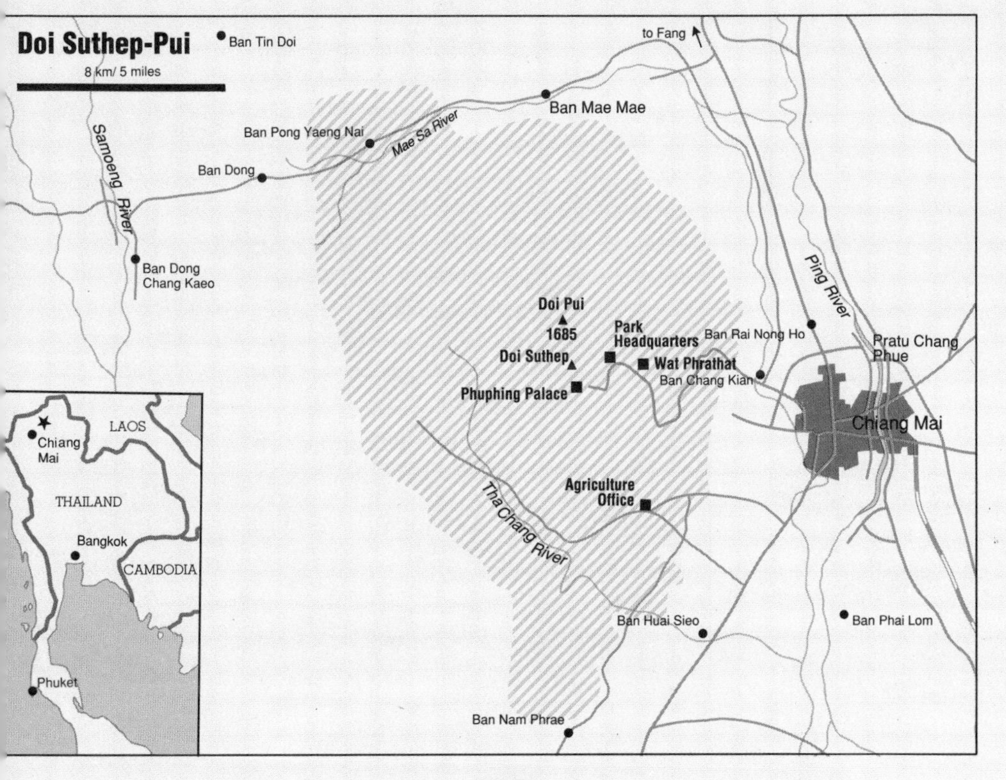

Public transportation: Village bus (pick-up truck) runs from Lan Sak market and will usually detour to the sanctuary headquarters; charges 20-25 baht.

Registration: Wildlife Conservation Division, Royal Forest Dept., Phahok Yothin Road, Bangkok 10900. Tel. 579-4847 or 579-1565.

Accommodation: Camping

Food: None; visitors to bring their own food or arrange with sanctuary staff to pay for food and for its preparation.

Transport in park: May be arranged.

Trails: Dirt road from HQ to Khao Nang Rum Research Station.

Further advice: Mosquito net advisable; also water bottle.

DOI INTHANON NATIONAL PARK

Location: 100 km southwest of Chiang Mai. Open all year, but the summit is army reserve and has to be left at night.

Size: 482 sq km

Access: Local buses from town of Chom Thong, 10 km outside park gate, to km 31, the headquarters. Local bus 15-20 baht.

Taxi fare: ca. 200 baht from Chom Tong, 1-2 day trip from Chiangmai at 1,200-2,000 baht.

Registration: Day visitors register at checkpoint at base of mountain. Campers may register at headquarters. Tel:579-4842.

Accommodation: Bungalows: 4 persons to a room at $20 per night, or 15 persons (2 large bedrooms) at $32 per night. They are located 1 km off the main road close to the headquarters.

Camping: Permitted; toilet facilities available.

Food: Park restaurant for larger groups. Local shops offer rice and noodle meals.

Transport in park: None provided.

Trails: Many dirt tracks; a road leads through the park. A 500-m trail exists on the summit.

Further advice: Can be very cool on the summit and at night; bring warm clothings. For overnight stays at the headquarter, bring a flashlight and food, since the foodstalls are difficult to locate in the evening.

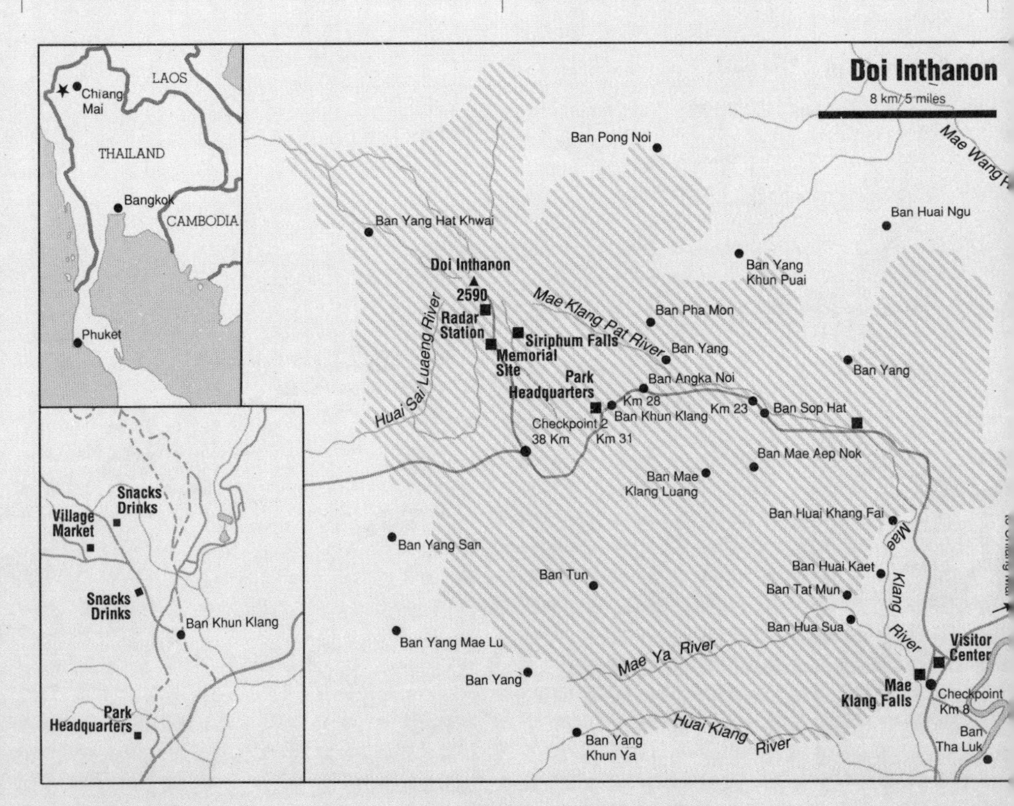

WAT PHAI LOM NON-HUNTING AREA

Location: 50 km north of Bangkok on the bank of Chao Phraya River in Amphoe Sam Khok, Pathum Thani Province. Open all year but storks are seen only from November to June.

Size: 0.05 sq km

Access: From Bangkok take highway no. 31 (Wipha Wadi Rangsit Rd.). Northeast (16 km), exit at Chaeng Wattana Rd. (no. 304) to the west to reach Pak Kret (9 km). Turn right to no. 306 to an intersection (6 km), cross Chao Phraya River at Nonthaburi Bridge (4 km), then turn right to no. 307 for about 8 km. Go north to no. 3111 for about 6 km to Amphoe Sam Khok. Wat Don (official name: Wat Surat Rangson) is about 1 km from Sam Khok town on highway 3111. At this port, long-tailed boats and ferries are available.

Registration: not necessary but for further information contact Wildlife Conservation Division, Royal Forest Department in Bangkok (Tel. 579 2776, 579 4847) or contact the forestry office at Wat Phai Lom.

Accommodation: Not available in the park.

Food: Available in town.

Transport in the area: Not necessary.

Trails: There are paths and trails under nesting colony.

Taxi fare: Approximately 300 baht (from Bangkok).

Public Transportation: Bangkok Mass Communication buses are available (No. 31 to 33) from Phra Mane Ground (Sanam Luang) to Pathum Thani. From Pathum Thani Bus Terminal, take a local bus Pathum Thani-Sam Khok.

Further advice: Since the paths and trails pass under the nesting colony, tourists are advised to bring hats to protect against stock droppings. Mites are abundant, particularly during the period February–April.

THALEBAN NATIONAL PARK

Location: 90 km from Hat Yai in the southwesternmost corner of peninusular Thailand, close to the Malaysian border.

Size: 101 sq km

Access: Hat Yai can be reached by plane from Bangkok, Kuala Lumpur or Singapore, or by bus from Bangkok or Phuket, by bus or train from the Malaysian west or east coast. A taxi from Hat Yai to Thaleban may charge 1500 baht. Cheaper to go first by bus from Hat Yai or Alor Star to Satrun, 37 km from Thaleban (ca 500 baht by taxi). The park entrance is 2 km from the border on highway 4184, on the east side of the road.

Accommodation: Available in the park. Dormitory for about 40 people. Also six bungalows for 6-10 persons each.

Food: Small canteen at headquarters sells drinks, but bring your own food.

Trails: Not well-marked. Advisable to get a local guide if you wish to explore the remote parts of the park.

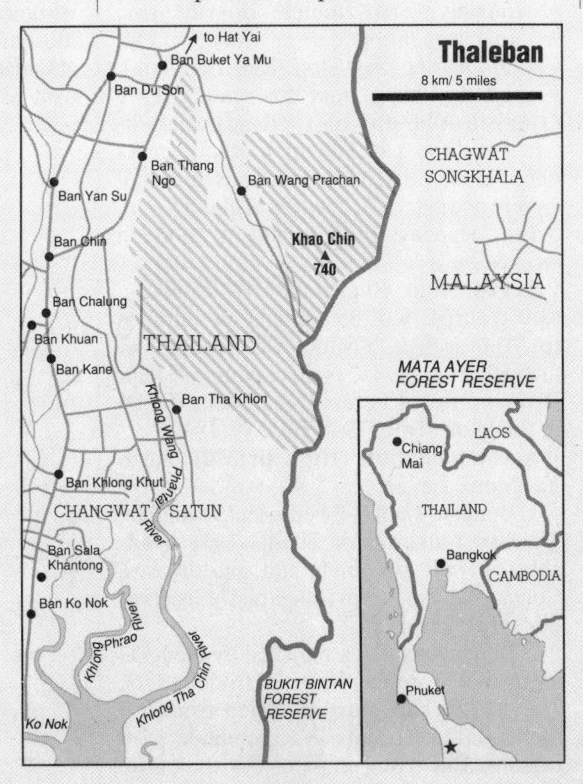

KHAO YAI NATIONAL PARK

Location: 160 km northeast of Bangkok. Open all year.

Size: 2,168 sq km

Access (car, taxi): From Bangkok take highway no. 1 or 31 north to Rangsit (about 30 km). Make a U-turn (200 m) at the intersection to northeast to get on highway no. 305 to reach Nakhon Nayok (73 km). From Nakhon Nayok take highway no. 33 southeast to reach Prachin Buri circus (20 km), turn left on to highway no. 3077 north to reach Khao Yai National Park (41 km).

Taxi fare: From Bangkok, taxis charge 1,000-1,500 baht.

Public transportation:

1. A direct air-conditioned bus to Khao Yai is available. The bus leaves at about 9 a.m. daily from the northeastern air-conditioned bus terminal and at about 2 p.m. from Khao Yai.

2. Regular air-conditioned buses (Bangkok to Nakhon Ratchasima) leaving approximately every 15 minutes from the same terminal as mentioned in 1). Get off at Pak Chong District (first stop from Bangkok). From Pak Chong, take an orange TAT (Tourism Authority of Thailand) truck directly to Khao Yai. The truck usually parks opposite the air-conditioned bus terminal (where one gets off). It leaves at about 5 p.m. during Mondays-Fridays and at noon on weekends.

Registration: Registration is not necessary. Tourists will be charged an entrance fee. Approach the Visitors' Centre for information, or otherwise contact National Park Division, Royal Forest Department (RFD) in Bangkok (Tel: 579-5269, 579-4842).

Accommodation (High priced): Khao Yai Motor Lodge.

1. Bungalows of 2-3 bedrooms owned by Tourism Authority of Thailand are available, as well as motels and guesthouses. Contact TAT office in Bangkok for reservation. (Tel: 282 1143-7).

2. Bungalows that are owned by the RFD are available around the park Headquarters. The size of bungalow is varied from 2-5 rooms and each room can accommodate 2-6 persons. But rental must be for the entire bungalow.

Accommodation (Low-priced): There are permanent camps (with wall and roof) near the park headquarters. No facilities are available except toilets. Cooking space is also available.

Camping: Camping ground is arranged for tourists with toilets.

1) RFD accommodation, is normally a available during the week but the facilities may be booked out on weekends. Prior arrangements through RFD in Bangkok are advisable.

2) Blanket and tents are available for rent.

Food: Restaurants are available at every tourist highlight spot. An expensive restaurant is located at KhaoYai Motor Lodge.

Transport in park: It is possible to hire a small pick-up car for sight-seeing (contact park headquarters).

Trails: Several well marked and beautiful trails start from the park headquarters and range from ½- to 3-hour walks. Trail map and descriptions can be obtained at visitor centre in the park.

Further advice: During December to January is sometimes quite cold. Insects, for instances mosquitoes including malaria mosquitoes, biting midges and leeches are abundant in the wet season. Ticks are abundant during the dry season.

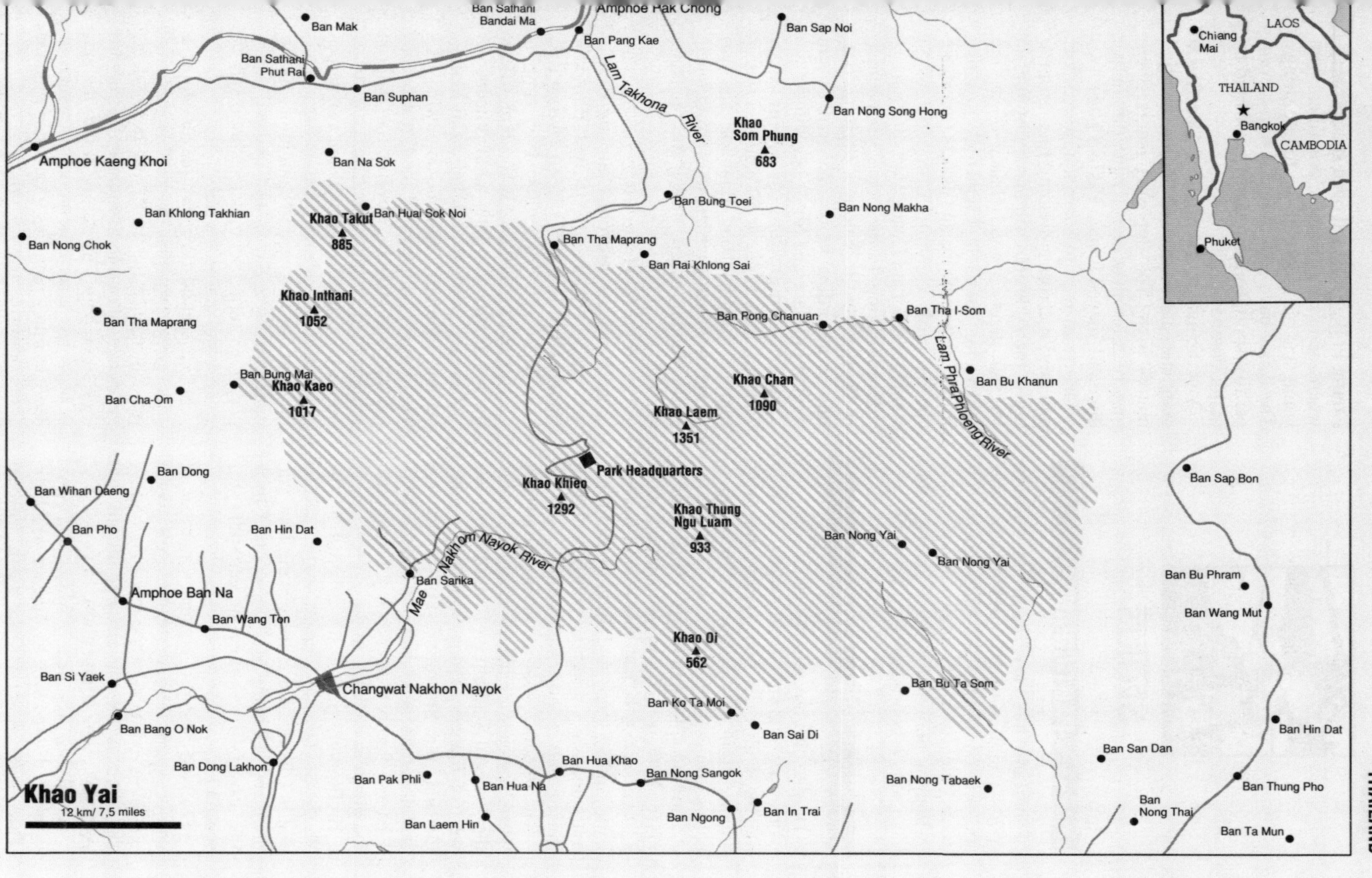
Ban Sathani Bandai Ma
Amphoe Pak Chong
Ban Mak
Ban Pang Kae
Ban Sap Noi
Ban Sathani Phut Rai
Ban Suphan
Lam Takhona River
Khao Som Phung
683
Ban Nong Song Hong
Amphoe Kaeng Khoi
Ban Na Sok
Ban Bung Toei
Ban Nong Makha
Ban Khlong Takhian
Khao Takut
885
Ban Huai Sok Noi
Ban Nong Chok
Ban Tha Maprang
Ban Rai Khlong Sai
Khao Inthani
1052
Ban Tha Maprang
Ban Pong Chanuan
Ban Tha I-Som
Lam Phra Phloeng River
Ban Bung Mai
Khao Kaeo
1017
Ban Cha-Om
Ban Bu Khanun
Khao Chan
1090
Khao Laem
1351
Park Headquarters
Khao Khieo
1292
Ban Dong
Ban Wihan Daeng
Ban Sap Bon
Khao Thung Ngu Luam
933
Ban Pho
Ban Hin Dat
Mae Nakhon Nayok River
Ban Nong Yai
Ban Nong Yai
Ban Sarika
Amphoe Ban Na
Ban Bu Phram
Ban Wang Mut
Ban Wang Ton
Khao Oi
562
Ban Si Yaek
Changwat Nakhon Nayok
Ban Bu Ta Som
Ban Ko Ta Moi
Ban Sai Di
Ban Bang O Nok
Ban Hin Dat
Ban San Dan
Ban Dong Lakhon
Ban Hua Khao
Ban Nong Sangok
Ban Pak Phli
Ban Hua Na
Ban Nong Tabaek
Ban Thung Pho
Khao Yai
12 km/ 7,5 miles
Ban Laem Hin
Ban Ngong
Ban In Trai
Ban Nong Thai
Ban Ta Mun
Chiang Mai
LAOS
THAILAND
Bangkok
CAMBODIA
Phuket

SPORTS

PARTICIPANT

Thailand has developed its outdoor sports facilities to a considerable degree and air-conditioned the ones played indoors. Nearly every major hotel has a swimming pool and a fitness centre; some have squash courts and jogging paths.

SPECTATOR

Despite the hot climate, Thai men and women are avid sports enthusiasts, actively playing both their own sports and those adopted from the West according to international rules.

The king of foreign sports is soccer and is played both by men and women. Following a close second is badminton with basketball, rugby, track and field, swimming, marksmanship, boxing, tennis and golf trailing only a short way behind.

The principal sports venues in Bangkok where one can watch the Thais in action are the National Stadium on Rama I Road just west of Mahboonkrong Shopping Centre; the Hua Mark Stadium east of the city next to Ramkamhaeng University; and the Thai-Japanese Sports Center at Din Daeng near the northern entrance to the expressway.

In Phuket, games are played at the Phuket Stadium on Vichaisongkhram Road and in Chiang Mai at the Chiang Mai Stadium and at Chiang Mai University. Check the English-language newspapers for schedules.

Thailand has also created a number of unique sports and these are well worth watching as much for the grace and agility displayed as for the element of fun that pervades every competition.

Kiteflying: The heat of March and April is relieved somewhat by breezes which the Thais use to send kites aloft. **Sanam Luang** in Bangkok and open spaces everywhere across the country are filled with young and old boys clinging to kite strings. The Thais have also turned it into a competitive sport, forming teams sponsored by major companies.

Two teams vie for trophies. One flies a giant star-shaped male Chula kite nearly 2 meters high. The opposing teams (there may be more than one) fly the diminutive diamond-shaped female Pakpao kites.

One team tries to snare the other's kite and drag it across a dividing line. Surprisingly, the odds are even and a tiny female Pakpao stands a good chance of pulling down a big lumbering Chula male (just like life). The teamwork and fast action make for exciting viewing. Competitions at Bangkok's Sanam Luang start at 2 p.m.

Horse racing: Horse racing is not unique to Thailand but it is no less exciting. Betting is according to the Western system with Win, Place and Show but without the various permutations of Quinella or Trifecta. Bets begin at 50 baht and run as high as 200,000 baht. Bangkok has two racecourses that alternate in offering meets every Sunday except during the monsoon season.

The **Royal Turf Club** is at 183 Phitsanuloke Road across from Chitrlada Palace, Tel: 282-3770; and the **Royal Bangkok Sports Club** is at No. 1 Henri Dunant Rd., Tel: 282-3770, 282-2008. Post time is 12:15 p.m.; the last race is at 6 p.m. Check the newspapers for dates and tipsheets.

Takraw: With close relatives in Malaysia and Indonesia, *takraw*, somewhat like Thai boxing, employs all the limbs except the hands to propel a woven rattan ball (or a more modern plastic ball) over a net or into a hoop. In the net version, two three-player teams face each other across a head-high net like that used in badminton. As the match heats up, it is not unusual for a player to turn a complete somersault to spike a ball across the net.

In the second type, six players form a wide circle around a basket-like net suspended high in the air. Using heads, feet, knees and elbows to keep the ball airborne, they score points by putting it into the net. A team has a set time period in which to score as many points as it can after which it is the opposing team's turn.

Tournaments are held at the **Thai-Japanese Sports Centre** (Tel: 465-5325 for dates and times) four times a year; admission is free. Competitions are also held in the northwest corner of Bangkok's **Sanam Luang** during the March-April kite contests. Free admission. During the non-monsoon months, wander into a park or a temple courtyard anywhere in the country late in the afternoon.

Thai Boxing: One of the most exciting and popular Thai sports is Thai Boxing. In Bangkok, **Rajdamnern Stadium** on Rajdamnern Nok Avenue next to the TAT, offers bouts on Monday, Wednesday, and Thursday at 6 p.m. and on Sunday at 4.30 and 8.30 p.m. The Sunday matinee at 4.30 p.m. is recommended as it has the cheapest seats. Ticket prices run between 500 and 1,000 baht for ringside seats (depending on the quality of the card), running downwards to 100 baht.

Lumpini Boxing Stadium on Rama 4 Road, 300 metres (330 yards) east of the Wireless intersection stages bouts on Tuesday and Friday at 6.30 p.m. and on Saturday at 1 and 6.30 p.m. Ticket prices are the same as at Rajdamnern. As above, weekend afternoon matinees are the cheapest.

There are also televised bouts on Saturday and Sunday and at 10.30 p.m. on some weeknights. For many visitors this will be sufficient introduction to the sport. The **Rose Garden** and Phuket's **Thai Village** offer short demonstrations of Thai boxing but these are played more for laughs than for authenticity.

Thai boxing bouts ceased in Chiang Mai several years ago but bouts are staged each Friday at 8 p.m. at the **Phuket Boxing Stadium** at Saphan Hin (to the right of the tin dredge memorial where Phuket Road meets the sea). Tickets are 30, 50 and 70 baht. Most large rural towns have their own boxing gyms and stage weekly bouts by young hopefuls.

PHOTOGRAPHY

Shimmering temples, beautiful scenery, people willing to be photographed, and no restrictions on photographing ceremonies and festivals make Thailand a photographer's delight. Pack far more film than you would for other destinations. If you don't, you will find all the major brands – Kodak, Fuji, Konica, Agfa – on sale in most towns. Slides can be processed within a day and there are mini-labs to process print film in as little as 23 minutes.

Thailand's tropical light tends to be rather harsh at midday, so pack a polarizing filter as well as a slight warming filter (81A) to take the blue tint out of shots, especially during the monsoon season. It is also advisable for slide photographers to set their ASA (ISO) meters one stop higher (i.e. instead of 64, set it at 80) to give a bit more colour saturation. A fill flash is also useful. Thais are not like silly tourists; when the sun is bright, they sit in the shade or wear hats. A flash will highlight faces that would normally be in shadowy areas.

LANGUAGE

Like the Thais, the Thai language was formed in the crucible of southern China with its host of tonal phenomena. It was only after the Thais crossed the mountains and came into contact with Indian influences, that it gained polysyllabic terms, a more formal structure and a written form. The result is a language which is grammatically simple but vocally complex (thanks to five tones) and difficult to read, especially as the words are all run together so that a single sentence looks like a single world.

The key is in the tones. Misinformed Westerners will claim that tones are not important. They soon discover that when they mispronounce, they aren't simply saying a word incorrectly, they are saying another word entirely, substituting confusion for communication.

From the languages of India have come the lexicon of literature and the terms used in the royal court. Thai names are among the longest in the world. Every Thai first name and surname has a meaning. By learning the meaning of the name of everyone you meet, you will acquire a formal, but quite extensive vocabulary.

There is no universal transliteration system from Thai into English which is why names and street names can be spelled three different ways. For example, the surname Chumsai is spelled Chumsai, Jumsai and Xoomsai depending on the family. This confuses even the Thais.

The way Thai consonants are written in English often confuses foreigners. An "h" following a letter like "p", and "t" gives the letter a soft sound; without the "h", the sound is more explosive. Thus, "ph" is not pronounced "f" but as a soft "p". Without the "h", the "p" has the sound of a very hard "b". The word Thanon (street) is pronounced "tanon" in the same way as "Thailand" is not meant to sound like "Thighland". Similarly, final letters are often not pronounced as they look. A "j" on the end of a word is pronounced "t"; "l" is pronounced as an "n".

Vowels are pronounced like this: "i" as in sip, "ii" as in seep, "e" as in bet, "a" as in pun, "aa" as in pal, "u" as in pool, "o" as in so, "ai" as in pie, "ow" as in cow, "aw" as in paw, "iw" as in you, "oy" as in toy.

In Thai, the pronoun "I" and "me" use the same word but is different for males and females. Men use the word *Phom* when referring to themselves; women say *chan* or *diichan*. Men use the word *Khrap* at the end of a sentence when addressing either a male or a female i.e. *Pai (f) nai, khrap (h)?* (Where are you going, sir?). Women append the word *Kha* to their statements i.e. *Pai (f) nai, kha (h)?*

To ask a question, add a high tone *mai* to the end of the phrase i.e. *Rao pai (We go)* or *Rao pai mai (h)?* (Shall we go?). To negate a statement, insert a falling tone *mai* between the subject and the verb i.e. *Rao pai* (We go), *Rao mai pai* (We don't go). "Very" or "much" are indicated by adding *maak* to the end of a phrase i.e. *ron* (hot), *ron maak* (very hot).

Here is a small vocabulary intended to get you on your way. The five tones have been indicated by appending letters after them viz high (h), low (l), middle (m), rising as when asking a question (r), and falling as when suddenly understanding something such as "ohh, I see" (f).

GREETINGS & OTHERS

Hello/goodbye	*Sawasdee* (a man then says *Khrap*; a woman says *Kha*; thus *Sawasdee, Khrap*)
How are you?	*Khun sabai dii, ruu (r)*
Excuse me	*Khaw (r) thoat (f)*
Well, thank you	*Sabai dii, Khap khun*
Thank you very much	*Khapkhun maak*
Please come in	*Chern khao (f)*
Please sit down	*Chern nang*
May I come in?	*khao dai (f) mai (h)*
May I take photo?	*Kaw thai roop (f) noi,* a *dai (f) mai (h).*
Never mind	*Mai (f) pen rai*
I cannot speak Thai	*Phom (r) (Chan) phuut Thai mai (f) dai (f).*
I can speak a little Thai	*Phom (r) (Chan) phuut Thai dai (f), nit (h) noi.*
Where do you live?	*Khun asai yoo thii (f) nai (r)*
What is this called in Thai?	*Nii (h), koo kaw riak away phasa Thai*
How much?	*Thao (f) rai*
When	*Mua (f) rai*

DIRECTIONS & TRAVEL

Go	*Pai*
Come	*Maa*
Where	*Thii (f) nai (r)*
Right	*Khwaa (r)*
Left	*Sai (h)*
Turn	*Leo*
Straight ahead	*Trong pai*
Please slow down	*Cha cha noi*

What street is this?	*Nii thanon arai*
What town is this?	*Nii muang arai*
How many kilometres to...?	*Kii kiilo by...*
Stop here	*Yood thii (f) nii (f)*
Fast	*Raew*
Hotel	*Rong raam*
Street	*Thanon*
Lane	*Soi*
Bridge	*Saphan*
Police Station	*Sathanii Dtam Ruat*
Market	*Talad*

USEFUL PHRASES

Yes	*Chai (f)*
No	*Mai (f) chai (f)*
Good	*Dii (m)*
Bad	*Mai (f) dii*
Do you have?	*Mii Mai (h)*
Expensive	*Phaeng*
Do you have something cheaper?	*Mii arai thii thook (l) kwa nii (h)?*
Can you lower the price a bit?	*Kaw long noi dai (f) mai (h)*
Do you have another colour?	*Mii sii uhn mai (h)*
Too big	*Yai kern pai*
Too small	*Lek kern pai*
Do you have bigger?	*Mii arai thii yai kwai nii (h) mai (h)*
Do you have smaller?	*Mii arai thii lek kwa nii (h) mai (h)*
Where is the toilet?	*Hong nam (h) yuu thii (f) nai (r)*

OTHER HANDY WORDS

Hot (heat)	*Ron (h)*
Hot (spicy)	*Phet*
Cold	*Yen*
Sweet	*Waan (r)*
Sour	*Prio (f)*
Delicious	*Aroy*
Coconut	*Ma-prao*

NUMBERS

One	*Nung (m)*
Two	*Song (r)*
Three	*Sam (r)*
Four	*Sii (m)*
Five	*Haa (f)*
Six	*Hok (m)*
Seven	*Jet (m)*
Eight	*Pat (m)*
Nine	*Kow (f)*
Ten	*Sip (m)*
Eleven	*Sip Et (m, m)*
Twelve	*Sip Song (m, r)*
Thirteen	*Sip Sam (m, r)*
Twenty	*Yii Sip (m, m)*
Thirty	*Sam Sip (r, m)*
100	*Nung Roi (m, m)*
1,000	*Nung Phan (m, m)*
10,000	*Muan*

DAYS OF THE WEEK

Monday	*Wan Jan*
Tuesday	*Wan Angkhan*
Wednesday	*Wan Phoot*
Thursday	*Wan Pharuhat*
Friday	*Wan Sook*
Saturday	*Wan Sao*
Sunday	*Wan Athit*
Today	*Wan nii (h)*
Yesterday	*Mua wan nii (h)*
Tomorrow	*Prung nii (h)*
Every day	*Thuk (h) wan*

GLOSSARY OF SIMPLE TERMS

***Bot*:** The ordination hall, usually open only to the monks. A bot is marked by six "bai sema" or boundary stones around the outside of the building which define the limits of sanctuary. Many *wats* do not have *bots*, only *viharns*.

***Chedi*:** Often interchangeable with stupa. A mound surmounted by a spire in which relics of the Buddha or revered religious teachers are kept.

***Chofah*:** The bird-like decoration on the ends of a bot or viharn roofs.

***Naga*:** A mythical serpent, usually running down the edge of the roof. In sculpture, it sheltered Buddha as he meditated.

***Prang*:** An Ayutthayan-style *chedi* that looks somewhat like a vertical ear of corn.

***Sala*:** An open-sided pavilion.

***Viharn*:** The sermon hall; the busiest building in a *wat*. A *wat* may have more than one.

***Wat*:** Translated as "temple" but describing a collection of buildings and monuments within a compound wall.

USEFUL ADDRESSES

TOURIST INFORMATION

Take the guesswork out of planning a trip to Thailand by contacting an overseas office of the Tourism Authority of Thailand (TAT), the Thai government's official tourism promotion organisation. At the offices listed below are brochures, maps and videotapes of the country's multitudinous attractions. Officials can answer questions regarding sights, facilities and services.

Additional tourism information can be found in the numerous travel magazines that give reliable information on current events in Bangkok, Pattaya, Chiang Mai, Chiang Rai, Phuket, Haad Yai, and Koh Samui. *The Nation's* newspaper, *Saen Sanuk* tourism newspaper, is issued each Friday as part of the morning edition.

TAT OFFICES IN THAILAND

Bangkok
No. 4 Rajdamnern Nok Ave., Bangkok 10100, Tel: 282-1143/4.

Chiang Mai
105/1 Chiang Mai-Lamphun Rd., Amphoe Muang, Chiang Mai 50000, Tel: (053) 235-334, 248-604/7.

Hat Yai
1/1 Soi 2, Niphat Uthit 3 Rd., Hat Yai, Songkhla 90110, Tel: (074) 243-747, 245-986.

Kanchanaburi
Saeng Chuto Rd., Amphoe Muang, Kanchanaburi 71000, Tel: (034) 511-200.

Nakhon Ratchasima
2102-2104 Mittraphap Rd., Amphoe Muang, Nakhon Ratchasima 30000, Tel: (044) 243-427, 243-751.

Pattaya
382/1 Chaihat Rd., South Pattaya 20260, Tel: (038) 428-750, 429-113.

Phitsanuloke
209/7-8 Surasi Trade Center, Boromtrailokanat Rd., Amphoe Muang, Phitsanuloke 65000, Tel: (055) 253-742, 252-743.

Phuket
73-75 Phuket Rd., Amphoe Muang, Phuket 83000, Tel: (076) 212-213, 211-036.

Surat Thani
5 Talat Mai Rd., Ban Don, Amphoe Muang, Surat Thani 84000, Tel: (077) 282-829, 281-828.

Ubon Ratchathani
264/1 Khuan Thani Rd., Amphoe Muang, Ubon Ratchathani 34000, Tel: (045) 243-770, 243-771.

GETTING THERE

BY AIR

Malaysia is well connected by airlines to all continents, and if you are coming from Europe or North America, you will enter the country by Subang Airport, Kuala Lumpur. If you are arriving from a nearby Asian country (Thailand, Indonesia or Singapore), it is possible to fly to some other Malaysian cities either directly or by connecting flights through Kuala Lumpur. Penang, Langkawi and Tioman can be reached directly from Singapore. For details of flights, contact your airline or travel office.

BY SEA

Cruise liners call at Malaysia on round-the-world trips, but most ports harbour cargo ships. There are boats to Penang from Phuket in Thailand and from Medan in Sumatra, Indonesia, to Penang. The latter operates twice weekly and details are available at travel agents in Penang. Another interesting route is with the Greek-built *MV Vigneswara* of the Greenseas Shipping Company, which plies the waters between Madras in South India, and Penang and Singapore twice a month. The ship is equipped with air-conditioning, casino, bar, TV lounge and a duty-free shop. Ask at travel agencies in Madras or Penang.

BY RAIL

An extensive railway network runs through Malaysia (for information on trains, see "Getting Around" section) from Singapore, with connections to Bangkok. The Thai-owned International Express leaves daily from Bangkok for Butterworth, and trains from Haadyai in the south of Thailand connect to trains on the Eastern Malaysian railway. The trip from Bangkok takes about two days, and you can travel in a first-class air-conditioned sleeper, a second-class non-air-conditioned sleeper, or in upright seats in third class. For reservations, contact the railway station in Bangkok (Hualamphong) or a local travel agent. Trains from Singapore to Kuala Lumpur and Kota Bharu run several times daily and take from 7 to 10 hours (Kuala Lumpur) or 12 to 15 hours (Kota Bharu) respectively.

BY ROAD

From Thailand, it is possible to get buses from Bangkok or Haadyai that cross the border at Padang Besar and travel to Penang or Kuala Lumpur, or at Sungei Golok on the East Coast. From Singapore, there are many buses which travel to all parts of peninsular Malaysia from the bus station at New Bridge Road, Singapore. Other buses run from Beach Road in Singapore. New Bridge Road buses may be booked by calling 221-6603. Long-distance taxis also run from Queen Street in Singapore, although you will get a better bargain by taking the bus to Johor Bahru across the Malaysian causeway and taking a taxi from there.

TRAVEL ESSENTIALS

PASSPORTS & VISAS

Valid passports and a health certificate of vaccination against yellow fever are required if travelling from an infected area. Citizens of Commonwealth countries, Ireland, Switzerland, the Netherlands and Liechtenstein do not need a visa to visit. The following countries do not need a visa for a visit not exceeding three months: Austria, Italy, Japan, South Korea, Tunisia, the United States, Germany, France, Norway, Sweden, Denmark, Belgium, Finland, Luxembourg and Iceland.

Citizens from communist countries are

granted visas from 7 to 14 days. Citizens from Israel and South Africa are not able to visit Malaysia, and Chinese, Kampuchean and Vietnamese citizens are only able to visit on an official basis.

Immigration requests that your passport be valid for at least 6 months. Bear in mind that Sabah and Sarawak are treated like other countries, and you will have to go through customs again there, both from peninsular Malaysia and between the two states.

On arrival, the most common visa will be for 30 days. If you wish to extend your stay, and are from one of the countries enjoying diplomatic relations with Malaysia, then you may do so at any of the following immigration offices:

Federal Territory
Blok 1, Tingkat 2-3, Pusat Bandar Damansara, Bukit Damansara,
50490 Kuala Lumpur
Tel: 03-7579063

Kuala Lumpur
Headquarters Office, Blok 1,
Tingkat 4-7, Pusat Bandar Damansara,
50490 Kuala Lumpur
Tel: 03-757-8155

Johor
Wisma Persekutuan Johor, Blok B,
Tingkat 1, Jln. Air Molek,
80550 Johor Bahru
Tel: 07-244255

Kedah
Tingkat 2, Wisma Persekutuan,
0500 Alor Setar, Kedah
Tel: 04-723302

Kelantan
Tingkat 2, Wisma Persekutuan, Jln. Bayam,
15550 Kota Bharu, Kelantan
Tel: 09-741644

Melaka (Malacca)
Tingkat 2, Bangunan Persekutuan,
Jln. Hang Tuah, 75300 Melaka
Tel: 06-224958

Negri Sembilan
Tingkat 2, Wisma Persekutuan,
Jln. Datuk Abdul Kadir,
70675 Seremban, Negri Sembilan
Tel: 06-727707

Pahang
Tingkat 1, Wisma Persekutuan,
Jln. Gambut, 25000 Kuantan, Pahang
Tel: 09-521373

Perak
Bangunan Persekutuan, Jln. Dato' Panglima Bukit Gantang, 30000 Ipoh, Perak
Tel: 05-540394

Perlis
Tingkat 1, Menara Kemajuan PKNP,
Jln. Bukit Lagi, 01000 Kangar, Perlis
Tel: 04-753535

Pulau Pinang (Penang)
Jln. Leboh Pantai, 10550 Pulau Pinang
Tel: 04-610678

Sabah
Tingkat 4 & 5, Bangunan Penerangan,
88550 Kota Kinabalu, Sabah
Tel: 088-51752

Sarawak
Peti Surat 639, 93908 Kuching, Sarawak
Tel: 082-20895

Selangor
Kompleks PKNS,
40550 Shah Alam, Selangor
Tel: 03-506061

Terengganu
Tingkat 1, Wisma Persekutuan,
Jln. Paya Bunya,
20200 Kuala Terengganu, Terengganu
Tel: 09-622457

MONEY MATTERS

Currency and Exchange: The Malaysian currency note is the *ringgit* or Malaysian Dollar, which is divided into 100 *sen*. The amount of Malaysian dollars you are allowed to bring in or take out of Malaysia is unlimited. At the current exchange, US$1 will give you M$2.70, A$1 – M$2.03, and £1 – M$4.24. Singapore and Brunei dollars are worth about 30 percent more than the Malaysian *ringgit*, and no longer circulate freely in Malaysia. You may be able to use Singapore dollars in the state of Johor, but they are counted as having the same value as the *ringgit*.

Banks and licensed money changers offer better rates than do hotels and shops, where a service charge may be levied (usually 2-4 percent). Make sure that you have enough cash before you leave for smaller towns or remote areas.

Travellers' Cheques and Credit Cards: In the more flashy quarters of the larger towns, in department stores, shops, first-class restaurants and hotels, travellers' cheques change hands easily. Have your passport ready when changing cheques. Off the beaten track, you may find it harder to change travellers' cheques. Established credit cards – Diners Club, American Express, Visa, Mastercard and Carte Blanche – are honoured in the major cities. Several hotel chains maintain their own credit card system. But when travelling extensively through Malaysia, nothing could be better than the coin of the realm.

Costs: Bear in mind that accommodation, transportation and food costs tend to be higher in East Malaysia than on the peninsula. But in general, costs in Malaysia are considerably lower than in Europe or North America (except for some imported goods), and higher than in neighbouring countries such as Thailand and Indonesia. Approximate prices for accommodation and some transportation are offered in this guide to give a general idea of costs. It is advisable to carry a mixture of monies, local and foreign currencies, as well as travellers' cheques and credit cards, and to store these separately in your luggage in case of theft.

Below is a budget guide published by the Tourist Corporation of Malaysia (TDC); you will find that you may possibly be able to exceed it or even undercut it!

HEALTH & EMERGENCY

Travellers have few worries in a country where the health standards are ranked amongst the highest in Asia.

Water in cities is generally safe for drinking, but it is safest to drink it boiled. Bottled drinks are also widely available. Avoid drinking iced water from roadside stalls. It is important to drink sufficiently to avoid dehydration; drink more than you would normally if you're coming from a cold country.

The sun is deceptively strong here: one hour of sunbathing a day for the first few days will get you a lasting tan without giving you sunstroke.

If you are visiting remote jungle areas, it is advisable to take malaria tablets; your doctor will know which type is suitable for the region. To help keep mosquitoes at bay, use insect repellents, mosquito coils and nets at night. If you intend to travel to Borneo, ask your doctor about outbreaks of cholera in the region. These are rare, but should there be any, you would be wise to have a vaccination before leaving home.

Treat open cuts and scratches immediately as infection in humid climates can delay healing, and at worst, cause tropical ulcers. If you are swimming in the sea near coral reefs, do not touch any of the interesting shells, snakes and other creatures you find there. Many of them are poisonous, but if you don't threaten them, you will be quite safe. To avoid getting sea urchin prickles in your feet, it is a good idea to wear plastic shoes while exploring the coral reefs.

Medical supplies are widely and readily available in Malaysia, and all large towns have government polyclinics as well as private clinics. Bring a small first-aid kit, or buy one from Kuala Lumpur or another large city. French and German-speaking doctors can be found by contacting the relevant embassies. Travel and health insurance, as well as documents concerning allergies to certain drugs should also be carried.

WHAT TO WEAR

In the delightful tropical climate of Malaysia, informal wear is most suitable and comfortable. However, since this is a predominantly Muslim and conservative country, observance of local customs is important. Men may wear tee-shirts or cotton shirts with short sleeves, and open sandals. Women should not wear dresses, skirts or shorts that are too short, and topless sunbathing is frowned upon. In cities, towns and villages, shorts are not a good idea – save them for the beach. In mosques, the legs should be covered to below the knee, and some mosques will provide scarves for the head and arms. When visiting government offices and passing through immigration points, long trousers and skirts are looked upon favourably.

WHAT TO BRING

There is very little need to worry about leaving something important behind when you visit Malaysia. Toiletries, medicines, clothes, photographic film, suntan lotion and straw hats are all readily available in most towns, and definitely in the large cities. In fact, the best advice is to take as little as possible so that you can travel lightly and comfortably.

If you are planning to visit the hill stations, a light sweater would be a good idea for the cooler evenings. If you're embarking upon the Mount Kinabalu climb, a lightweight plastic raincoat is a must, as are a warm hat and gloves, but all these items can be found in Kuala Lumpur, Singapore, or Kota Kinabalu. Camping gear is often available for hire in national parks, but it is also under heavy demand; so it may be best to bring a lightweight tent with you. If you intend to go jungle-trekking, the Taman Negara National Park issues a list of contents for the average backpack.

In more remote areas, you will not have the luxury of a shaving point, but disposable razors are sold widely in towns and cities, or you can buy a battery-operated razor at reasonable prices. Sanitary protection for women is also available in larger centres. Cheap clothes are everywhere – batik shirts are colourful and cool, and tee-shirts with interesting slogans are also a good buy. You may even decide to adopt the multi-purpose sarong as skirt, towel or sheet! So...just bring your camera. Or buy one in Singapore!

CUSTOMS

Import duties seldom affect the average traveller, who may bring in 250 grammes of tobacco or cigars, or 200 cigarettes, and a one-quart bottle of liquor duty-free. Other duty-free items include food items not exceeding $75, 3 pieces of clothing (shirts, scarves, ties etc.), one pair of shoes, and any gift item not exceeding $25. Used portable articles are normally exempted from import tax. Pornography, weapons and walkie-talkies are strictly prohibited. Possession of narcotics and other illegal drugs carries the death sentence, and firearms are subject to licensing.

The above duty-free items are not available to you if you are travelling on a domestic flight, or from Singapore.

ON DEPARTURE

Departure Tax: Airport tax is collected at all airports. For domestic flights, the tax is $3; for flights to Brunei and Singapore, $5; for all other international flights, the departure tax is $15.

GETTING ACQUAINTED

GOVERNMENT & ECONOMY

Malaysia is the official name of the former British protectorates of Malaya, British North Borneo and Sarawak. Independent since 1957, the Malaysian government is regulated by the Parliament comprising the *Yang di-Pertuan Agong*, King or Supreme Sovereign, and two Houses: the House of Representatives and the Senate. The executive functions of the government are carried out by the Cabinet, led by Dato' Seri Mahathir Mohammed who became Prime Minister in 1981.

The population of 17.36 million comprising of Malays, Chinese, Indians, Pakistanis and other indigenes, is spread over 13 states. The capital city of Kuala Lumpur alone has a population of approximately 1 million.

Petroleum oil, natural gas, tin, timber, pepper, palm oil and rubber are the main exports. Its main trading partners are Japan and the United States.

CLIMATE

A tropical sun and clouds laden with the makings of a sudden downpour compete for the skies of Malaysia, with the odds on the sun. Malaysia's seasons follow the monsoon winds, which splash rains inland from September to December on the west coast of the peninsula, only to be overtaken by sunshine within the hour. Rains arrive later, between October and February, on the east coast of peninsular Malaysia and in Sabah and Sarawak. Malaysia's weather, however, is generally warm, humid and sunny all year round, with temperatures wavering between 32°C during the day and 22°C at night. The highlands, both during the day and at night, and the lowlands in the evening, are comfortably cooler, which is why Malaysia's nightlife is liveliest outdoors.

CULTURE & CUSTOMS

The customs, religions and language of many nations converge in Malaysia. With everyday etiquette relaxed and straight-forward, visitors behaving courteously stand little chance of unintentionally giving offence. It is beneficial, however, to learn something of how Malaysians behave towards one another so that you will become more integrated in the culture.

Seniority is much respected. The oldest male member of a family is greeted first, often sits in the best and highest seat, and is consulted first on any matter. Pointing with the finger is considered very rude and a whole hand is best used to indicate a direction (but not a person). For those interested in learning more about Malaysian customs, have a look at the Times Editions guidebook *Culture shock! Malaysia and Singapore*.

Temples & Mosques: Removing one's shoes before entering a mosque or an Indian temple has been an unspoken tradition for centuries. Within, devotees do not smoke. Neither of these customs generally apply to Chinese temples where more informal styles prevail. Visitors are most welcome to look around at their leisure and are invited to stay during some religious rituals. While people pray, it is understood that those not participating in the service will stand quietly to one side. A polite gesture would be to ask permission before taking photographs; unless there is a sign indicating otherwise, this request is seldom if ever refused. Moderate clothing, rather than short skirts or shorts should be worn. Most temples and mosques have a donation box for funds to help maintain the building. Contributing a few coins before leaving is customary.

Tipping: Tipping is not common in Malaysia, especially in more rural areas. In most hotels and large restaurants, a 10 percent service charge is added to the bill along with 5 percent government tax. In large hotels, bellboys and porters usually receive tips from 50 *sen* to $2 depending on the service rendered. Outside these international establishments, however, simply a thank you (*terima kasih*) and a smile will do.

WEIGHTS & MEASURES

Malaysia is fast converting from the English Standard System to the metric system, though it will be some time before it is complete. Road distances are always given in kilometres, but if you ask a *kampong* dweller for directions, he will give distances in both miles and kilometres. There is a similar confusion over weights.

ELECTRICITY

Do not expect to plug an American electric razor into the communal circuit of a clapboard coconut palm-shaded *kampong* house. The Malaysian current is 220 volts, 50 cycles, although most first-class hotels can supply an adaptor for 110-volt, 60 cycles appliances. Bargains in electrical appliances can be found in Singapore.

BUSINESS HOURS

In an Islamic nation with a British colonial past, weekly holidays vary. In the former Federated States which were united under the British – Selangor, Melaka, Penang, Perak, Pahang and Negri Sembilan – there is a half-day holiday on Saturday and a full-day holiday on Sunday. The former unfederated states, which remained semi-autonomous under British rule – Johor, Kedah, Perlis, Terengganu and Kelantan – retain the traditional half-day on Thursday and full-day holiday on Friday; Saturday and Sunday are treated as weekdays. The workday begins at 8 a.m. and ends at 4.30 p.m. with time off on Friday from noon until 2.30 p.m. for communal *Jumaat* prayers at the mosque. Most private businesses stick to the nine-to-five routine. Shops start to close at 6 p.m., unless they are attached to a night market, and department stores like *Klasse* and *Metrojaya* in Kuala Lumpur keep their cash registers ringing past 9 p.m.

Banking: The nationwide network of 42 commercial banks has all the facilities to cope with simple as well as more complex transactions. The various banks operate more than 550 offices throughout the country and have connections with the major financial centres of the world. Banking hours are Monday through Friday 10 a.m. to 3 p.m., and Saturday 9.30 a.m. to 11.30 a.m.

COMMUNICATIONS

MEDIA

Television: This is the most popular communication medium in Malaysia, watched in international hotel rooms and longhouses with the same enthusiasm! Programmes are highly cosmopolitan and British and American sit-coms and documentaries are shown alongside Indonesia's hottest film and Koran reading competitions from Kuala Lumpur. British football is also given wide coverage, and sports in general have a generous slot of transmission time. The news is given in English on Channel RTM 1 at 6 p.m. and on TV 3 at 6.30 p.m.

Radio: Radios can also be heard everywhere, blaring out a wild assortment of different sounds; a flick of the dial will tune you into Indonesian pop, Malay rock and heavy metal, Indian pop or classical music, Chinese theatrical, John Williams sci-fi music scores or the number one sound in Britain or America. Soap operas and programmes of daily events around the country will also familiarise you with the Malay culture. For the news in English, tune into Blue Network (280 MW) at 7 a.m., 8 a.m., 1.30 p.m., 5 p.m. or 7 p.m., or to "Beautiful Malaysia" (97.2 FM) between 6 and 7 p.m.

The Press: Malaysia's newspapers come in a variety of languages, from *Bahasa Melayu* (Malay, the national language), English and Chinese, to Tamil, Punjabi and Malayali. The *New Straits Times* and the *Star*, masters of the English press, arrive every morning crammed with national and world news, occasional eye-opening letters to the editors, and "Peanuts" and local cartoonist Lat comic strips. The *Malay Mail*, an afternoon paper, is less formal and more chatty, and entertains its readers by focussing more on local news and entertainment. The *Sabah Times, People's Mirror, Sarawak Tribune* and *Borneo Post* hail from East Malaysia with international news appearing next to events from the remote jungle. Foreign newspapers and magazines can also be purchased in large cities.

POSTAL SERVICES

Malaysia has one of the most efficient postal services in Asia. There are post offices in all state capitals and in most cities and towns. Except for the General Post Office in Kuala Lumpur, which opens from 8 a.m. to 7 p.m. (Monday to Saturday), all post offices are open from 8 a.m. to 5 p.m.

An aerogramme to any country costs 40 *sen*. Postcards to other Asian countries cost between 20 and 25 *sen*, to Australia and New Zealand 30 *sen*, to Europe 40 *sen* and North America 55 *sen*. Most international hotels provide postal services; and stamps and aerogrammes are often sold at the small Indian sweet and tobacco stalls on street corners. Malaysian stamps are very attractive, often picturing the country's beautiful flora and fauna.

TELEPHONE & TELEGRAM

Local calls in Malaysia cost 10 *sen*, and public phones can be found in most towns, often located in front of restaurants. Long-distance and international calls can be made from the post office or from a hotel.

On Penang Island, different area codes have been given to each suburban district – Balik Pulau 898, Batu Ferringhi 811, Batu

Uban 883, Bayan Lepas 831, Penang Hill 892, and Tanjong Bungah 894. Inter-state calls require the following prefix codes:

Ipoh	05
Johor Bahru	07
Kota Bharu	09
Kota Kinabalu	088
Kuala Lumpur	03
Kuala Terengganu	09
Kuantan	09
Kuching	082
Melaka	06
Penang	04
Seremban	06
Sungai Petani	04
Taiping	05

For international calls, dial "108" to book the call. If you dial "104", an operator will take your telegram dictation over the phone. International Direct Dialling is available in most international hotels, as are secretarial and facsimile services.

EMERGENCIES

In an emergency, the charge-free number to call is 999 for ambulance, fire or police services. Most hotels have security services and many of them have safes for valuables. Never carry too much cash on your person, and keep travellers' cheque numbers apart from the cheques. In the event of loss, you can call the Tourist Police Unit at the following numbers:

Johor Bahru	07-232222
Kuala Lumpur	03-241-5522
	03-243-5522
Melaka	06-222-2222

GETTING AROUND

ORIENTATION

Malaysia is one of the easiest Asian countries to travel around in. Transportation ranges from an Orang Asli dugout canoe up remote rivers, to a trishaw in Kuala Terengganu, to funicular railways and fast modern jets. There are almost always several alternatives to travelling to a place. Travelling in Malaysia can be as exciting and as adventurous as you want to make it. To get a feel of the country, you should try several of the different modes of transport.

FROM THE AIRPORT

Buses, public and private, taxis and even limousine services operate from major airports in Malaysia. Many airports have a taxi desk, where taxis can be booked and paid for and where the price is fixed. Where there is no such service, inquire at the information desk about how far the town is from the airport, and how much you should pay a taxi driver. Otherwise, you might fall prey to an unscrupulous taxi driver who takes advantage of you by overcharging you for a very short distance. Taxi fares from airports are in general much higher than around the town.

DOMESTIC TRAVEL

By Air: Malaysian Airlines System (MAS) runs an extensive network of airways over the entire nation. In remote jungle areas, the Fokker 27s and Northland Islanders operate buses linking out-of-the-way places to national centres. Singapore Airlines, Royal Brunei and Thai International have flights to Malaysian destinations other than Kuala Lumpur.

By Rail: Malaysian Railways or Keretapi Tanah Melayu (KTM), runs right from the heart of Singapore's business centre, through the Malay peninsula and on into Thailand in the north, calling at major cities and towns including the capital Kuala Lumpur. Another line, the East Coast line, branches off the main one at Gemas, plunges through the central forests and emerges eventually at Tumpat, near the border to Thailand. Malaysian trains are generally comfortable, and travelling by train gives one an excellent idea of Malaysia and its varying countryside. Passengers can choose from air-conditioned first class coaches on the day trains and first-class air-conditioned twin berth cabins on the night trains. In the second class are fan-cooled sleepers, and sleeperettes are in the third class coaches. First and second class tickets may be purchased 30 days in advance, third class tickets 10 days in advance. Passengers holding tickets for distances over 200 kilometres are allowed to "break journey" at any station – one day for every 200 km or part thereof, in addition to time occupied by the journey. Passengers taking advantage of this scheme must remember to have their tickets endorsed by the station master at the alighting station immediately upon arrival.

For foreign tourists, KTM offers a railpass which entitles the holder to unlimited travel in any class and to any destination for a period of 10 or 30 days. The Railpass costs M$85 for 10 days and M$175 for 30 days. The cost of the pass does not include sleeping berth charges, and for these you would be wise to book in advance.

There are a number of different train services available. The passenger can either choose the normal train which stops at most stations, or the express which only stops at major towns. Most stations have left luggage services. Charges are a reasonable 50 *sen* a day.

WATER TRANSPORT

Traditionally, transport in Malaysia, particularly in the west, was by water. In Pahang, on the Endau River, and of course, in Sarawak, water transport still has some importance, but in general, roads have taken over as the main means of transport.

On rivers in peninsular Malaysia, you may find boats for hire, but often the best way is to find out which boats are going where and hitch a ride. Boat rental can sometimes be phenomenally expensive, especially, for example, if you want a ride downriver. The boatman will be reluctant, because he has to motor all the way upriver again afterwards. He might prefer to sell the boat to you!

In Sarawak, there is a lot of traffic on the rivers inland throughout most of the year as roads are still few and far between and mostly in a poor condition. On the Rejang River, regular boats run between Sibu and Kapit, and if the waters are high enough, all the way to Belaga. Boats to smaller rivers and to remote longhouses can be fearfully expensive, and it is best to go down to the jetty and see where all those women with baskets are heading for and change your plans accordingly.

Other regular ferry and boat services include boats to the islands Pangkor, Penang and Langkawi. Boats to Pangkor stop running at 7 p.m., but ferries to Penang run 24 hours, and boats to Langkawi till 6 p.m.

Boats out to islands on the east coast are slightly less regular, especially in the monsoon season when several services may stop altogether. There are services to the Perhentian Islands, Kepas, Redang and Tenggol in the north, and many boats run out to the islands off Mersing. Mersing boats can be a little confusing as there is a wide choice. Boats to Tioman range from fishing boats that charge $15 per person, or catamarans and ferries that cost $25-$30 and take 2½ hours, to the hydrofoil which takes 75 minutes and costs around $25. Some of the other islands also have ferries (e.g. Rawa and Sibu) or you can hire a fishing boat to get there. The price of the boat is the same whether you are one person or twelve (maximum capacity).

Besides ferries to Penang, there are several leisurely cruises which allow one to see Malaysian shores on the peninsula and also in Borneo. Introduced in August 1986, Feri Malaysia operates cruise *Muhibah*, a holiday cruise ship, between Singapore, Kuantan, Kuching and Kota Kinabalu. The ship offers air-conditioned cabins and comfortable suites, as well as facilities such as restaurants, a discotheque, gymnasium, cinema, swimming pool and golf putting green. Stopover packages and shore excursions are available and further information can be obtained at Feri Malaysia Sdn. Bhd., Ground Floor, Menara Utama UMBC, Jln. Sultan

Sulaiman, 50000 Kuala Lumpur, Tel: 03-238-8899, Telex: FM KUL MA 50055.

PUBLIC TRANSPORT

By Taxi: Taxis remain one of the most popular and cheap means of transport, especially on a shared basis. You can hail them by the roadside, hire them from authorised taxi stands, or book them by phone, in which case, mileage is calculated from the stand or garage from which the vehicle is hired.

Although most taxis are fitted with meters, they are only used in major towns such as Kuala Lumpur, Johor Bahru and Ipoh.

At the start of the journey, the meter reads $1.00 and turns over in 10 *sen* lots every 200 metres. Check that your driver turns his meter on only when you have got into the car. Apart from the amount stated on the meter, you also have to pay 30 percent of the price extra for air-conditioning, regardless of whether you want it or not! If you are travelling in the early morning or late at night, taxi drivers are reluctant to use their meters and prefer to fix a price for the destination: this is where you'll need your bargaining skills! You can also negotiate a day price with town taxis.

Outstation taxis are a popular way of getting to another city or town, and if your bargaining powers are good, it can be very economical. The taxis usually operate on a shared cost basis, with four passengers, and many drivers will not leave until that quota has been reached. If you want to pay four times as much, then you can charter a taxi, which will mean you don't have to wait around for other passengers to turn up.

There are no meters in taxis in Sabah or Sarawak. Find out how much other transport cost to your destination and calculate accordingly. Transportation costs are generally much higher in East Malaysia than on the peninsula.

By Bus: There are three types of buses that operate in Malaysia: the non-air-conditioned buses plying between the states, the non-air-conditioned buses that provide services within each state, and the air-conditioned express buses connecting major towns in Malaysia. All prices are reasonable though bus departures do not always adhere to the schedule.

By Trishaw: If you want to see the city at your own pace, you can still find trishaws at your service. These are very popular in cities such as Melaka, Kota Bharu, Kuala Terengganu and Georgetown, Penang. For short trips, they are better than taxis as their slow pace allows you to see points of interest along the way.

Except in Penang, where passengers sit in a sun-hooded carriage in front of the cyclist, a trishaw is a bicycle with a side carriage. In Penang, trishaw drivers will warn you to hold on to your bags firmly for fear of snatch-thieves on motorcycles. Incidences of this kind, however, are rare nowadays.

It is important that you fix the price before proceeding in a trishaw. A little bargaining is necessary, and you should inform the peddler if you wish to stop somewhere along the way, as stopping time also has to be calculated. In some places, especially in Penang, you can rent trishaws by the hour, which can be economical should you wish to see several places.

PRIVATE TRANSPORT

Car Rental: Having your own transport gives you the freedom to explore places off the beaten track at your leisure. The principal car rental firms are listed here. Most have branches in the main towns throughout Malaysia including those in Sabah and Sarawak. Check at the main office.

Cars are usually for rent on an unlimited mileage basis. The daily rates vary from $125 for economy cars per day to $300 for cars in the super luxury class. Weekly rates are also available. Four-wheel drive is advisable in Sabah, Sarawak, and the central regions of the Malay peninsula.

MOTORING ADVISORIES

An international driving licence is required by visitors who wish to drive in Malaysia. National driving licences are only acceptable upon endorsement by the Registrar of Motor Vehicles. Travel insurance must also be taken out.

From the causeway connecting Singapore and peninsular Malaysia, the main trunk road runs up the west coast to the Thai border. From this road, two main highways cross the peninsula to the east coast. In the north, the East-West Highway connects

Butterworth with Kota Bharu, while in the central part of the peninsula, the Kuala Lumpur-Karak Highway cuts through the Main Range and joins a road leading to Kuantan on the east coast. In Sabah and Sarawak, motorways run along the coast connecting major towns. The distances between towns in kilometres is given in the chart below. Roads leading to more remote areas inland are often unpaved or rough, and a four-wheel drive is advisable for these routes.

Driving is on the left-hand side of the road. International traffic signs are used, along with a few local ones such as "*Awas*" meaning "caution", "*ikut kiri*" meaning "Keep left", "*kurangkan laju*" meaning "slow down", and "*jalan sehala*" meaning "one-way street" in the direction of the arrow. Where compass points are given, "*Utara*" is north, "*Selatan*" south, "*Timur*" east, and "*Barat*" west.

The speed limit in towns is 50 km/h. Outside towns, the familiar speed limit signs are displayed where limits have been imposed. The wearing of seat belts by drivers and front seat passengers is compulsory. Monsoon rains can cause hazards for motorists. Drive slowly and be prepared for delays on smaller roads as whole roads are sometimes washed away entirely.

For safety, local drivers have developed a few signals of their own. There are individual variations on these, so watch to see what the other motorists do. If the driver in front flashes his right indicator, he is signalling to you not to overtake. This is usually because of an oncoming vehicle or a bend in the road, or he himself might be about to overtake the vehicle in front of him. If he flashes the left indicator, this means to overtake with caution. A driver flashing his headlamps at you is claiming the right of way. At roundabouts or traffic circles, the driver on the right has the right of way.

Petrol is inexpensive and petrol stations are to be found in or on the fringes of towns. Very few of them operate 24 hours (24 *jam*), so be sure to fill up your tank by 6 or 7 p.m. if you intend to drive at night.

The Automobile Association of Malaysia (AAM) is the national motoring organisation and has offices in most states. Tourists who are members of motoring associations affiliated to AAM are given free reciprocal membership. The head office is located at: 22/23 The Arcade, Hotel Equatorial, Jln. Sultan Ismail, KL, Tel: 03-242-0042.

HITCHHIKING

Hitchhiking is fairly common in Malaysia, and Malaysian students and budget travellers use this form of transport. Malaysians are helpful people, and will sometimes offer to take you where you want to go. This is another good way of getting to know the Malaysians.

WHERE TO STAY

Accommodation in Malaysia encompasses many different styles. You can choose anything, from youth hostel and crash pads to top-class international hotels with saunas, jacuzzis and tennis courts. To sample all of Malaysia, it may be well worth your while to try out accommodation from several different categories. The smaller and more homely establishments will be more likely to bring you closer to the Malaysians.

Most small Malaysian towns, in any case, do not offer the cosmopolitan facilities of a worldwide hotel chain; instead, they provide the personal touch, simplicity and cleanliness of a wayside inn. A typical urban street is dotted with small budget hotels renting simply furnished rooms for between M$10 and M$40 (depending on whether the room has air-conditioning or a simple ceiling fan).

If you intend to visit one of Malaysia's national parks, it is necessary to go to the national park's office in the nearest town in order to book accommodation. In some cases, a deposit must be paid towards the accommodation and permit.

FOOD DIGEST

The different people that comprise Malaysia's multi-racial population provide the country with enough flavours to please every palate. Variety in food is not just restricted to taste either, but extends to the many dining environments you can find yourself in. These range from plush restaurants with air-conditioning, a formal setting and attentive waiters, to the Chinese coffee shop, Malay *kedai makan* (eating shop) and Indian *roti canai* (Indian bread), to the open-air foodstalls to be found in every Malaysian village, town or city.

During your stay in the country, you should eat at the roadside stalls at least once, for it is there that some of the country's most famous and tastiest foods are cooked. And very often, you will find the stall holders as attentive as the waiters of the hotels.

The most popular cuisines are those of the Malays, the Chinese and the Indians. Thai food is also well represented, especially in the north-east Malaysian states. Western food is now ubiquitous, and American fast food outlets are springing up in the smallest of towns.

Malay food is generally rich and spicy, though not as scorchingly hot as Thai or some Indian food. Although each state has its distinctive style of preparation and taste, ingredients are common to all. White steamed rice (*nasi*) is the staple grain. Seafood, chicken and meat (except of course, pork) are cooked in a variety of ways. Coconut forms the basis for many dishes. The "milk" is a popular drink, while the meat is usually grated and squeezed to provide the juice for a tasty sauce. Perhaps the best known of all Malay dishes is *satay*, tender slivers of meat on wooden skewers, which are barbecued over charcoal, and served with a peanut sauce. Depending on the regional recipe, the sauce can either be sweet or spicy. *Satay* is traditionally served with sliced onion, cucumber and *ketupat* (rice cakes steamed in woven palm leaf packets).

Nasi Padang is a variation of Malay food which hails from Padang in Sumatra, and which has the reputation of being extremely hot, though it isn't always, if the local cooks favour sweetness. Dishes are usually displayed on stalls, so you can point to what you want. Other tasty dishes include *tahu goreng*, fried cubes of soya bean curd with fresh bean sprouts; *gado gado*, a salad of raw vegetables topped with a spicy peanut sauce; *laksa*, a type of spicy soup made of fine noodles and fish stock; *mee rebus*, boiled noodles, and *mee siam*, Thai-style noodles.

Besides cakes and sweets mostly made of sago and coconut, there is the popular *gula melaka*, a dessert treat of sago swimming in coconut milk topped by a syrup of palm sugar, a delight for those with a sweet tooth.

Chinese cuisine is everywhere in Malaysia, and you'll find it has influenced many of the Malay dishes. Therefore Malays cook noodles and the Chinese use chilli – it's a mutual exchange, except for the forbidden pork, so loved by the Chinese. Hainanese chicken rice (rice cooked in chicken stock and served with steamed chicken and chilli, the latter optional) makes a good lunch. Other dishes are Hakka *yong tau foo* (beancurd stuffed with meat), Hokkien fried *mee* (noodles fried with pieces of meat and seafood) and Chinese *laksa*, which differs from the Malay version. Teochew rice porridge should also be tried; often written as *congee*, the porridge is served with numerous side-dishes of meat and vegetables. *Nonya* food spans both Malay and Chinese cuisines, has many delicacies and tends to be sweeter than Chinese food.

Indian cooking is characterised by its complex and generous use of spices, not all of them hot. Spices include cardamom, clove, anise and cinnamon, as well as the hotter chilli, curry leaves, cumin and turmeric. Many claim that Indian food in Malaysia is better than that in India. There are three traditions of Indian food represented here: North Indian food, which has rich, creamy sauces and which uses the *tandoor* oven; Muslim Indian food, which serves spicy foods such as *rojak* (a selection of eggs, cuttlefish, potatoes and prawns served with a sweet and hot gravy); and South Indian Hindu food, which, near the local temple, is purely vegetarian. The banana leaf curry and fish head curry are two Indian favourites. Indian breads served with a spiced gravy make an excellent breakfast. *Roti canai* (an unleavened bread tossed and cooked on a griddle, with or without egg) is the most popular. Others are *naan* and *chapati*.

FRUITS OF THE LAND

Malaysia is a veritable garden of Eden of fruits all year round. There are the usual tropical fruits – golden pineapple, rosy papaya, and endless types of bananas – as well as a host of other fruits you may not have seen before. Many are seasonal so you'll need to find out what's "in". For certain, at least one or two of them will become your favourite fruits, which you will hanker after when you've left Malaysian shores.

Durian is the king of fruits in South-East Asia. Mixed with a very sweet taste and a texture like blancmange is a subtle flavour of onions. You must try it at least once!

Jambu Batu is also known as *guava*. Both the green skin and the apple-like flesh can be eaten.

Rambutans have a marvellous hairy red-tinged skin. The flesh is similar to lychees and is very refreshing.

Mango comes long or rounded, with yellow or white flesh.

Mangosteen is dark purple and somewhat forbidding on the outside. But squeeze the fruit in your palm, and it opens to reveal white, juicy segments of delicious sweetness. Excellent for travelling with to cool you down on the road. The mangosteen ripens at the same time as the *durian*, and they are traditionally eaten together. Chinese believe that the "heatiness" of the *durian* is balanced by the "coolness" of the *mangosteens*. Beware of the purple juice which stains clothes.

Nangka or jackfruit is the huge green/brown fruit which you often see on trees in villages covered with sacks or plastic bags to protect them from the hungry birds and insects. The pulp is juicy and chewy at the same time. The *chempedak* is a smaller, sweeter version. Jackfruit and *chempedak* are often deep-fried – a delicious snack, and even the nutty seed inside is edible.

Starfruit, yellow and shiny, is a good thirst-quencher. It is cut horizontally into star-shaped slices and dipped in salt before being eaten.

Pomelo looks and tastes like a sweet, overgrown and somewhat dry grapefruit.

Buah Duku. To open a *duku*, just squeeze the top gently. The flesh is sweet and rather like the rambutan, but with a sour tinge.

Buah Susu, literally translated "milk fruit", is better known as passion fruit. There are many different varieties, all equally delicious. Crisp-skinned and orange from Indonesia, purple from Australia, New Zealand and California, but the local ones have soft, velvety yellow skins. The grey seeds inside are sweet and juicy.

Buah Durian Blanda (literally Dutch *durian*) or soursop, resembles a chunky durian without the smell or the prickles. The soft, creamy flesh inside has just the hint of a sour tang to it.

DRINKING NOTES

In general, alcohol is expensive in Malaysia. A glass of wine may cost as much as a tot of brandy, and the heat of the tropics does not always guarantee a good flavour. Alcohol is forbidden to Muslims, so if you want to indulge in alcoholic beverages, head for the hotels and the Chinese liquor stores. The latter have a fascinating range of familiar and strange bottles on their shelves. Tiger and Anchor beer and Guinness Stout are the most popular and the cheapest.

But far more refreshing in the steamy atmosphere of the tropics are the fruit juices, to be found at small stalls and night markets. Choose your combination: pineapple and orange, starfruit and water melon. If you don't want sugar, ask them not to add the sugar syrup; in Kelantan, salt is often added to cut fruits and juices, which may not be to your taste – ask them for plain fruit or juice. There are a variety of very sweet coconut and soya bean drinks for sale at small stalls, with alarmingly bright colours added to them. Young coconuts produce a refreshing milk which can be drunk straight from the coconut with ice added and a straw stuck into a hole made in the top. Mineral water is sold in grocery stores and in supermarkets.

PARKS & RESERVES

PENINSULAR MALAYSIA

PANTAI ACHEH FOREST RESERVE

Location: Northwest corner of Penang Island; 40 km from George Town. Open to public all the year round.

Size: 20 sq km

Access: From George Town take the northern coastal road to Telok Bahang Village. At Telok Bahang village roundabout, take Jalan Hasan Abas to Telok Awak (1½ km). The trail into the Pantai Acheh Forest Reserve starts from behind Ah Sim Restaurant at the fishermen's jetty.

Public Transportation: Take the Hin Bus Company bus no. 93 at the cross channel ferry terminal (referred to as "Jetty" locally) or at the KOMTAR bus interchange at downtown George Town.

Registration: Not necessary.

Accommodation: Nil. There are a number of beach hotels along the Telok Bahang route and a beach hotel within walking distance from the reserve. There are also moderately priced hotels in George Town.

Camping: Possible at the beaches and near streams.

Transportation within the reserve: Nil.

Trails: Well defined and often used. Very steep near the beaches.

PENANG GINTING BEE-EATER NESTING GROUNDS

Location: 3 km south of the town of Balik Pulau on the western side of Penang Island. About 45 minutes drive from George Town.

Size: About 3 sq km

Access by car: Take the round-the-island road from George Town. If you start your journey in a southerly direction, you will see signpost "Ginting" on the left. If you are taking the northern coastal road, Ginting is about 3 km after the town of Balik Pulau on your right. Turn into Ginting and take the straight road for about 1½ km. Turn right into the dirt road beside the school. You can park along the dirt road after about 200 metres.

Public Transportation: Take the Penang Yellow Bus Company bus service no. 66 to Balik Pulau. Alight at Ginting village and walk down the road till you reach the school.

Registration: Private land but freely accessible to the public.

Accommodation: There are moderately priced hotels in George Town.

Trails: Open land; free to walk anywhere although there are some paths to follow.

Further advice: It gets extremely hot during the day so hat and drinks are necessary.

PENANG HILL (BUKIT BENDARA)

Location: 10 km from downtown George Town. Open all year.

Access by car: Take the road to Air Itam. At Air Itam roundabout take the Hill Railway Road (2 km). Park at the parking bays at the Hill Railway Lower Station. Parking fees have to be paid.

Public Transportation: The Hin Bus Company and the City Council run bus services from downtown George Town to the township of Air Itam. Take the Hill Railway bus from Air Itam.

Access up the hill is by cable cars on rails. Take the train to the Upper Station. Passengers have to change trains at the Middle Station to take the train to the Upper station.

Registration: Nil.

Accommodation: There is a small hotel on top of the hill.

Camping: Not possible.

Food: There are food stalls and a hotel restaurant.

Transportation on top of the hill: Not always available and restricted to some areas only.

Trails: There is a tarred road on top of the hill up to Western Hill. There are also small marked and tarred tracks and walking trails.

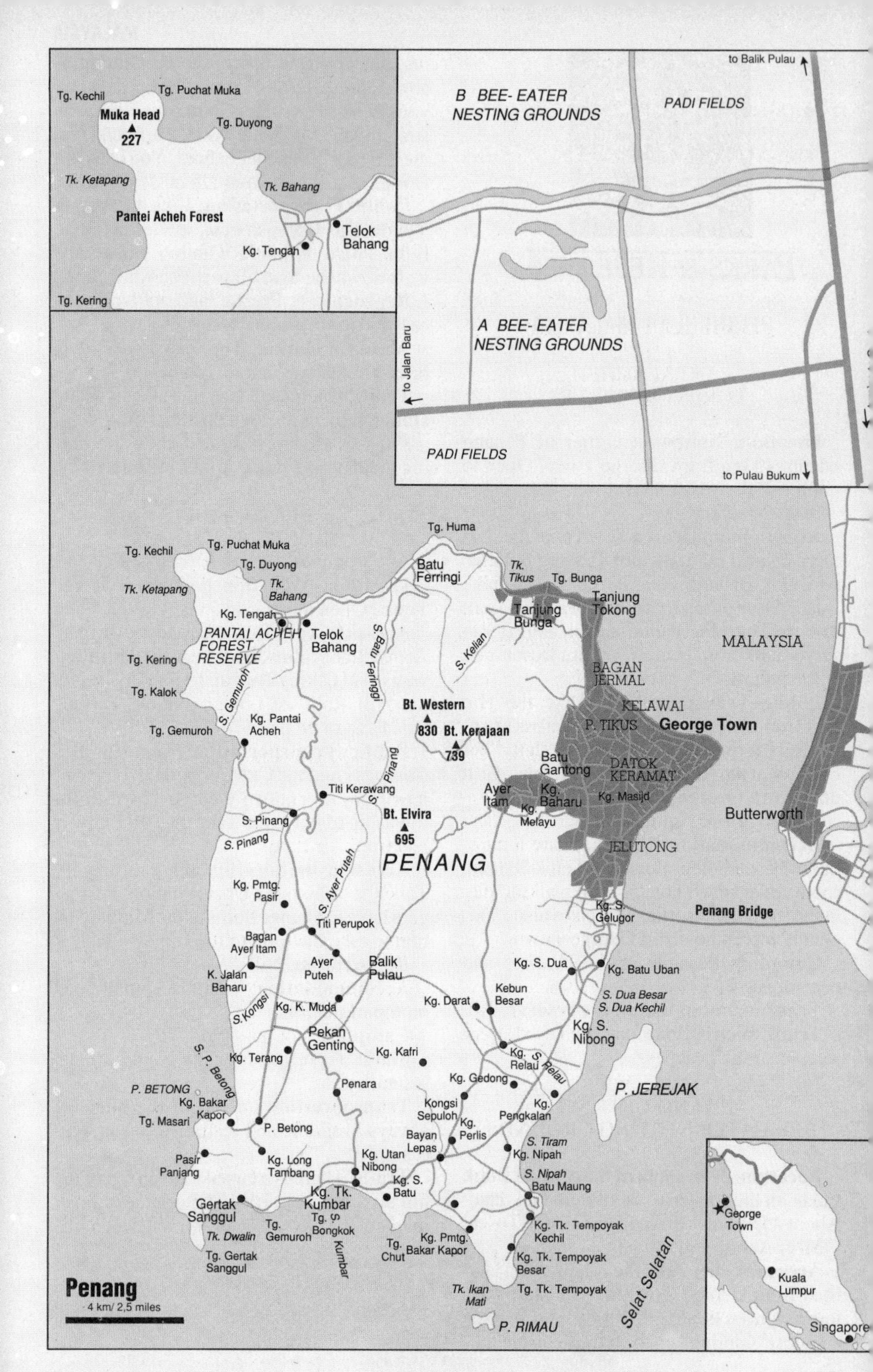

Tg. Kechil
Tg. Puchat Muka
Muka Head
227
Tg. Duyong
Tk. Ketapang
Tk. Bahang
Pantei Acheh Forest
Telok Bahang
Kg. Tengah
Tg. Kering
B BEE- EATER NESTING GROUNDS
PADI FIELDS
to Balik Pulau
A BEE- EATER NESTING GROUNDS
to Jalan Baru
PADI FIELDS
to Pulau Bukum
Tg. Huma
Tg. Kechil
Tg. Puchat Muka
Tg. Duyong
Tk. Ketapang
Tk. Bahang
Batu Ferringi
Tk. Tikus
Tg. Bunga
Tanjung Tokong
Tanjung Bunga
Kg. Tengah
Telok Bahang
PANTAI ACHEH FOREST RESERVE
S. Batu Feringgi
S. Kelian
MALAYSIA
Tg. Kering
BAGAN JERMAL
Tg. Kalok
S. Gemuroh
Bt. Western
KELAWAI
P. TIKUS
George Town
Kg. Pantai Acheh
830
Bt. Kerajaan
739
Tg. Gemuroh
S. Pinang
Batu Gantong
DATOK KERAMAT
Titi Kerawang
Ayer Itam
Kg. Baharu
Kg. Masjid
Kg. Melayu
Butterworth
S. Pinang
Bt. Elvira
695
S. Pinang
JELUTONG
S. Ayer Puteh
PENANG
Kg. Pmtg. Pasir
Kg. S. Gelugor
Penang Bridge
Titi Perupok
Bagan Ayer Itam
Ayer Puteh
Balik Pulau
Kg. S. Dua
Kg. Batu Uban
K. Jalan Baharu
Kebun Besar
S. Dua Besar
S. Dua Kechil
S. Kongsi
Kg. K. Muda
Kg. Darat
Kg. S. Nibong
Pekan Genting
S. P. Betong
Kg. Terang
Kg. Kafri
Kg. Relau
S. Relau
Penara
Kg. Gedong
P. JEREJAK
P. BETONG
Kg. Bakar Kapor
Kongsi Sepuloh
Kg. Perlis
Kg. Pengkalan
Tg. Masari
P. Betong
Bayan Lepas
S. Tiram
Kg. Nipah
Pasir Panjang
Kg. Long Tambang
Kg. Utan Nibong
S. Nipah
Batu Maung
Kg. Tk. Kumbar
Kg. S. Batu
Gertak Sanggul
Tg. Gemuroh
Tg. S. Bongkok
Kg. Tk. Tempoyak Kechil
Tk. Dwalin
Tg. Chut
Kg. Pmtg. Bakar Kapor
S. Kumbar
Tg. Gertak Sanggul
Kg. Tk. Tempoyak Besar
Selat Selatan
Tk. Ikan Mati
Tg. Tk. Tempoyak
P. RIMAU
George Town
Kuala Lumpur
Singapore
Penang
4 km/ 2,5 miles

LANGKAWI ISLANDS

Location: Off the coast of the northern states of Kedah and Perlis.

Size: A group of about 100 small islands with four main Islands.

Access: Most visitors arrive at Langkawi by plane from Penang, Kuala Lumpur or Singapore.

Access by car: Take the turning to Kuala Perlis along the main north-south trunk road just before Alor Setar town. The road leads right up to Kuala Perlis. Park the car near the jetty.

Public Transportation: There are express bus services from the Butterworth Bus Terminal to Kuala Perlis. The ferry ride from Kuala Perlis to Langkawi takes just over one hour. The cost of the ferry ticket is M$10 one way. It is also possible to fly from either Kuala Lumpur or Penang to Langkawi.

Transportation at Langkawi: There are cars and bicycles for hire at Langkawi. Taxis can also be hired. The bus service is not extensive and limited to certain areas of the main island only.

Registration: Not necessary for entering forest reserve but the Forestry Department in Langkawi must first be informed. Tel: 04-788835.

Accommodation: There are expensive to moderately priced resorts, as well as budget accommodations. New budget accommodations are being planned and some are nearing completion. Get in touch with the Tourist Development Corporation (TDC) office at Langkawi for cost and locations of the new accommodations.

CAMERON HIGHLANDS

Location: 220 km north of Kuala Lumpur, situated in the Main Range. Open all year.

Size: 712 sq km

Access (car, taxi): From Kuala Lumpur, drive north on the North-South Highway to Tapah (158 km) in the state of Perak. From there drive 59 km northeast to Tanah Rata. The drive from Tapah to Tanah Rata takes about an hour. Taxis from Kuala Lumpur to Cameron Highlands are available at Puduraya Bus Terminal.

Taxi fare: From Kuala Lumpur to Tanah Rata at M$21 per person or M$84 if taxi is chartered. Taxis at Tapah charge much lower rates.

Public Transportation: An express bus leaves the Puduraya bus terminal for Tanah Rata at 8.30 am daily. The fare is M$9.20. From Puduraya, there are also regular bus services plying the Kuala Lumpur-Tapah route. From Tapah there are regular bus services to Tanah Rata. From Tanah Rata, there are taxi services to Brinchang.

Registration: Not necessary but it is advisable to book accommodation in advance especially during holiday season.

Accommodation (High Priced): The Merlin Inn Resort, tel: 05-941205/941211/941313; Ye Olde Smokehouse Hotel, tel: 05-941214/941215.

Accommodation (Moderately Priced): BALA's The Holiday Chalets. Tel: 05-941660. There are several hotels at both Tanah Rata and Brinchang. Contact The Cameron Highlands Tourist Promotion Association Office, tel: 05-941266.

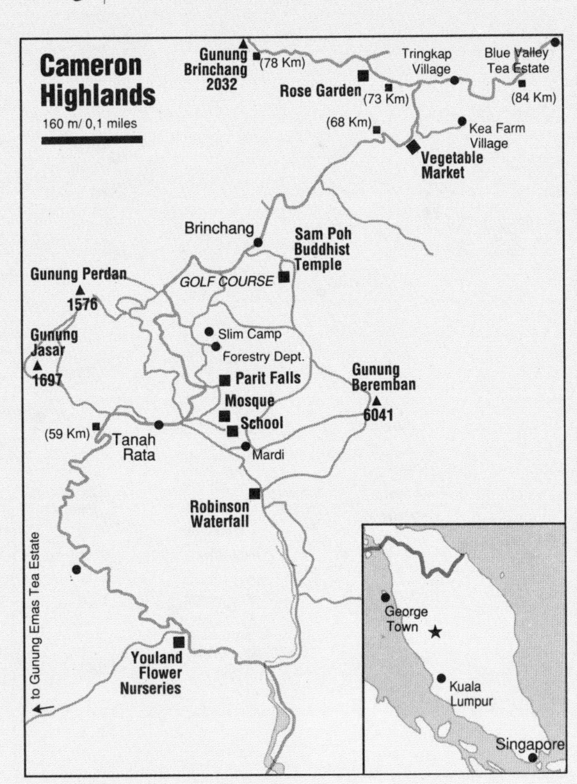

Food: There are many restaurants serving both local and western food at both Tanah Rata and Brinchang.

Transport in Cameron Highlands: By car, taxi or on foot.

Trails: There a number of marked jungle trails leading through scenic forest to Parit and Robinson waterfalls and also some interesting peaks – Mount Beremban, Mount Jasar and Mount Perdah. Contact Bala (the proprietor of BALA's The Holiday Chalets) who is a Tourist Development Corporation (TDC) licensed guide at Cameron Highlands. Tel: 05-941660.

Further advice: Bring along adequate warm clothing as it can be cold at night. Wear leech socks and comfortable footwear for long walks along trails. Visitors are also advised not to stray away from marked trails.

FRASER'S HILL

Location: 103 km north of Kuala Lumpur, situated in the Main Range in Pahang and Selangor. Open all year. The best time to visit this resort is February–October.

Size: 64 sq km (The area from The Gap to Jeriau under the jurisdiction of the Fraser's Hill Development Corporation). The montane forest (forest reserve) surrounding Fraser's Hill is much larger.

Access (car, taxi): From Kuala Lumpur drive north the North-South Highway to Kuala Kubu Baharu (55 km). From there drive east to The Gap (40 km) then up 8 km along the one-way road to Fraser's Hill. Control gates are opened from 6.30 a.m. to 7 p.m. with uphill traffic limited to odd hours and downhill traffic limited to even hours. There is no regular taxi service to Fraser's Hill but taxis may be chartered for M$60 (one way) at Puduraya bus terminal. Taxis may also be hired at Kuala Kubu Baharu.

Public Transportation: There are two bus services leaving Kuala Kubu Baharu at 8 a.m. to noon, and going up the hill by the 9 a.m. and 1 p.m. gates respectively. Alternatively visitors can catch the 8.30 a.m. Kuala Lumpur–Kuala Lipis express bus from the Jalan Tun Razak bus terminal (Kuala Lumpur) which passes The Gap at 10.30 a.m. and allows you to catch the 1 p.m. bus up. There are regular bus services to Kuala Kubu Baharu from Kuala Lumpur.

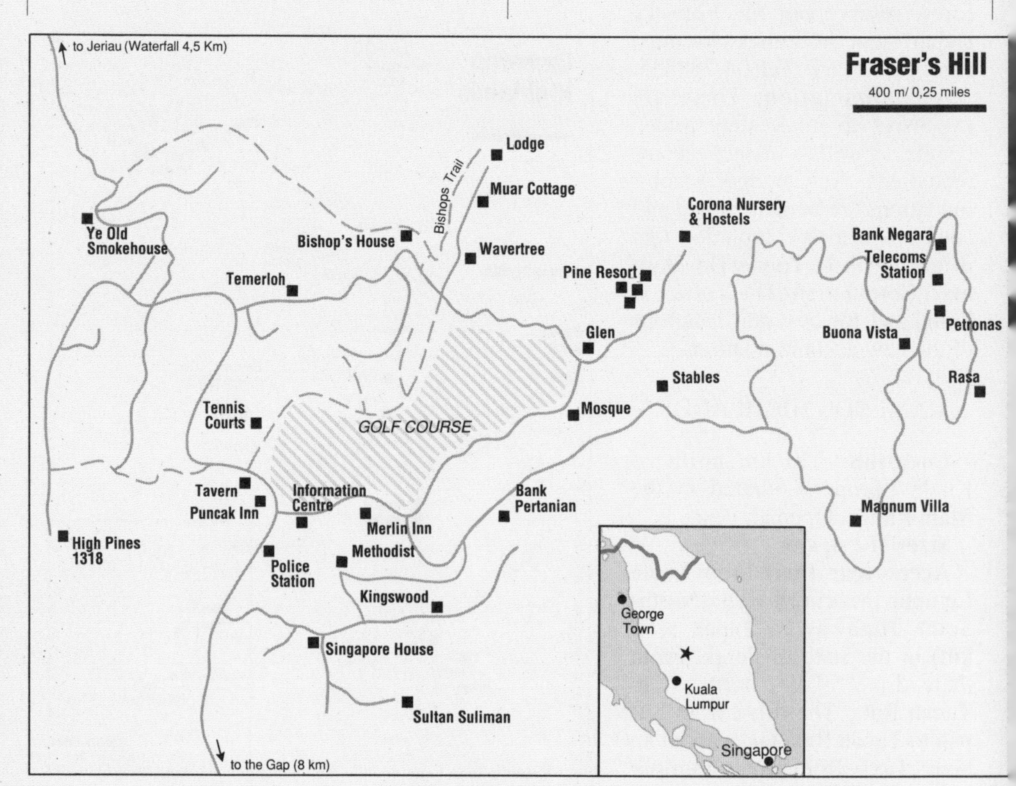

Registration: No registration is required but during weekends, public holidays and school holidays, it is always advisable to book accommodation in advance.

Accommodation (High Priced): Merlin Inn Resort. Rooms from M$85, tel: 09-382300; Ye Olde Smokehouse Hotel. Rooms and suites from M$100–M$380 per night, tel: 09-382226; Fraser's Pine Resort. Condominiums from M$90 (1 room; max 3 persons) to M$240 (3 rooms; max 10 persons), tel: 03-7832810 (for reservations).

Accommodation (Moderately Priced): The Fraser's Hill Development Corporation (FHDC) runs bungalows and chalets which range from M$40–M$55 per night, tel: 09-382201/382248; Puncak Inn has 28 rooms located in the town centre, from M$40, tel: 09-382055; The Hulu Selangor District Office runs The Gap Rest House and the Seri Berkat Rest House at Fraser's Hill. Room rates are M$21 per night (Gap) and M$32 per night (Seri Berkat). For reservations, contact: The District Officer, Hulu Selangor District Office, 44000 Kuala Kubu Baharu, Selangor. Tel: 03-8041026/8041027/8041030.

Accommodation (Low Priced): The Corona Flower Nursery runs a hostel at M$8 per person per night.

Camping: Not available.

Food: The FHDC bungalows and chalets as well as The Gap and Seri Berkat Rest Houses have catering facilities. The Merlin Inn Resort Coffee House and Ye Olde Smokehouse Hotel (and its branch, Ye Olde Tavern) offer local and western dishes but are highly priced. In the town centre, there are four restaurants serving Chinese, Malay and western food at moderate prices.

Transport in Fraser's Hill: On foot or own vehicle.

Trails: There is a good system of well maintained nature trails running through montane forest.

Further advice: There is no petrol station at Fraser's Hill. Visitors are reminded that the Jeriau waterfalls is a security area and can only be visited from 7 a.m.–7 p.m. daily. Warm clothings are essential as it can be very cold at night. Leech socks and insect repellents are necessary when walking along the moist jungle trails.

KUALA SELANGOR NATURE PARK

Location: Peninsular Malaysia, west coast 65 km from Kuala Lumpur. Open all year.

Size: 26 sq km

Access (car, taxi): One-hour ride from Kuala Lumpur.

Taxi fare: M$30 from Kuala Lumpur, for 4 persons.

Public transportation: M$5 from Kuala Lumpur to Kuala Selangor Town; bus-stop is walking distance from park entrance.

Registration: Reservation of chalets through: MNS Kuala Lumpur Office, tel. 03-7912185.

Accommodation: 8 chalets available, M$18 for a 4-room, M$12 for a 2-room. Bukit Melawati Hotel in town offers rest-house at M$20, and air-conditioned room at M$28.

Camping: No facilities.

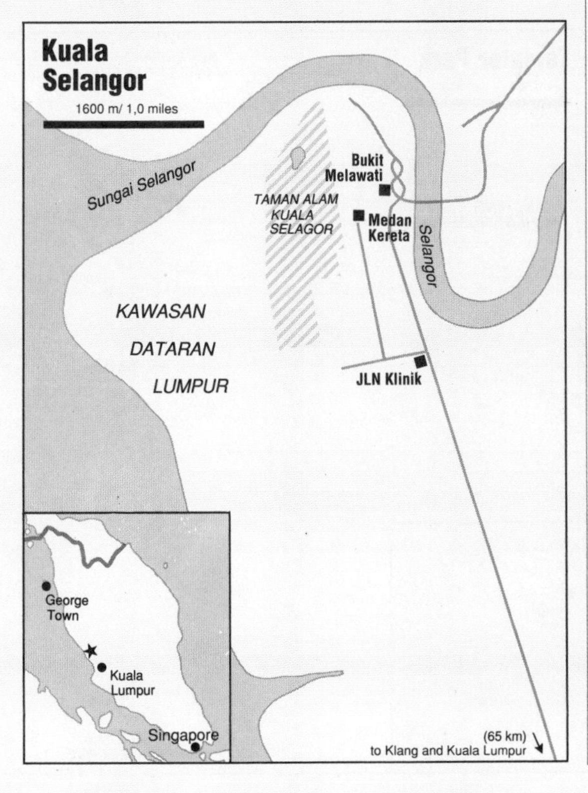

Food: MCE, a tranquil restaurant on hilltop patio outside resthouse. Usual local fare from foodstalls in town.

Transport in park: You walk!

Trails: Well-marked trails to lookout points and hides. Small park, perfect for a day-trip from Kuala Lumpur.

Further advice: Bring along insect repellent as there are plenty of mosquitoes during the evening.

TEMPLER PARK

Location: 21 km north of Kuala Lumpur along the North-South Highway in the state of Selangor. Open all year.

Size: 12 sq km within park boundaries.

Access (car, taxi): From Kuala Lumpur drive north along the North-South Highway and after the 20th km, turn right into Templer Park. Visitors can also take any of the Rawang-bound taxis from Puduraya Bus Terminal.

Taxi fare: From Kuala Lumpur to Rawang, the fare per person is M$6. Templer Park is before Rawang.

Public Transportation: Buses for Rawang, Kuala Kubu Baharu and Tanjong Malim leave regularly from Puduraya Bus Terminal. Visitors can take any one of these buses and alight at Templer Park.

Registration: Nil.

Accommodation: There are no accommodation facilities at Templer Park. Since Kuala Lumpur is close by, visitors could stay in town at any of the hotels (a wide variety) or at any of the youth hostels.

Camping: There are no campsites provided. Those wishing to camp may do so if camping gear is brought along. Campers are advised not to litter in the park.

Food: A canteen opposite the park serves food and drinks. No substantial meals are available so visitors are advised to bring along a packed lunch.

Trails: There are well-kept paths along which visitors could walk and explore the park.

Transport in the park: On foot.

Further advice: Bring along insect repellants or leech socks as the trails when wet are frequented by leeches.

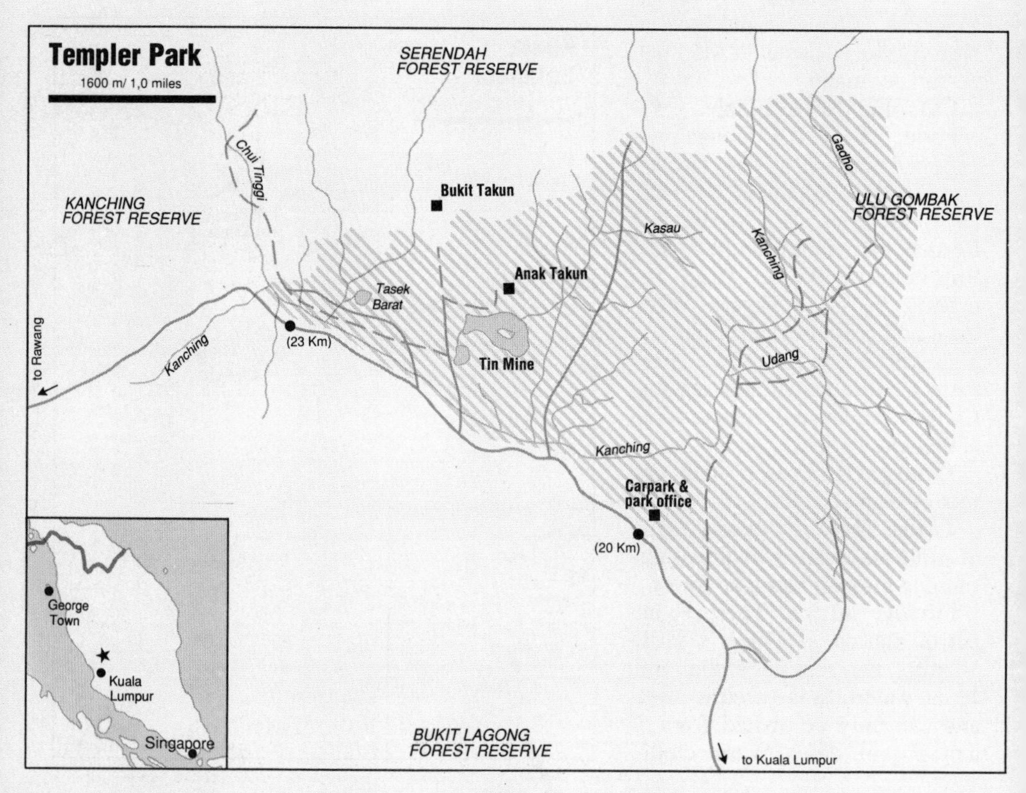

TAMAN NEGARA

Location: 300 km northeast of Kuala Lumpur in the states of Pahang, Kelantan and Trengganu. The park is closed from 15 December to 15 January every year.

Size: 4,343 sq km within park boundaries.

Access (car, taxi): From Kuala Lumpur, drive east to Mentakab on the Kuala Lumpur -Karak Highway (120 km). Three km after Mentakab town, turn left at the junction to Kuala Krau and head north for Jerantut (50 km). From Jerantut, drive to Kuala Tembeling (15 km) and head for the Taman Negara jetty. Park your vehicle in the jetty car park, report to park officials there before boarding boat for Kuala Tahan (the park headquarters), which leaves at 2 p.m. Taxis from Kuala Lumpur to Jerantut are available from Puduraya Bus Terminal. Visitors going by taxi should leave Kuala Lumpur not later than 10 a.m. From Jerantut there are local taxi and bus services to Kuala Tembeling.

Taxi fare: From Kuala Lumpur to Jerantut, M$14 per person or M$56 if taxi is chartered. From Jerantut to Kuala Tembeling, the local taxi fare is M$3 per person. Bus fares are much cheaper.

Boat fare (from K. Tembeling to K. Tahan): M$30 per person (return trip).

Public Transportation: Buses from Temerloh leave the Putra World Trade Central terminal, Kuala Lumpur. Visitors are advised to leave by 7 a.m. or 8 a.m. Journey to Temerloh takes 2½ hours. From Temerloh, take another bus to Jerantut (1½ hours) and from thereon there is the option of either bus or taxi service to Kuala Tembeling.

From Singapore, night trains depart around 8 p.m. and 10 p.m. and arrive at the Tembeling Halt at 6 a.m. and 8 a.m. the following morning. From here it is a ½-hour walk to the jetty. For an updated schedule enquire at the Singapore railway station.

Registration: Prior bookings have to made through Taman Negara Booking, Malaysia Tourist Information Complex (MATIC), No 109 Jalan Ampang, 50450 Kuala Lumpur. Tel: 03-2434929 ext 108.

Bookings should be accompanied by M$1 per person (entry permit) and a deposit of M$30 per person.

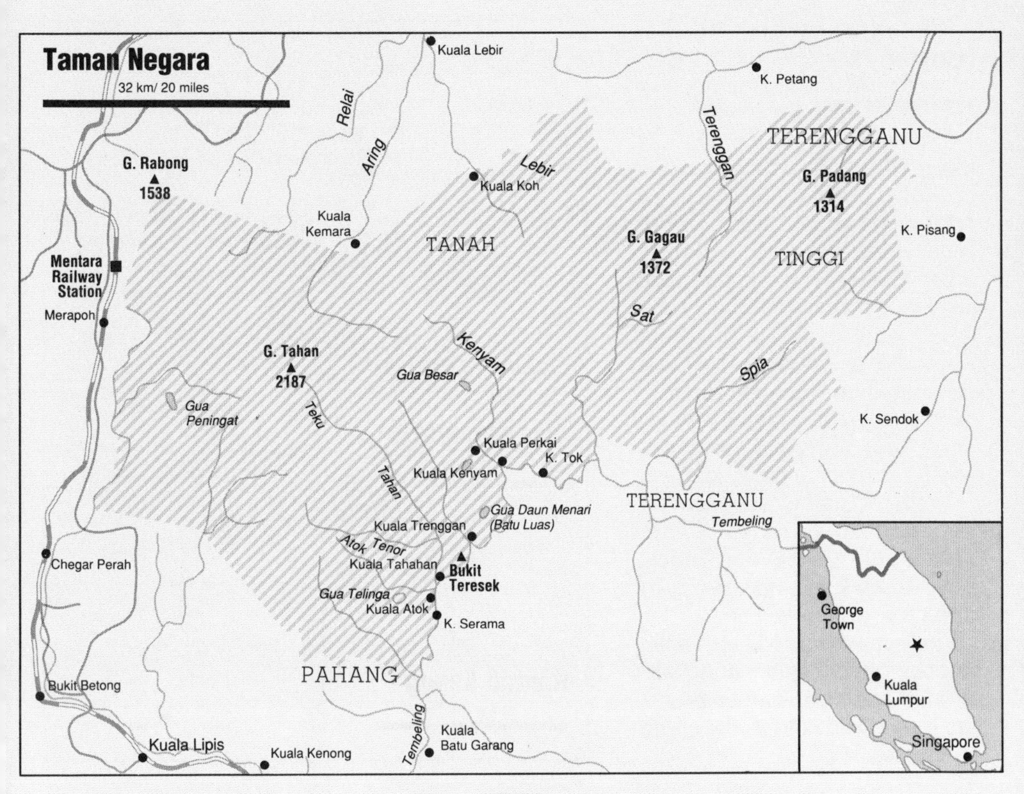

Accommodation: At Kuala Tahan there is a modern 12-room resthouse and 11 chalets of 2 rooms each, with 2 beds and attached bathrooms. Resthouse rooms are M$38 per night and chalet rooms are M$30 per night.

An 8-room hostel with 64 beds, bathrooms and communal cooking facilities is also available for M$10 per person per night. A camp site at Kuala Tahan is available for M$1 per person per night. Tents may be hired at Kuala Tahan.

Food: Two restaurants serve local and western cuisine daily. Food and drinks are moderately priced.

Observation Hides: Reservations for hides (Kumbang, Belau, Yong, Tabing and Cegar Anjing) can only be made at park headquarters at M$5 per person per day. All have bunks and toilets.

Forest/Fishing Lodges: Two forest lodges are located at Kuala Terengganu and Lata Berkoh and for anglers, two fishing lodges are available at Kuala Kenyam and Kuala Perkai. These are available at M$8 per person per day. There are no catering facilities so visitors must do their own cooking.

Photography/Fishing Licenses: A camera license of M$5 must be obtained. Anglers pay M$10 per tackle per month.

Transport in the park: On foot or by boat which may be booked at park headquarters.

Mount Tahan Climb: Those undertaking this climb will have to pay M$400 per week, per guide and M$50 for each additional day thereafter. The trek begins from Kuala Tahan, taking 5 days at a casual pace to reach the summit, the return journey taking 4 days. There are campsites along the way for visitors to break journey.

Trails: There is a system of well defined trails to Mount Tahan, the observation hides, Bukit Teresek and other places of scenic interest in the park.

Further advice: Mount Tahan climbers should bring along waterproof and warm clothings as it can be wet and cold as the climb progresses.

RANTAU ABANG TURTLE SANCTUARY

Location: Rantau Abang, Terengganu, peninsular Malaysia. Open all year but from October to March there are no turtles.

Size: 18 km long (coastal strip)

Access (car, taxi): Kuala Terengganu, 70 km north; Kerteh, 50 km south.

Taxi fare: M$45 from K.T. airport, and M$30 from either Kuala Terengganu town or Kerteh, for 1-4 persons.

Public transportation: Regular buses available from most points in peninsular Malaysia.

Registration: Not necessary.

Accommodation At nearby Tanjong Jara. There are several resthouses charging economical rates in immediate vicinity.

Camping: Possible, but no facilities.

Food: Several restaurants and foodstalls.

Transport in park: None.

Trails: Can walk on beach in daytime.

Further advice: Visit "Turtle Information Centre" for more information to see turtle exhibits and souvenir shop.

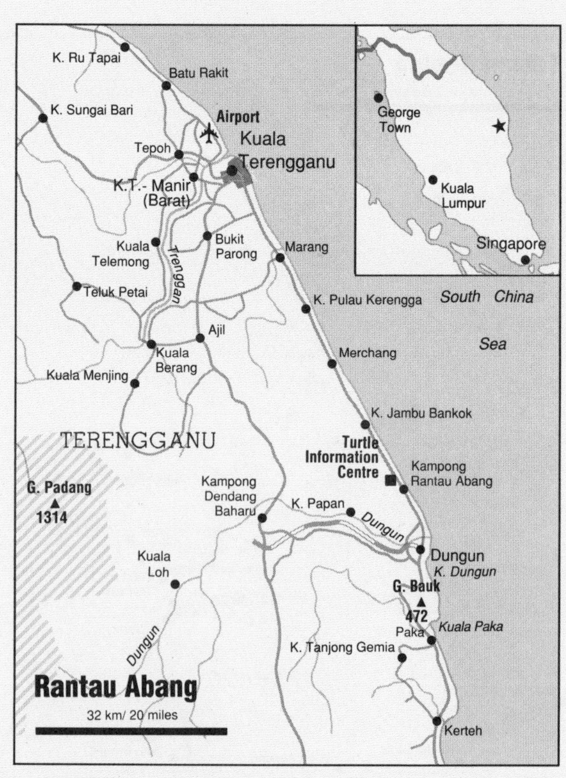

PULAU TIOMAN

Location: East of Mersing, southeastern coast of peninsular Malaysia. Open all year.

Size: Approximately 100 sq km

Access: There are daily flights from Kuala Lumpur and Kuantan by Malaysia Airways and from Singapore by Tradewinds and Malaysia Airways (M$100 one way).

There are daily bumboat trips between Mersing and Tioman that can be arranged on an ad hoc basis, or booked by travel agents. Speedboat services from Mersing and Singapore were introduced in 1989.

Registration: Tioman is not a national park and can be entered without formalities.

Accommodation: Tioman Island Resort – room rates vary, depending on standard and season. Beach chalets at Tekek Village.

Camping: Permitted.

Transport in park: Boat tours around Tioman and to neighbouring islands at M$80-120 per boat.

Trails: There are many trails that are normally difficult to find and follow. Easily accessible is the trail from Tekek to Juara.

ENDAU–ROMPIN

Location: 40 km from Kuala Rompin in southeastern peninsular Malaysia. Open all year. (See map on next page.)

Size: 800 sq km

Access (car, taxi): From Kuala Rompin through Kampung Mok to banks of Sungai Kinchin; or from Kluang in Johore to Kahang, then to Bukit Cantik and Kampung Peta, followed by a 45-minute boat ride.

Public transportation: By bus to Kuala Rompin or Kluang and Kahang; thereafter own arrangements are necessary.

Accommodation: None.

Food: Bring your own.

Transport in park: On foot; boat hire can be arranged from local villagers.

Trails: Many trails from camp-sites at Kuala Jasin and Sungai Kinchin, especially those leading to Mount Janing, Mount Keriong and Buaya Sangkut.

Further advice: The trip is best arranged through a tour company with accompanying guide.

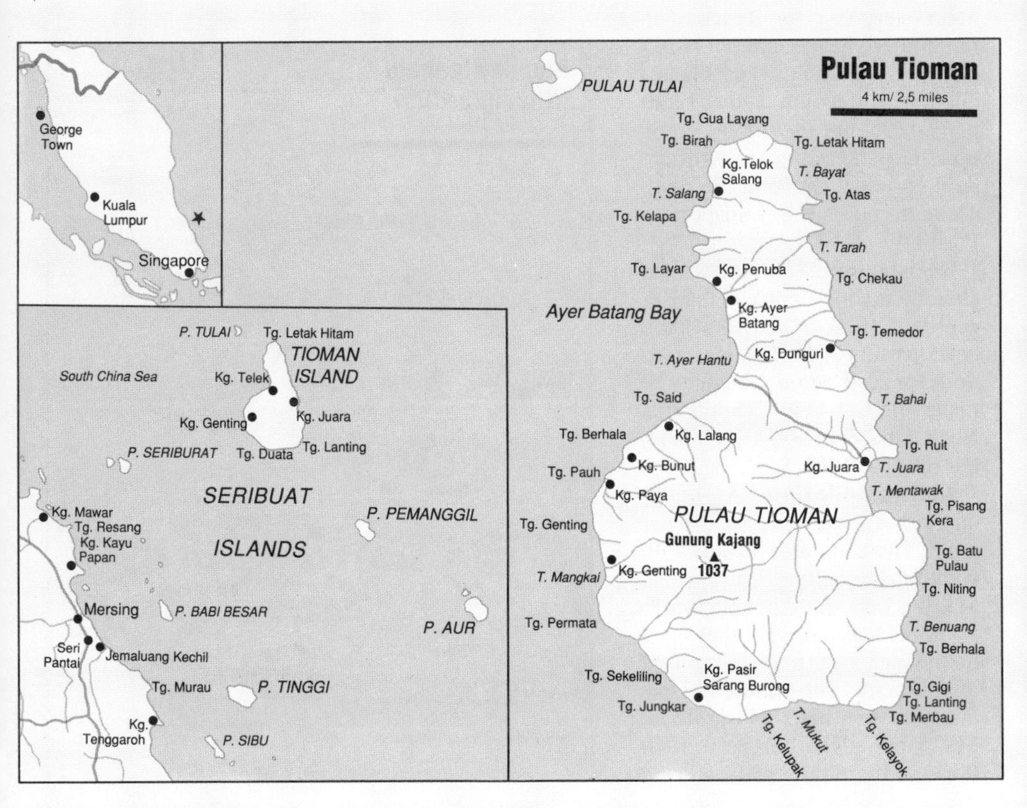

EAST MALAYSIA

MOUNT KINABALU PARK

Location: 90 km northeast of Kota Kinabalu, Sabah, Malaysia. Open all year round.

Size: 754 sq km

Access (car, taxi): From Kota Kinabalu take the Tuaran road north to the Tamparuli junction. Turn right on to the Ranau road, to reach park headquarters in 2–2½ hrs. The Ranau road continues another 16 km to Ranau town, and another 27 km beyond that to the Poring Hot Springs Ranger Station. All these roads are sealed.

Taxi fare: A shared 4-person taxi from Kota Kinabalu to Ranau costs about M$15 per person. Fares to Poring from Ranau are negotiable. A trip by taxi from Kota Kinabalu to park headquarters costs M$70–100.

Public transportation: Buses leave Kota Kinabalu 8 a.m. and noon every day for Ranau. The fare to park headquarters is M$5 one way. Return buses pass headquarters at 8.20 a.m. and at 12.20 p.m. sharp, but they don't come up to the park headquarters office, so it is best to wait on the road. There is no public transport to Poring.

Registration: All visitors must sign the visitor's book at the Reception Office on arrival at park headquarters or Poring. Reservations for accommodation must be made through the Sabah Parks Head Office in Kota Kinabalu, tel: (088) 211881.

Accommodation: At headquarters – chalets at M$30 per night. There are also hostels with communal cooking facilities at M$10-15 per night per person.

On the summit trail, there is a resthouse with a restaurant at 3,352 metres for climbers. Dormitories are also available at M$25 per night per person. In Poring, there are self-catering

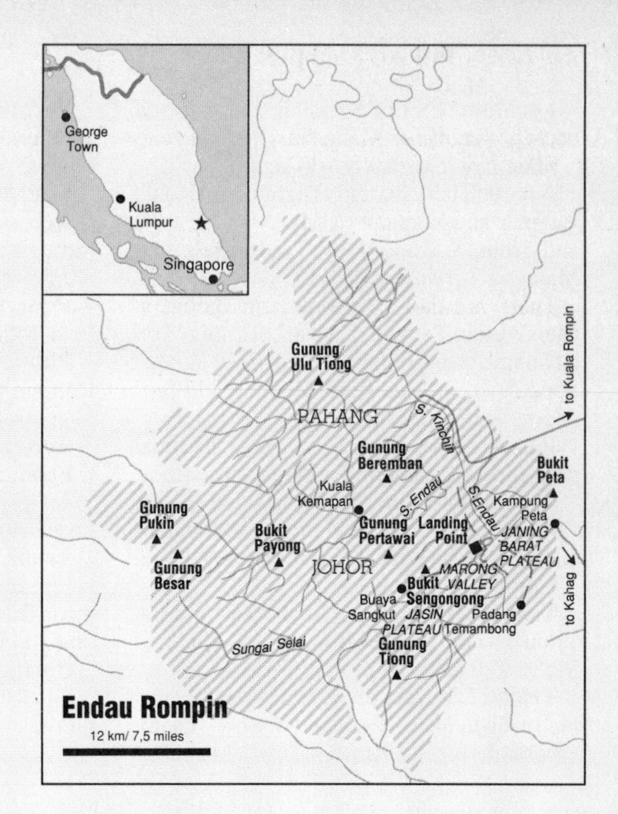

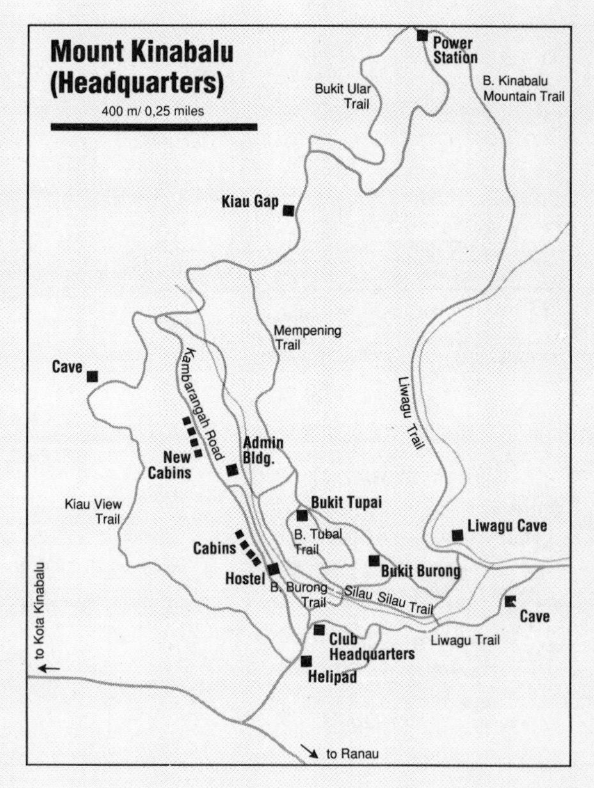

chalets, sleeping 4-6 persons and hostel with communal cooking facilities.

Camping: Not permitted.

Food: Two restaurants at park headquarters; bring your own to Poring.

Transport in park: There is a 5½-km road at the start of Mt. Kinabalu Summit Trail. Climbers can hire vehicles at park headquarters.

Trails: The Summit Trail is the most popular and takes 2 days. There is an 11-km network of walking trails at headquarters and one 2½-hour trail to a waterfall at Poring.

Further advice: Bring warm and waterproof clothes. Climbers *must* hire a guide; porters can be hired at park headquarters.

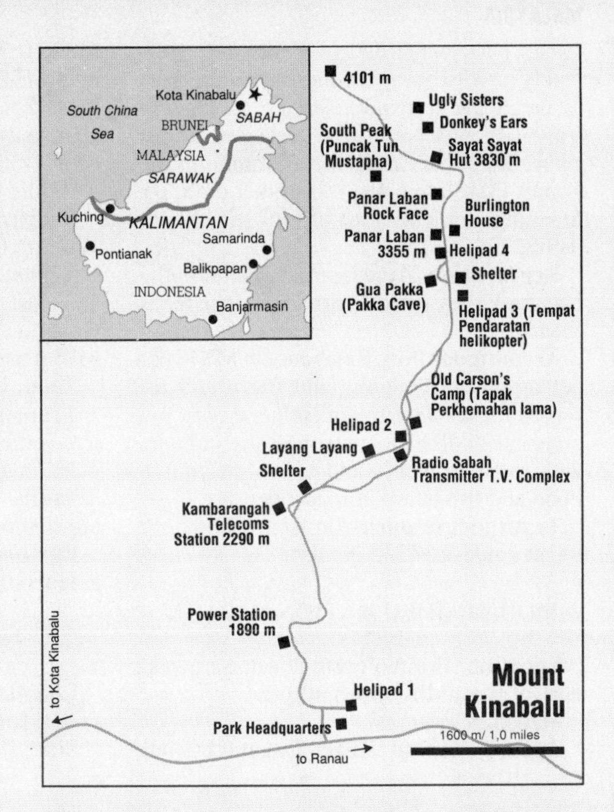

SEPILOK

Location: 24 km from Sandakan, in eastern Sabah, Borneo. Open all year.

Access (car, taxi): Taxi at Sandakan airport, 11 km from Sepilok.

Public transportation: Bus is available from Sandakan by Labuk Road Bus Co., tel: 215106. Sandakan is best reached by Malaysia Airways from Kota Kinabalu; can also be reached by a long bus ride (ca. 12 hours).

Registration: Entrance is free; reserve opens 9 a.m.– 4 p.m.

Accommodation: No accommodation in Sepilok, but there are about 12 hotels in Sandakan.

Food: At the reception area.

Transport in park: On foot.

Further advice: The orangutans are usually fed at 9.30 a.m.

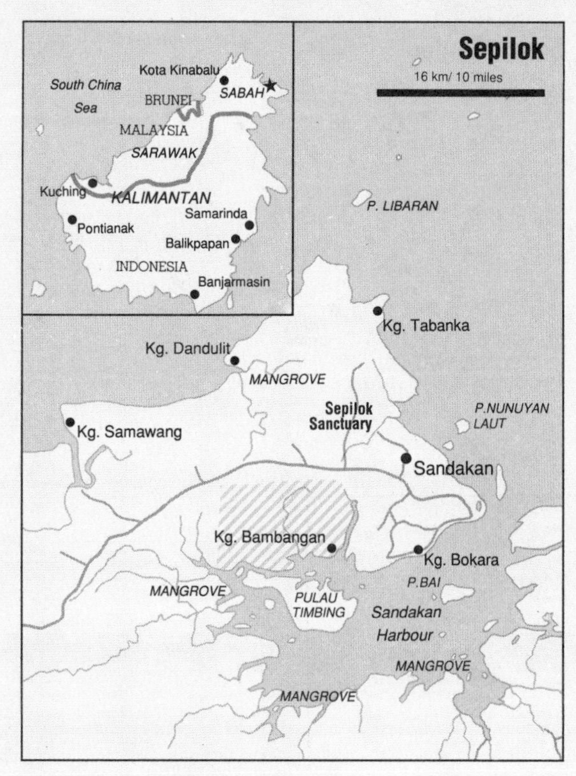

DANUM VALLEY CONSERVATION AREA

Location: Eastern Sabah (Borneo). Open all year.

Size: 438 sq km

Access: Visitors can fly into

Lahad Datu with Malaysia Airways, preferably on Mondays, Wednesdays or Fridays to catch the transport service to Danum Valley Field Centre. The transport at the centre also operates in the afternoon to the airport (M$30 one way). On other days, try to arrange for a minibus trip for M$130, tel 089-82324.

Registration: At reception counter in the Danum Valley Field Centre. Entrance fee is M$15.

Accommodation: Resthouse at M$45 per person (popular among and frequently reserved for resident scientists).

Hostel at M$30 per person (2-bed cubicle, with separate men's and ladies' sections). Food and full board at M$30 per day.

Transport in park: On foot. Obligatory forest guide at M$30 per day.

MOUNT MULU NATIONAL PARK

Location: Borneo north coast, Sarawak; upriver from Miri. Open all year.

Size: 528 sq km

Access: Miri to Kuala Bakam by bus M$2.10; Kuala Bakam to Marudi by express boat M$12; Marudi to Kuala Apoh by express boat M$18; Long Terawan to park headquarters (Long Pala) by chartered longboat – minimum M$160 per boat.

Public transportation: None.

Registration: Through National Parks & Wildlife Office, tel. Miri 085-36637.

Accommodation: Private lodges available just outside park at M$60 per room; Government bungalows at headquarters at M$5 per person or M$25 per room.

Food: Food/meals available.

Transport in park: Boats can be chartered otherwise you walk with a government guide at all times – M$20 per day.

Trails: Sometimes hard to follow; rivers are shallow.

Further advice: Package tours are available from travel agents in Miri.

NIAH NATIONAL PARK

Location: Borneo north coast, Sarawak; 96 km south of Miri. Open all year.

Size: 78 sq km

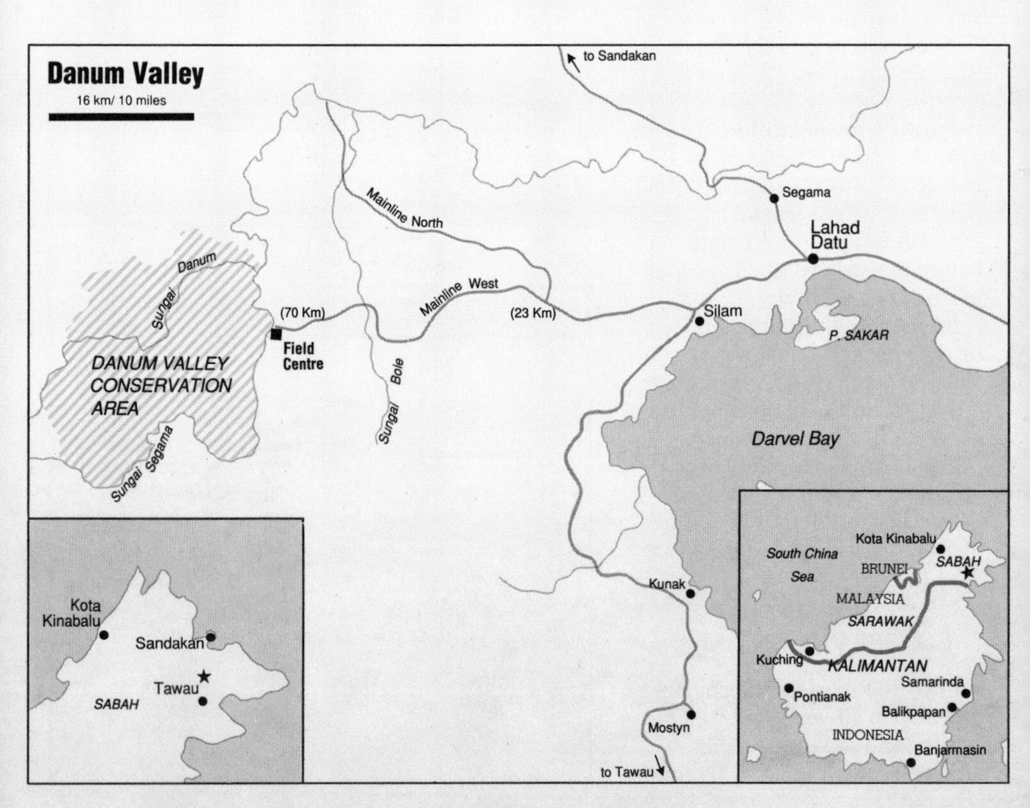

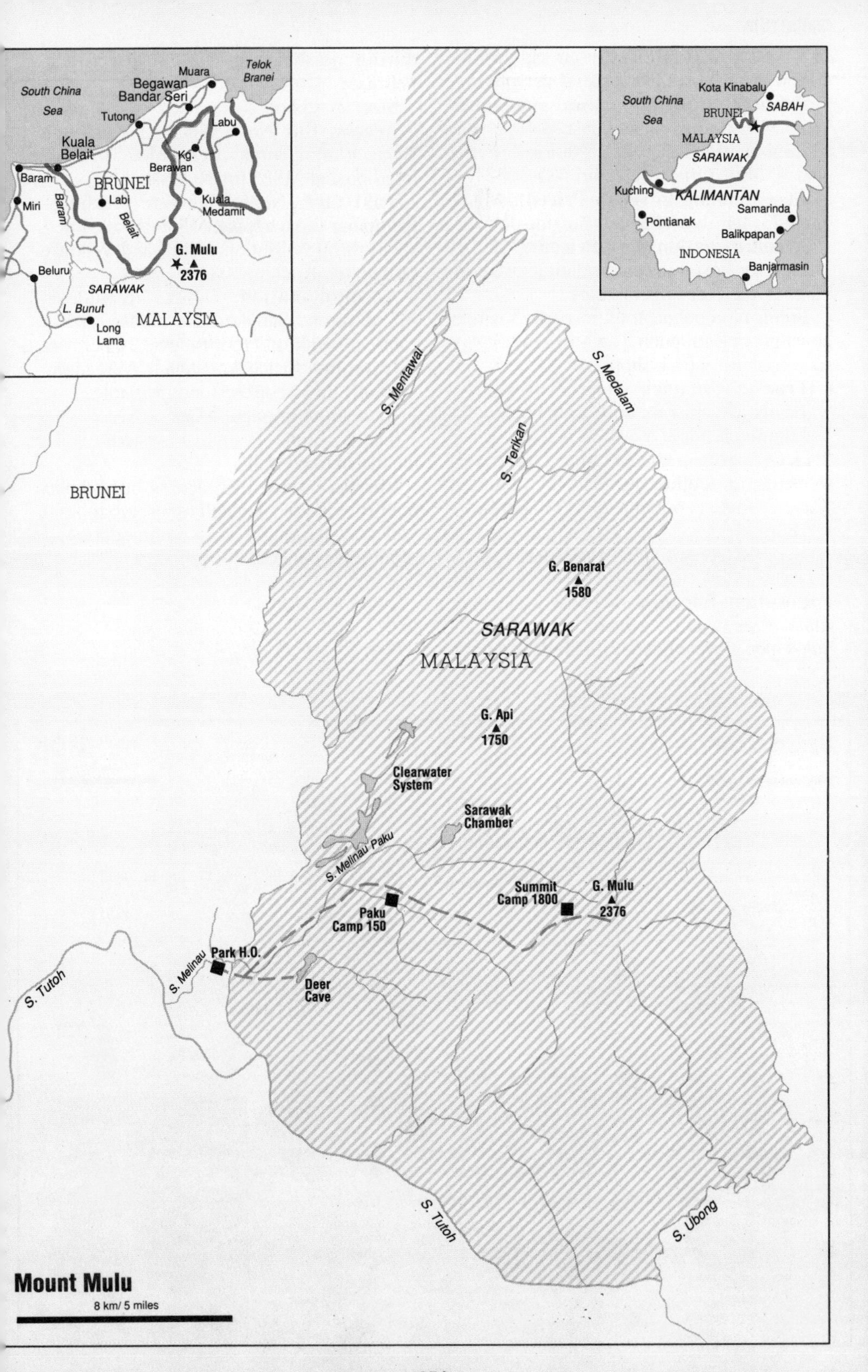
South China
Sea
Muara
Telok
Branei
Begawan
Bandar Seri
Tutong
Labu
Kuala
Belait
Kg.
Berawan
Baram
BRUNEI
Miri
Labi
Kuala
Medamit
Baram
Belait
G. Mulu
2376
Beluru
SARAWAK
L. Bunut
MALAYSIA
Long
Lama
Kota Kinabalu
South China
Sea
BRUNEI
SABAH
MALAYSIA
SARAWAK
Kuching
KALIMANTAN
Samarinda
Pontianak
Balikpapan
INDONESIA
Banjarmasin
S. Mentawai
S. Medalam
S. Terikan
BRUNEI
G. Benarat
1580
SARAWAK
MALAYSIA
G. Api
1750
Clearwater
System
Sarawak
Chamber
S. Melinau Paku
Summit
Camp 1800
G. Mulu
2376
Paku
Camp 150
Park H.O.
S. Melinau
S. Tutoh
Deer
Cave
S. Tutoh
S. Ubong
Mount Mulu
8 km/ 5 miles

Access (car, taxi): A 2-hour trip costing M$60 from Miri (Taxi takes 4 persons).

Public transportation: Buses run 4 times daily Miri-Batu Niah and cost M$9 person.

Registration: Through National Parks and Wildlife Office, tel. Miri 085-36637.

Accommodation (High Priced): M$32 night for family chalet on near side of river.

Accommodation (Low Priced): M$2.50 person for hostel accommodation. Visitors must cross river to get to park.

Food: No cooked food in park. Visitors must go to Batu Niah Town (3 km) or buy groceries from park shop.

Transport in park: None.

Trails: Broad walk trails to main caves.

Further advice: You can register at the Park Office in Pangkalan Lubang on arrival. A permit is required to visit the "Painted Cave". Guide is optional at M$30 per day

BAKO NATIONAL PARK

Location: Borneo north coast, Malaysian state of Sarawak, 37 km northeast of Kuching. Open all year but be prepared for delays during northeast monsoon November-February.

Size: 27 sq km

Access: Bus No 6 Kuching–Kampung Bako, 40-min journey, ticket M$2.10; chartered boat at M$25 for 10 persons.

Taxi fare: Negotiable fare Kuching–Kampung Bako approx M$30 (takes 4 persons) through: National Park and Wildlife Office, tel. Kuching: 082–248088.

Accommodation: Deluxe Resthouse, M$32 room; Standard Resthouse, M$21 room, 6-bed hostel rooms, M$2.10 per bed.

Camping: Beach tent/shelter, M$1 night.

Food: Basic cooked food available.

Transport in park: None.

Trails: Very extensive system of 16 clearly marked trails.

Further advice: Easy access, buses/boats readily available, but call to reserve room if you stay overnight as accommodation is often fully booked.

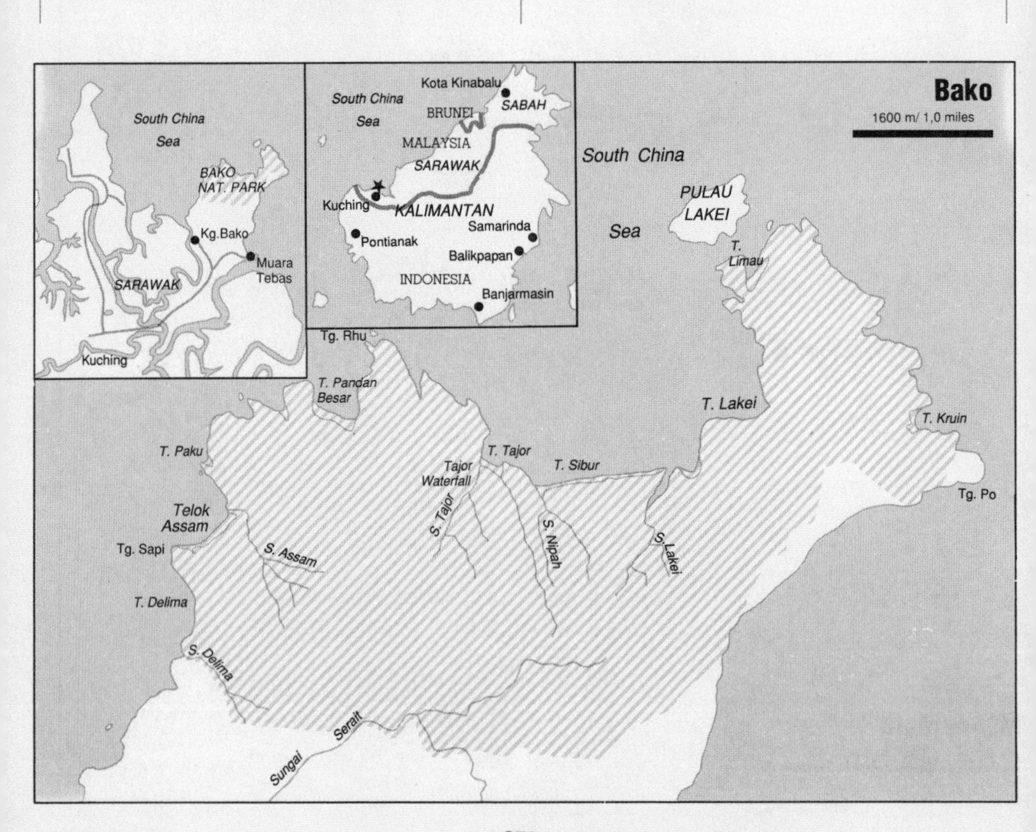

SPORTS

Malaysia is a land of sports. A visitor can enjoy a game of football at an urban field or witness a top-spinning contest in a rural village. Courts for squash, tennis and badminton are found in most international hotels, as are gymnasiums. Many hotels have swimming pools.

Badminton nets are erected on almost every vacant lot available, and not having the right equipment will never deter the sporty Malays from playing the game with improvised rackets and shuttlecocks. If you wish to participate in a sport, there are abundant facilities throughout the country.

There is good **angling** to be had in the numerous river tributaries which indent the country's coastline. Marine game fishes such as barracuda, shark and Spanish mackerel may be sighted off the east coast of the peninsula from May to October.

Hunting is restricted in many areas and is subject to a licence from the Department of Wildlife and National Parks in Kuala Lumpur, Tel: 03-941272. There are hundreds of protected animal species in Malaysia. Obtain more information from reception of the Parks Office.

Scuba-diving is catching on in Malaysia, and you will find scuba centres in most developed beach resorts such as Tioman, Langkawi, and Desaru, to name a few. Some centres also teach scuba-diving and you can gain an internationally recognised certificate. Equipment can be hired for scuba-diving, sailing, wind-surfing, snorkelling and water skiing. Check with a resort by phone first. Snorkelling equipment can be bought cheaply in local shops.

Spectator sports include horse racing as well as popular local pastimes such as kite-flying and top-spinning (*gasing*) and *sepak takraw*. Also popular is *silat*, the Malay equivalent of the Chinese *kung fu*, *silat* is the art of self-defence. The origin of this art is accredited to the famous Hang Tuah of old Melaka, who did not hesitate to draw his sword, and even to strike to kill, for justice's sake. Youths today regard *silat* as a form of physical exercise in an artistic form. Demonstrations at weddings and other feasts are given to the rhythmic beat of gongs and drums. It is also part of the school curriculum for boys.

PHOTOGRAPHY

Professionals working in the tropics have one big suggestion for good results in colour: beware of the heat. Exposure of film or camera equipment to hot sun causes changes in the chemical emulsions of the film, which detract from natural colour. Whenever possible, store your camera and film in a cool place; if not in an air-conditioned room, at least in the shade. Experienced photographers also recommend buying film in the cities rather than in the countryside where proper storage facilities for colour films are not guaranteed. Also, get your films processed as soon as possible, either in Malaysia or in Singapore.

Humidity can be another tropical hazard, particularly with jungle photography. The solution here is to carry equipment and film in a closed camera bag containing silica gel, a chemical that absorbs moisture. For suitable tones and rich colour, the best times to photograph are before 10.30 a.m. or after 3 p.m. Few films take noontime sunlight well. Pictures often lose subtle gradations in colour because the light is too strong. In the early morning or late afternoon, sidelights give softer contrasts and deeper colour density. You perspire less as well!

Most Malaysians are more than amiable about having their pictures taken. It usually takes a gang of schoolchildren about 15 seconds before they merrily begin jabbing peace signs in front of your 20 mm lens. Mosques and temples are rightly more reserved about photographers posing their subjects in front of altars. Whatever the situation, you should always ask first for permission, especially with tribal people, who may have an aversion to having their photo taken. Keep a respectful distance from religious ceremonies. If you can bear to carry one, a zoom lens will enable you to photograph interesting groups of people without interfering with them.

Film processing is offered everywhere in Kuala Lumpur. Komal in Petaling Jaya develops Kodak colour film only. Black and white normally needs 24 hours. Kodakcolour and Ektachrome take one or two weeks, as they are sent off to Australia for developing. General colour film can take as little time as half an hour in a developing booth in major shopping centres in Kuala Lumpur.

LANGUAGE

Malay, mother tongue of more than 150 million Asians, is as ancient as a Grecian urn, and nearly as practical. A man can travel from the tip of the Malay peninsula, through the southern Philippines and all along the island-hopping trail that zigzags across the Indonesian archipelago – speaking Malay. New nations have adopted the old language to their own ends, lending it a variety of sophisticated nuances in grammar, spelling and scientific terms. But all countries with official letterheads in Malay trace them back to the trade fairs of antiquity when merchants bargained over gold dust and rhinoceros horn in a tongue similar to today's "Bazaar Malay", the language of the marketplace.

Though formal Malay is a complex language demanding some time of serious study, the construction of "Basic Malay" is fairly simple, with many things about the language conducive to learning. Malay is written in the Latin alphabet and, unlike some Asian tongues, is not a tonal language. There are no articles in Malay – *buku* means "the book" or "a book", *anak* means "the child" or "a child". Plurals are made simply by doubling the noun – *buku-buku* means books. To denote time, a few key adverbs are used: *sudah* (already) shows past time, *belum* (not yet), *akan* (will) for the future, and *sedang* (in the process of doing, e.g. *Saya sedang makan* = I am eating) for a present action, being performed at the moment of speaking.

When speaking Malay, you need a few basic rules. Adjectives always follow the noun. *Rumah* (house) and *besar* (big) together as *rumah besar* means "a big house" and so on. When constructing a sentence, the order is subject-verb-object: *Dia* (he) *makan* (eats) *nasi* (rice) *goreng* (fried). *Dia makan nasi goreng* = He eats fried rice. The traditional greeting in Malay is not "Hello!" but rather *Ke mana*? – "Where are you going?" The question is merely a token of friendliness which does not require a specific answer. One simply returns the smile by replying *Tak ada ke mana* – "Nowhere in particular" – and passes on.

Below are some very general guidelines for the pronunciation of Malay, or *Bahasa Malaysia* as it is known here. No written descriptions of the phonetics can replace the guidance of a native speaker, but once you've tried pronouncing a few words, Malaysians are quick to understand and their response is the best way to pick up a feeling for the language.

a	is pronounced short as in *matter* or *cat*. *apa* – what; *makan* – to eat
ai	is pronounced like the sound in *aisle*. *kedai* – shop; *sungai* – river
au	sounds like the *ow* of *how*. *pulau* – island; *jauh* – far
c	is pronounced like *ch* as in *chat*. *capal* – sandal, *cinta* – love
e	is very soft, hardly pronounced at all. *membeli* – to buy; *besar* – big
g	is pronounced as in *go*, never as in *gem*. *pergi* – go; *guru* – teacher
gg	is pronounced as *ng* plus a hard *g* sound: i.e. sing-ging. *ringgit* – Malaysian dollar; *tetangga* – household
h	is pronounced as in *halt*. *mahal* – expensive; *murah* – cheap

i sounds like *i* in *machine* or *ee* in *feet*. *minum* – to drink; *lagi* – again
j sounds like the English *j* in *judge*. *Jalan* – Street; *juta* – million
ng a single *g* in a word is pronouced like the *ng* in *sing*, not with a hard sound. *sangat* – very; *bunga* – flower
ny is similar to *ni* in *onion* or *n* in *news*. *harganya* – price; *banyak* – a lot
o is most similar to the *o* in *hop*. *orang* – human being; *tolong* – help
u is pronounced as *oo* in *pool*. *tujuh* – seven; *minum* – to drink.
y sounds like *y* in *young*, never as in *why*. *wayang* – opera; *kaya* – rich.

There is no specific syllabic stress in Malay as in English (i.e. **na**-tion, not na-**tion**), nearly all syllables are given equal stress; however, the Malays add to their speech a sing-song intonation which often gives more emphasis to the final syllable of a word, especially the last word in an utterance. This has led to the widespread use of the appendage – *lah* to the important word. This can either charm or irritate the visitor! Its purpose is purely emphatic, and it is now used generally in Malaysia, whether the speaker is talking in Malay, Chinese, Tamil or even English! Perhaps the most famous example of this is the phrase "Cannot-lah!", uttered when you have asked something the speaker considers impossible!

USEFUL ADDRESSES

TOURIST INFORMATION

The Tourist Development Corporation (TDC) has various offices throughout Malaysia (see the following list). Offices vary in the amount of literature available, but there are usually brochures on local places of interest and some of the staff (especially in KL) are very knowledgeable. The TDC also has a few offices overseas. In smaller places, there are other tourist offices that service the area. Besides these, it is always a good idea to ask several local people for their opinion of a place. You'll get varying reports and will have to make up your own mind as to whether the place is worth a visit!

Head Office
24-27th Floor, Menara Dato' Onn, Putra World Trade Centre, 45 Jln. Tun Ismail, 50480 KL
Tel: 03-293-5884

East Coast Region
2243 Tingkat Bawah, Wisma MCIS,
Jln. Sultan Zainal Abidin,
20000 Kuala Terengganu, Terengganu
Tel: 09-621433

Northern Region
10 Jln. Tun Syed Sheikh Barakbah,
10200 Pulau Pinang
Tel: 04-619067

Sabah
Block L, Lot 4 Bandaran Sinsuran,
Mail Bag 136, 88700 Kota Kinabalu, Sabah
Tel: 088-211723

Sarawak
2nd Floor AIA Bldg.,
Jln. Song Thian Cheok, 93100 Kuching
Tel: 082-246575

Southern Region
No. 1, 4th Floor, Kompleks Tun Razak,
Jln. Wong Ah Fook,
80000 Johor Bahru, Johor
Tel: 07-223590

GETTING THERE

BY AIR

Touch down at Singapore's Changi airport , which was host to 12.6 million passengers in 1988, projects an aura of grandeur. These passengers were travelling on the national carrier, Singapore Airlines, or on the planes of the 48 other carriers which, among them, provided 1,500 flights a week to and from 101 cities in 53 countries. If visitors are impressed with Terminal 1 (T1), then the second terminal (T2), which opens in November 1990 is an eye-opener. It will double Changi's capacity, enabling it to cope annually with 24 million passengers. The two terminals will be linked by the "Changi Sky Train", an automated miniature rapid transit train system with a track length of 600 metres. The system is the first of its kind outside the USA and UK.

Finding your way about Changi is a breeze. A comprehensive information system combines computerised TV screens, illuminated signposts and flip boards.

Horizontal travelators and escalators whisk you down one level to the arrival area where you can complete health and passport formalities, and grab duty-free buys from the emporium. If your airplane is parked away at one of the remote bays from the terminal, a shuttle bus will take you to a bus station. From there, you can proceed directly to the arrival area.

Immigration and customs clearance are fairly speedy, though it is easiest to enter Singapore well dressed. After clearing immigration and customs, you come to eight conveyor belts carrying baggage. Look for the track which states your flight number; claim your baggage; present it for clearance at any one of the 12 counters and enter the Arrival Meeting Hall. Banking, money changing and postal facilities; hotel reservation and tourist information desks; convention bureau, car rental and airport information counters; and meeting services are all housed here.

Both transit and transfer passengers stopping over at Changi Airport proceed straight to the Departure/Transit Waiting Lounge on the second level. Transfer passengers must remember to reconfirm their onward flights at the transfer counters immediately upon landing. Television monitors in the Departure Hall inform passengers of their boarding time and check-in counters.

Departing passengers are advised to keep some Singapore dollars handy fo rthe Passenger Service Charge. Alternatively, the Charge tickets can be purchased at nearly all hotels.

BY SEA

Three hundred shipping lines flying the flags of some 80 countries and serving 300 other ports call regularly. In fact, a ship arrives at and leaves every 10 minutes, making Singapore one of the busiest ports in the world.

Cruise and passenger ships arrive from Australia, Europe, North America, India and Hong Kong. Entry formalities comply with standard immigration laws and, for quick orientation, the Tourist Promotion Board is on hand to greet arrivals with a mobile information service on the sights of the town.

Singapore, surrounded by waters on all sides, is connected by road and by rail with the peninsular Malaysia.

OVERLAND

The railway brings travellers to Singapore from Bangkok via Haadyai, Butterworth and Kuala Lumpur. The International Express departs from Bangkok Monday, Wednesday and Saturday at 4.10 p.m., stopping at Haadyai at 10.28 a.m. the next day and arriving at Butterworth at 6.45 p.m. From Butterworth, the Express Rakyat departs at 8.15 a.m. and arrives at Kuala Lumpur at 2.35 p.m. The last link to Singapore is also provided by the Express Rakyat which leaves Kuala Lumpur at 3.05 p.m. and arrives at Singapore at 9.35 p.m.

The fare from Butterworth to Singapore is $50 in an air-conditioned coach and $30 in one without air-conditioning.

Six other daily services run from Singapore to Malaysia. The morning trains depart at 6.30 a.m. and 7.30 a.m. The afternoon service leaves Singapore at 3 p.m. while the evening trains leave at 8 p.m., 8.30 p.m. and 10 p.m.

For detailed fares and schedules, call the railway station at Keppel Road (tel: 222-5165). Reservations are accepted with advance payment, though arriving early for your train is advisable.

TRAVEL ESSENTIALS

PASSPORTS & VISAS

Visitors entering Singapore must possess valid national passports or internationally recognised travel documents. A certificate of vaccination against smallpox is necessary for those coming from infected countries within the preceding 14 days. Visas are not necessary for a stay up to 14 days, provided the visitor has confirmed onward passage and has adequate finances. Collective passports or travel documents are permitted for group travel in tours of five to 20 people. However, these regulations do not apply to nationals from communist countries, and to those from Taiwan and South Africa, who need visas to enter Singapore.

Tourist visas are usually issued and extended up to a maximum of three months at the Immigration Office, Pidemco Centre (for enquires, tel: 532-2877).

MONEY MATTERS

The Singapore dollar was valued at approximately 1.9 to the U.S. dollar or 3.2 to the British pound sterling in 1990. Brunei dollars are equivalent to the Singapore dollar. Malaysian dollars are about 25 percent lower in value. There is no limit on the amount of Singapore and foreign currency notes, traveller's checques and letters of credit which may be brought in or taken out of the country.

Make no haste to change big amounts of money at luxurious hotels. Singapore's ubiquitous money-changers at Raffles Place and most shopping complexes are government licensed. They invariably give better rates than those of leading hotels. Traveller's checques generally get a slightly better rate of exchange than cash. It is best to deal directly with banks and/or licensed moneychangers. Banking hours are Monday through Friday 10 a.m. to 3 p.m. and Saturday 9.30 a.m. to 11.30 a.m. Some local banks now open on Sundays from 11 a.m. to 4 p.m.

If you wish to deal with banks for your traveller's checques and other foreign currency transactions, it is advisable to do so on weekdays. Some banks do not handle such transactions on Saturdays while others conduct them in small amounts only, based on Friday's rate.

Credit cards commonly accepted are American Express, Diner's Club, Carte Blanche, Asia Card, Visa Master Charge Card and those of international hotel chains and airlines.

HEALTH

Inevitably, the third word belonging to Singapore after "clean" and "green" is "healthy." Travellers have no worries about drinking water straight from the tap and eating food by the streetside. They can even participate in the hygiene campaigns like "Keep Singapore Pollution Free." The Republic lives up to this comforting reputation with efficient contemporary medical facilities in numerous hospitals and clinics. The closest anyone can come to a physical ailment originating in Singapore is over-eating and, fortunately, there is a quick cure for this: shopping for a few hours.

The daytime trend is pure casual comfort. Light summer fashions, easy to move in, are the right choice for a full day out in town. Wear a white shirt and tie for office calls. Evening dress is a more subtle combination of fads and formality. Only few plush nightclubs and exclusive restaurants favor the traditional jacket and tie. Most hotels, restaurants, coffee houses and discos accept more casual wear of shirt-and-tie (or even shirts without ties) for men and pleasant dress suits for the ladies. Jeans and tee-shirts are taboo at most discos.

To avoid embarrassment, it is best to call in advance to check an establishment's dress policy.

CUSTOMS

Singapore is essentially a free port. Personal effects (including cameras and radios), 1 litre spirits, 1 litre wine and 200 cigarettes or 50 cigars or 250 grams tobacco may be brought in duty-free. Narcotics are strictly forbidden, as are firearms and weapons, subject to licensing.

GETTING ACQUAINTED

GEOGRAPHY

Singapore, at the tip of the Malay Peninsula, is about 80 miles (128 km) north of the equator and covers 238 sq miles (616 sq. km). The Republic city-state consists of diamond-shaped Singapore island and 57 other islets of which fewer than half are inhabited. The Government, not to be deterred by limited land space, is constantly clearing swamps and jungles and reclaiming land from the surrounding waters for housing and industrial uses.

About two-thirds of the island is less than 50 feet (15 metres) high and the highest point (Bukit Timah, literally "Tin Hill") soars to 580 feet (170 metres).

TIME ZONE

Singapore Standard Time is eight hours ahead of Greenwich Mean Time. Aside from time variations made in certain countries during specific seasons, if it is noon in Singapore, it is 11 a.m. in Bangkok, and 9.30 a.m. in New Delhi, 5 a.m. in Bonn and Paris and 1 p.m. in Tokyo.

CLIMATE

Temperatures teeter between 87°F (30.6°C) at noon and 75°F (23.8°C) at night, the daytime heat cooled by sea winds and ample air-conditioning. Asia's subtle seasons follow the monsoons, and from November to January high winds turn buckets over Singapore. For 10 minutes the city comes to a standstill under hawker stalls, arcades and umbrellas. Then, as suddenly as they come, the rains vanish, leaving the pavements wet-washed, the lawns refreshed, and the umbrellas behind for sunshades.

LOCAL LAWS & CUSTOMS

The customs, religions and languages of nearly every nation in the world have converged in Singapore at some time in history. Adjectives beginning with "multi" are common sounds on the Singapore scene, and a cosmopolitan tolerance is part of the city's character. With everyday etiquette relaxed and straightforward, visitors behaving courteously stand little chance of unintentionally giving offense. Some ceremonies and special occasions, however, recall inherited traditions and a familiarity with certain customs will set everyone at ease.

TEMPLES & MOSQUES

Removing one's shoes before entering a mosque or an Indian temple has been an unspoken tradition for centuries. Within, devotees do not smoke, though neither of these customs generally applies to Chinese temples where more informal styles prevail. Visitors are most welcome to look around at their leisure and are invited to remain during religious rituals.

While people pray, it is understood that those not participating in the service will quietly stand aside. A polite gesture would be to ask permission before taking photographs: the request is seldom, if ever, refused. Modest clothing, rather than brief skirts or shorts, is appropriate for a visit.

Most temples and mosques have a donation box for funds to help maintain the building. It is customary for visitors to contribute a few coins before leaving.

PRIVATE HOMES

The hospitality of a Singaporean friend is a good feeling. In private homes, visitors are received as honored guests.

Without hesitation, the hostess prepares some drinks, the best of whatever is available in the house whether it be rare imported tea or an iced soft drink. Wives pride themselves in serving good food any time a guest arrives, and when returning the visit they bring a small gift of fruit or cakes, as is the custom. Though not everything served is expected to be eaten, nothing pleases a hostess more than knowing her guests enjoy her cooking.

Malays, Indians and Chinese remove their shoes at the door so as not to bring dirt into the house. No host would insist his visitors do so, but it is the polite way to enter a home.

SHARING A MEAL

As every taste has its flavor, every food has its style. Chinese food is eaten with chopsticks, most Malay and Indian food with the right hand (never the left), Indonesian and Thai food with a large spoon and fork. A gourmet would no more eat a Chinese meal with knife and fork than a filet mignon with chopsticks.

Asian meals are usually served in large bowls placed in the center of the table, with each diner helping himself to a little from each bowl. Piling up your plate with food is not only impolite but unwise. With more dishes to follow, by taking a little you can always help yourself to more. Local people are inwardly pleased if you join them in their styles of dining, for a simple reason: they know it tastes better that way.

TIPPING

Smiles follow tips everywhere. Singapore is no exception especially in fashionable places such as nightclubs, friendly bars and expensive restaurants. Here the magic number is 10 percent. Leading hotels are kind enough to do the tipping for you by adding a 10 percent service charge to every bill; a second time around is not neccessary though neither is it refused. Bellboys and porters receive from $1 upwards, depending upon the complexity of the errand. Yet, beyond the international thoroughfares, tipping is exceptional. In small local restaurants, food stalls and taxis, the bill includes the service, and with thank you (*terima kasih* in Malay), simply a smile will do.

WEIGHTS & MEASURES

The transition from imperial to the metric system of weights and measures was a gradual one which first started almost a decade ago. Now, except for smaller shops which might still be using the imperial system, most transactions are metric.

ELECTRICITY

Singapore is on a 220-240 voltage and 50 Hz system. Most hotels provide transformers to step down electrical appliances to the suitable voltage. Check with your concierge.

BUSINESS HOURS

Government offices are open 9 a.m. to 5 p.m. Monday through Friday, and to 1 p.m. on Saturday. Banks are open to the public 10 a.m. to 3 p.m. weekdays and 9.30 a.m. to 11.30 a.m. Saturday. Some are also open on Sundays from 11 a.m. to 4 p.m.

Post offices operate from 8.30 a.m. to 4.30 p.m. weekdays and to 1 p.m. on Saturday. The post office at Orchard Point and the Changi Airport operate from 8 a.m. to 9 p.m. every day including Public Holidays.

Business hours of restaurants, hotel coffee shops, coffee houses and music lounges differ widely. Many open at 10 a.m. and close at 10 p.m Some offer 24-hour service: there is always some place to eat and relax.

Most shopping complexes close by 9 p.m. although larger chain stores and department stores stay open till 10 p.m.

COMMUNICATIONS

MEDIA

The Press: Singapore's oldest newspaper is *The Straits Times*, an English daily which began in 1845 on the merry note "Good morning to you, kind reader!" It has been the venue for local and foreign news ever since.

Business Times is popular among businessmen and offers details on ships calling at the local port and their schedules.

The most widely read local Chinese-medium newspaper is the *Lian He Zao Bao*, first published in 1923. Singapore also supports four other Chinese dailies, a Malay and a Tamil daily.

POSTAL SERVICES

Despite low postage rates, Singapore's communication system is advanced and fast. An aerogram to anywhere but the moon costs 35 cents while an airmail postcard to similar destinations costs 30 cents. Letters weighing not more than 10 grams to the USA, Europe, Africa and Middle East countries cost 75 cents while to Australia, New Zealand, Japan and all countries in Southeast Asia 35 cents. The fee for a registered item is $1.20. An express mail service is available to more than 40 countries: cost for any item under 0.5 kg sent by this service to the USA is $40.

Most hotels handle mail service or you may post letters and parcels personally at the General Post Office, Fullerton Building (Tel: 4493377) which operates from 8 a.m. to 6 p.m. on weekdays and to 4 p.m. on Saturdays (closed on Sundays) or at any of nearly 60 branch offices. Branch offices open at 8.30 a.m. and closes at 5 p.m. on weekdays and at 1 p.m. on Saturdays (closed Sundays): half of these remain open on Wednesdays until 8 p.m. while half-a -dozen are open until 8 p.m. on all working days.

Not known to many is the fact that the annex to the GPO and the Comcentre, 31 Exeter Road (near Orchard Road) never closes and at both of these most postal transactions can be carried out.

TELEPHONE & TELEX

Local calls can be made from public telephone booths whose phones are invariably in working order. Clearly-marked public phones can be used for world-wide calls. Phone cards can be purchased for $10, $20 and $50 and can be used for both local and overseas calls. Local calls cost 10 cents for the first three minutes and 10 cents for every subsequent three minutes, for a maximum of nine minutes.

The easiest way to the right number is through your hotel operator, the directory, or by dialling Enquiries at 103. You may dial a ship in port by ringing 105, or radio-telephone a ship at sea by ringing 107.

Most hotels provide cable, telex and telephone services day and night. All forms of telecoms can be filled at the GPO, Fullerton Building; Telecoms (Comcentre branch), 31 Exeter Road; Telecommunication Building, 35 Robinson Road and Telecoms (Hill Street Branch), 15 Hill Street. All open at 8 a.m. on weekdays and close at 6 p.m. except the Hill Street branch and 24-hour service are both available at the GPO annexe and the Telecoms at Exeter Road.

EMERGENCIES

MEDICAL SERVICES

Pharmacies: The long waiting time at clinics for consultation of a minor ailment is something that proves too much for an average Singaporean with his innate hurry-scurry trait. This accounts for the popularity of pharmacies or drug stores or medical stores. They are known by a string of names here but they mean the same thing. Their shelves stock a whole range of drugs; from linctus for dry coughs and expectorant for wet coughs to anti-histamine for colds and antipyretic for fevers. These stores are registered and all of them have a qualified pharmacist on duty.

PRIVATE PRACTITIONERS

There is no shortage of these in Singapore, all of whom speak at least English and one other language. They are professionally trained either here or abroad and an average cost per visit varies somewhere between $10 to $35 for a practitioner and $35 to $70 for a first consultation by a specialist. Ask your embassy to recommend a specialist or private practitioner.

HOSPITALS

Government hospitals are numerous all over the island. They compare favorably with those in the West. The hospitals here are advanced and well-equipped to cope with the most complicated and difficult operations. Most of their services are made available to non-citizens who pay a slightly higher rate than the citizens.

GETTING AROUND

TOURIST PROMOTION BOARD

In the inner circles of the travel trade, Singapore's Tourist Promotion Board is highly esteemed for the crisp efficiency characteristic of the place it represents. Through its steady campaigns to add extra sheen to the Republic's facilities, the Board convinces taxi drivers that it pays to have tidy upholstery. It rewards service stations if they look nice. It encourages shops through every incentive, including a metal plaque to keep their premises clean, display their items attractively, and serve their customers with honesty and courtesy.

Tour operators and travel agents are gleaned with a fine comb to make sure their services rank high by international standards. And to see that Singapore's new arrivals are well received, the Board goes down to the grass-roots level by organising "Courtesy Courses" for employees stationed at the island's principal points of entry. "So they will know the meaning of PR and how to exercise it," a Board member explains.

Since the Board's numerous services command such diligent attention, they are bound to be professionally tailored to a visitor's needs. And the traveler who appreciates smooth organisation and quick answers will find they work well as a cushion for feeling at home.

The Board is so insistent upon the capabilities of the people it selects as guides that an applicant must pass an intensive three-month training course before he is allowed to make his first professional contact with a visitor. Since quality is handled with such care, a visitor is assured of the service he deserves when touring with one of the Board's registered guides. For half a day, English-speaking guides charge $40. Foreign-language guides charge $70. Guides with cars may charge higher.

All guides speak English and at least one other major language, and though the Board is somewhat shy about its lovely female guides, it certainly suggests they accompany ladies on a shopping spree. The Board also recommends guides who specialise in subjects such as special ceremonies or places. By law, only registered guides are allowed to practise.

For the less costly pleasures of launching out unguided, the Board's receptionists pave the way by furnishing maps, transportation routes and schedules, suggestions for walking tours, and a library of pertinent brochures as an instant means for getting a footing in "Instant Asia." They suggest good restaurants for tasting different styles of food, pass on what is currently "in" after sundown and, armed with a notepad directed to the boss upstairs, handle any complaints which may arise.

It is safe to say that just about every practical tip a traveler needs to get around in Singapore has been considered and recorded in the Board's booklets. The STPB's *Official Guide to Singapore* has the advantage of clear, condensed information in the form of one small booklet. You may follow the numbers in the Map of Singapore for orientations to the city plan. The Board keeps the visitor informed daily with a calendar of current affairs and religious ceremonies, Chinese operas and art exhibits in its *Singapore This Week*, distributed to all leading hotels. There is only one publication which isn't free and it is remarkably helpful; *City Guide: Singapore*, a coffee table-sized colour panorama with many views of a city of many moods, which sells for $32.90 at the Board's offices and at bookshops.

Publicity writers, journalists, photographers, or editors, whether on assignment to feature Singapore in travel publications or seriously considering the idea, are welcome to take a short cut through the Tourist Promotion Board. By writing in advance, complimentary tours, suggested itineraries, a bundle of literature and 9,000 slides, conveniently filed in special subjects for articles, is at your disposal. And if research demands more depth than the gloss of a tourist brochure, the Board refers writers to specific publications and library references, or to officers heading its departments who personally handle the subject.

FROM THE AIRPORT

The Airport is linked to the city center by the East Coast Parkway (20 minutes traveling time) and to the rest of Singapore by the Pan-Island Expressway.

There are three types of transport from the airport – private car, taxi or public bus. Inclined travelators in the Arrival Hall descend to a short tunnel that leads to the Passenger Crescent (the curbside for private cars) and the Arrival Crescent (the pick-up point for taxis). A surcharge, $3 more than the fare shown on the taxi meter, is charged if you board a taxi at the Airport. Several public shuttle bus services run between the Airport and nearby bus interchanges. For service numbers and routes, check the basement passenger terminal.

AIR TRAVEL

There are no domestic air flights in the compact island of Singapore. The closest you can come to is a bird's eye view of the island from a conducted tour on a helicopter.

The helicopter takes a maximum load of five passengers including the pilot, who is also the tour guide and who will give commentaries and explanations of the places he flies over. The flight, costing about $750 on a per-helicopter per-flight basis, takes you round the island (taking off at the Seletar secondary airport, down to the waterfront, round Sentosa island, up west to Jurong and back to Seletar) in 40 minutes. By law, cameras are not allowed but as long as you do not make it too obvious, the security guards will not object. Call Helicopter Services at 481-5711 to make arrangements.

MRT

The $5 billion, 41-mile (66-km) Mass Rapid Transit (MRT) system was offically opened in early 1988. It has 66 air-conditioned trains running over a north-south and east-west line. The six-car trains, each of accommodates 1,800 persons, travel at 45 kph (30 mph) and stop at each of the 41 stations for 20-30 seconds. About one-third of the stations are underground and must rate as among the most handsome underground stations in the world. A journey aboard the MRT, especially on the underground section, which runs through the heart of tourist and business Singapore, is a must for all visitors.

Fare collection is automatic: magnetically coded plastic cards costs between 50 cents for 3.5 kms (two miles) and $1.40. Stored-value tickets costs $10.

Depending on the station the first train rolls out between 6 a.m. and 6.40 a.m., Mondays to Saturdays, and between 6.45 a.m. and 7.25 a.m. on Sundays and holidays. Last trains are between 11 p.m. and midnight: 15 minutes earlier on Sundays and public holidays.

BUSES

Nearly 250 bus services ply the paved roads of Singapore and connect every corner of the island.

Buses (single and double-deckers) run from 6.15 a.m. to 11.30 p.m. on the average, with an extension of about a half hour for both starting and ending times on weekends and public holidays. Fares are cheap (minimum 50 cents, maximum 90 cents), and the amount payable is structured according to fare stages (0.5 miles/0.8 km is equivalent to one fare stage). You pay 50 cents if traveling over four or fewer fare stages; 60 cents for five to seven stages; 70 cents for eight to 10 stages; 80 cents for 11 to 13 stages; and 90 cents for 14 or more stages. Too much to remember? Then just have some change ready and ask the bus driver on boarding.

For good value, purchase the Singapore Explorer ticket and get the special Explorer Route Map: $5 for a one-day ticket or $12 for a three-day ticket. Holders can hop on and off, as frequently as they wish, the buses run by the Singapore Bus Services and the Trans Island Bus Service. Tour the island by following the color-coded and clearly indicated major routes and visit points of interest on the map. Tickets are available from travel agents and major hotels. Contact the SBS public relations office, tel: 287-2727 (8 a.m. to 4.30 p.m. weekdays, 8 a.m. to 12.30 p.m. Saturdays) for further information.

An especially helpful source, available at most bookstores and newsstands, is the Bus Guide. This booklet gives complete details of all bus routes and contains a section on bus services to major tourist spots. Or call Singapore Bus Service, Tel: 284-8866.

TAXIS

One of the rarest sights in Singapore is a main street without a taxi. More than 10,000 taxis are on the roads and everyone uses them. It is the fastest and easiest way to move around in comfort. The vehicles are clean and kept in tip-top condition (the cleanest taxis get a prize in Singapore). Black and yellow, solid blue, green and white or red and white – all with "SH" or "SHA" on their license plates – taxis run by meter. Each taxi may carry a maximum of four passengers. The fare is $2.20 for the first mile (1.5 km) and 15 cents for each 0.15 mile (0.25 km) thereafter. Most taxi stands are found just outside shopping centers and other public buildings. You may join the queue at these or you may hail one from any curbside (except those marked with double yellow lines).

Most drivers speak or understand English. Still, it is better to be sure the driver knows exactly where you want to go before starting. Tipping is purely optional and is discouraged by the government.

Taxis on radio call are available 24 hours at tel: 452-5555, 533-9009 and 250-0700.

TRISHAWS

A direct descendant of the historical rickshaw – covered carriage pulled by man on foot – is the trishaw, a bicycle with a sidecar. This is a vanishing mode of transport among locals due to its slow speed, its lack of sophistication and the unending hassle over the fare with the rider – often a stubborn grumpy man in his 60s.

However, it is fast becoming a hot favorite among visitors. Its selling point lies in the fact that it goes at a speed slow enough for its passenger to absorb what goes on around but fast enough to cover most of the picturesque sights of downtown within an hour. Full information of itinerarized trishaw tours at standard prices is available at your hotel's tour desk.

A word of caution: for a ride by a freelance rider, be sure you agree upon a fare before getting on. Licensed riders are distinguished from these by colored badges.

PRIVATE TRANSPORT

Rent-A-Car: In the free spirit of independent travel on and off the main roads of Singapore, rent-a-car services provide the wheels if you provide the valid driver's license. Self-drive cars cost from $60 to $350 a day plus mileage, depending on the size and comfort of your limousine. Contact any of the many companies renting cars through your hotel or the Yellow Pages – they will be the first to remind drivers to bring along their passports.

COMPLAINTS

The Singapore Tourist Promotion Board has full authority to protect the interest of visitors whether they are here to shop, to trade or just to see. Stringent measures, including imposition of heavy fines and suspension of licences, have been taken against shopkeepers, taxi drivers or trishaw pullers who tried to fleece a visitor. Since then, shams are rare. However, if you have a legitimate complaint, bring it to the attention of the board which will assist you promptly and efficiently.

Note: The Singapore Tourist Promotion Board (STPB) has a telephone number with assistance on the receiving end: 339-6622. Or pay a visit to its office in Raffles City Tower (with a superb waterfront view from the 37th floor). They are open from 8.30 a.m. to 5 p.m. Mondays through Fridays, from 8.30 a.m. to 1 p.m., Saturdays.

WHERE TO STAY

Hotels in Singapore are well-maintained with courteous and helpful staff. Most have restaurants and some, especially those in the tourist belt, have a wide range of amenities available – discotheque, conference rooms, banquet rooms, swimming pool, business centre, fitness centre, health centre and tennis courts. Listed below in alphabetical order are a selection of the hotels.

Amara Hotel
165 Tanjong Pagar Road, S.0208. Tel: 224-4488

Apollo
405 Havelock Road, S.0316. Tel: 733-2081

Asia
37 Scotts Road, S.0922. Tel 737-8388

Boulevard
200 Orchard Blvd., S.1024. Tel: 737-2911

Cairnhill
19 Cairnhill Circle, S.0922. Tel: 734-6622

Carlton Hotel Singapore
76 Bras Basah Road, S.0718. Tel: 338-8333

Chateau at Scotts
Scotts Road, S.0922. Tel: 732-5885

Cockpit
115 Penang Road, S.0923. Tel 737-9111

Crown Prince
271 Orchard Road, S.0923. Tel: 732-1111

Dai-Ichi
81 Anson Road, S.0207. Tel: 224-1133

Dynasty
320 Orchard Road, S.0923. Tel: 734-9900

Goodwood Park
22 Scotts Road, S.0922. Tel: 737-7411

Grand Central
22 Orchard Road/Cavenagh Road, S. 0923. Tel: 737-9944

Hilton International
581 Orchard Road, S. 0923. Tel: 737-2233

Holiday Inn Park View
Cuppage Road/Cavenagh Road, S.0922. Tel: 733-8333

Hyatt Regency
10 Scotts Road, S.0922. Tel: 733-1188

Imperial
1 Jln Rumbia, S.0923. Tel: 737-1666

The Mandarin Singapore
333 Orchard Road, S.0923. Tel: 737-4411

Marina Mandarin
6 Raffles Boulevard, S.0923. Tel: 338-3388

Le Meridien Singapour
100 Orchard Road, S.0923. Tel: 733-8855

Ming Court
1 Tanglin Road, S.1024. Tel: 737-1133

New Otani
177A River Valley Road, S.0617. Tel: 338-3333

Omni Marco Polo
247 Tanglin Road, S.1024. Tel: 474-7141

Orchard
442 Orchard Road, S.0923. Tel: 734-7766

Oriental
6 Raffles Boulevard, S. 0923. Tel: 338-2266

The Pan Pacific Hotel Singapore
Marina Square, 7 Raffles Boulevard, S. 0103. Tel: 336-8111

Phoenix
Orchard Road/Somerset Road, S.0923. Tel: 737-8666

(**Raffles**, as much an institution as a hotel, is currently closed for restoration. Note the word "restoration" for, when it opens at the beginning of 1991, Raffles will have all the state-of-the-art facilities although these will be cleverly hidden behind a mid-20th century ambience. Thus, hotel guests will be met at Changi airport by vintage cars. The hotel will consist of suites only – 104 of them.)

River View
382 Havelock Road, S.0316. Tel: 732-9922

Royal Holiday Inn
25 Scotts Road, S.0922. Tel: 737-7966

Shangri-la
22 Orange Grove Road, S.1025. Tel: 737-3644
Sheraton Towers
39 Scotts Road, S.0922. Tel: 732-0022
Tai-Pan Ramada
101 Victoria St., S.0718. Tel: 336-0811
Westin Plaza
2 Stamford Road, S.0617. Tel: 338-8585
Westin Stamford
2 Stamford Road, S.0617. Tel: 338-8585
York
21 Mt. Elizabeth, S.0922. Tel: 737-0511

FOOD DIGEST

Food has always been the main source of entertainment and topic of conversation on this island. A Ministry of Environment survey records more than 20,000 cooked food stalls housed in hawker centres which are scattered throughout public housing estates and the central business district. Let your tastebuds lead you, be it Chinese cuisine – Cantonese, Fukien, Peking, Teochew, Hainanese, Hakka, Szechuan – or Malay, Indian, Thai, Korean, Japanese, Vietnamese, or even Western cuisine – Italian, French, English, Swiss, American… and don't leave Singapore without experiencing the very popular local dish, chilli crab!

PARKS & RESERVES

Singapore has more than bright lights, skyscrapers and shopping centres. If you prefer to get away from the hustle and bustle, or just want to soak in the quiet soothing clean environment, here are a few not-so-trodden remnants of nature to immerse yourself in.

BUKIT TIMAH NATURE RESERVE

Location: In the geographic centre of Singapore, about 10 km from Orchard Road (the principal hotel district). Open all year.

Size: 0.75 sq km reserve and primary forest, to the east lie about 25 sq km of secondary forest.

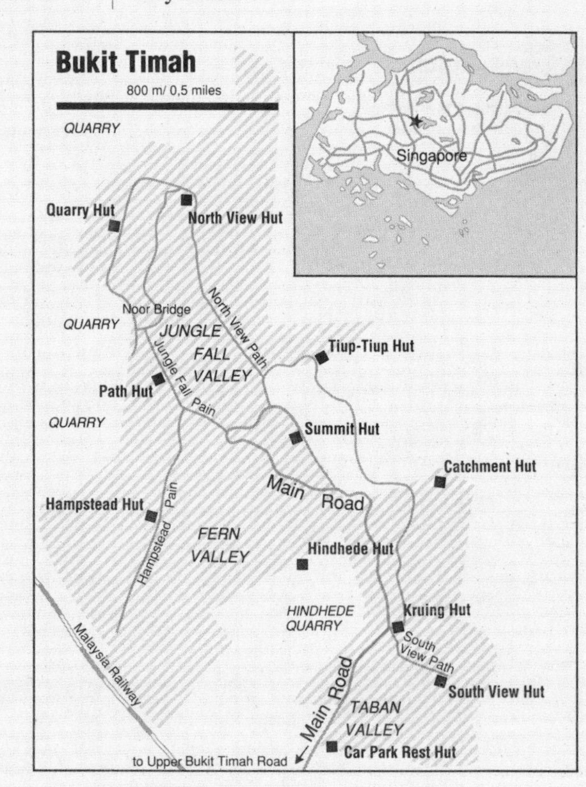

Access (car, taxi): Taxi drivers normally don't know the reserve by name and tend to confuse it with Bukit Timah district. Alight at Courts Furniture, Upper Bukit Timah Road. Cross the road and walk the last 400 metres towards the hill.

Taxi fare: S$9 from Orchard Road.

Public transportation: From Scotts Road, SBS bus service no. 171 or 173. Alight at Bukit Timah Shopping Centre.

Registration: Not necessary.

Accommodation/Food: No accommodation or food in the reserve.

Transport in park: None.

Trails: The park has a broad entrance and an extensive network of small trails.

SUNGEI BULOH BIRD SANCTUARY

Location: On northwest coast of Singapore. Open all year.

Size: 85 hectares of orchards, ponds and mangroves.

Public Transportation: SBS bus service nos. 172 and 206 end their journeys at Lim Chu Kang Bus Terminal. Enter the sanctuary along a track opposite the canteen at the terminal. In 1991, a visitor centre will connect the reserve's eastern boundary with Kranji dam.

Food/Transport: None in the sanctuary.

SINGAPORE BOTANIC GARDENS

Location: Lies just outside Orchard Road. Open all year: weekdays, 5 a.m. to 11 p.m.; Saturdays, Sundays and Public Holidays, 5 a.m. to midnight.

Size: 23 hectares

Public Transportation: Take any of SBS bus service nos. 7, 14, 105, 106 or 174 from Orchard Boulevard. As bus turns into Napier Road, alight at the second stop along the road. You will easily notice the entrance gate across the road.

Food: Opposite the entrance is a hawker centre. Inside the gardens, there is a cafeteria which serves light snacks and drinks.

Trails: There are well-paved walkways throughout the gardens.

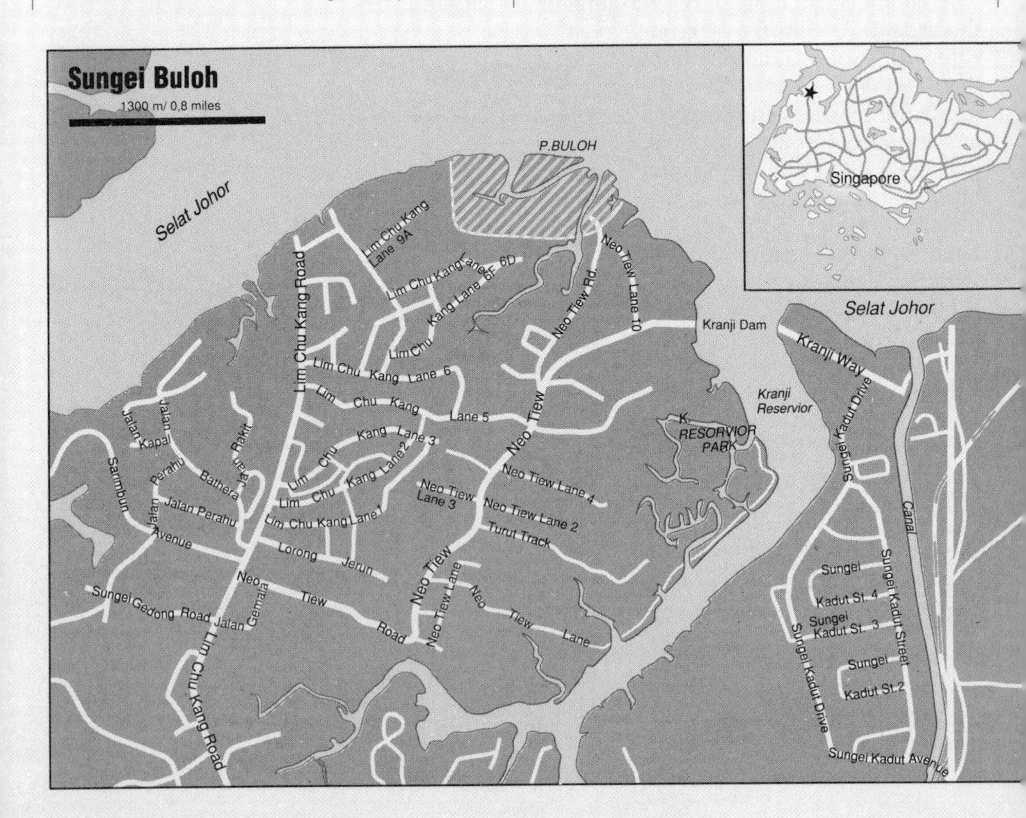

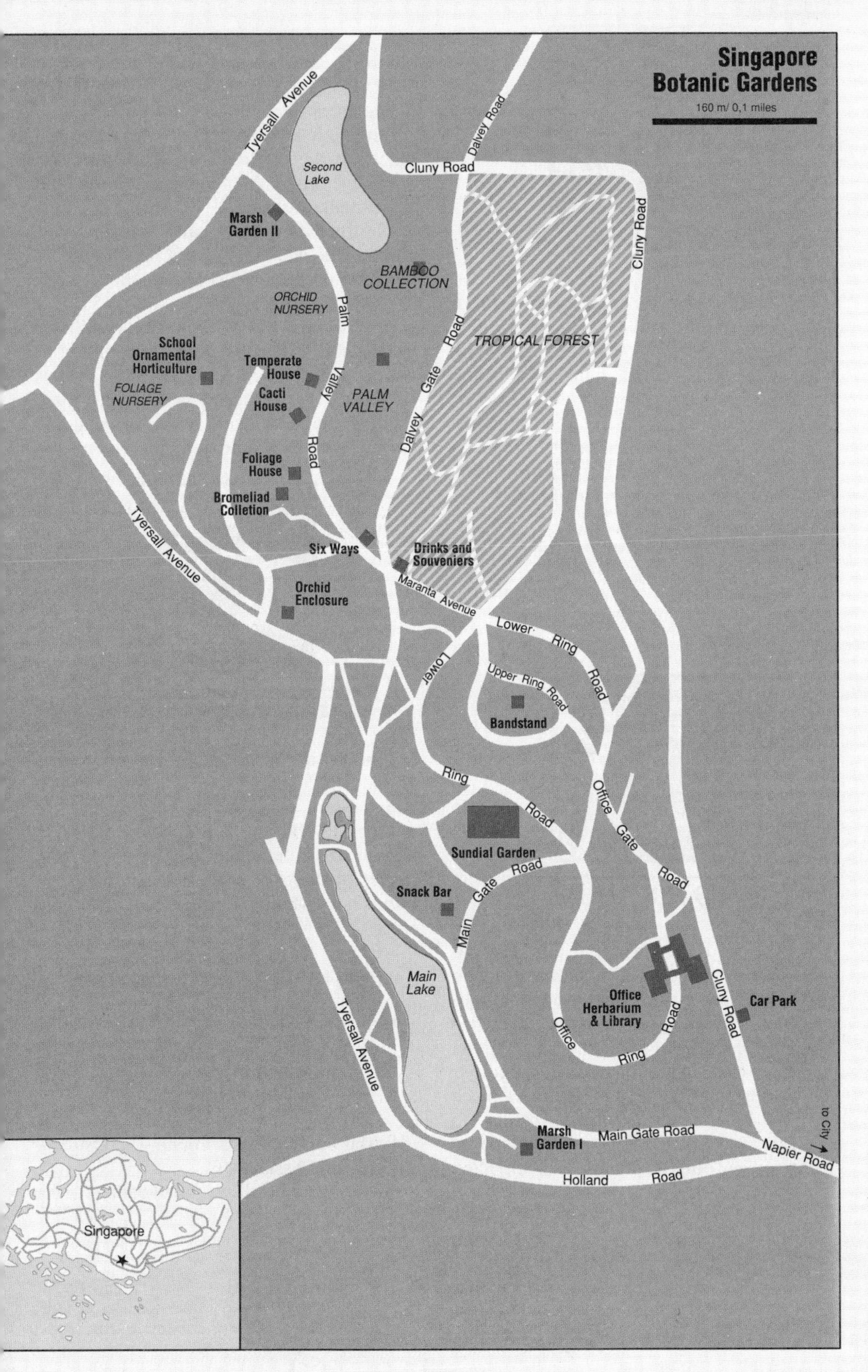
Singapore
Botanic Gardens
160 m/ 0,1 miles
Tyersall Avenue
Second Lake
Cluny Road
Dalvey Road
Marsh Garden II
BAMBOO COLLECTION
Cluny Road
ORCHID NURSERY
Palm Valley Road
Dalvey Gate Road
TROPICAL FOREST
School Ornamental Horticulture
Temperate House
FOLIAGE NURSERY
Cacti House
PALM VALLEY
Foliage House
Bromeliad Colletion
Six Ways
Drinks and Souveniers
Tyersall Avenue
Orchid Enclosure
Maranta Avenue
Lower Ring Road
Lower
Upper Ring Road
Bandstand
Ring
Road
Office Gate Road
Sundial Garden
Main Gate Road
Snack Bar
Main Lake
Tyersall Avenue
Office Herbarium & Library
Office Ring Road
Car Park
Cluny Road
to City
Marsh Garden I
Main Gate Road
Napier Road
Holland Road
Singapore

SPORTS

The enthusiastic sports person will find facilities catering to almost every form of organised physical exertion. Check with your hotel. Many hotels here have fitness centres with saunas, weights and aerobics, and even tennis and squash courts. Staying at one of the bigger establishments will often permit a guest to play golf or to swim at some of the island's exclusive "members only" country clubs.

If not, Singapore has facilities for badminton, bowling, canoeing, cricket, cycling, flying, golf, horseracing, horseback riding, squash, tennis, waterskiing, windsurfing even ice-skating.

PHOTOGRAPHY

In tropical places better results are obtained by protecting your camera and film from excessive exposure to heat and humidity. Do not leave a camera in the sun or in a hot car because heat gives the film a green overtone. Use a haze filter for better colour and store your equipment in a dry cool place such as an air-conditioned hotel room. Do not leave unexposed films in luggage as incoming and outgoing baggage is subjected to a liberal use of X-ray. For the same reason do not mail unprocessed films.

Some oldtimers believe that a photograph may carry away one's soul and will protest if someone tries to invade their world with a snapshot. In this case, lies in a telephoto lens; the scene is more spontaneous that way, too.

Processing of both monochrome and colour films can be done within the hour, if urgent, and in several weeks for Kodachrome.

Singapore is undoubtedly one of the best places in the world in which to purchase leading-brands of cameras, lenses and equipment duty-free.

Members of clubs and associations abroad who wish to give talks on Singapore upon their return, can complement the story of their trip with duplicate slides ($50 per slide) from the STPB photo library. Photographs portraying all sightseeing attractions which appear in brochures, plus various subjects in the Singapore scene – pretty girls, orchids, National Day, and a many-in-one culture – are available at their office.

USEFUL ADDRESSES

The Yellow Pages of Singapore's Telephone Directory is a first-rate listing of just about every product and service in Singapore. Its Buying Guide comes with a handy cross-referenced fast-find index, with names of items of interest to tourist translated into Malay, French, Spanish and Dutch.

USEFUL TELEPHONE NUMBERS

Fire, Ambulance	995
Police	999
Flight Information	542-5680
Meterological (forecast) Office	542-7788
Postal Service Department	533-0234
Directory Assistance	103
Assistance in Calling (local calls)	100
Time Announcing Service	1711
Singapore Bus Service	287-2727
Singapore Tourist Promotion Board	235-6611
Overseas Call Booking	104

GETTING THERE

BY AIR

Coming from outside the Indonesian archipelago you have two main airports at Cengkareng 20 km (12.5 miles) west of Jakarta or through Ngurah Rai Airport near Denpasar on the neighbouring island of Bali with connecting flights to Yogyakarta. Airport tax on international flight departures is Rp 11,000. Halim international airport will continue to serve only special flights including government and state guests as well as *hajis*, and some connecting flights to Java and Kalimantan. A highway links Cengkareng with Jakarta and until a railway system is completed buses will operate at regular intervals to Gambir – itself the location of a railway station and only a few minutes by road from the city.

A majority of visitors arrive in Jakarta from Singapore. Garuda and Singapore Airlines have five to eight flights daily from Singapore to Jakarta at between US$170 and $190 for round-trip, one-month excursion fares. A one-month excursion fare is available from Singapore to Bali with stops in Jakarta and Yogyakarta for approximately US$300.

BY SEA

If you're one of the lucky ones with plenty of time (and money), an ocean cruise to Indonesia should not be missed. Luxury cruise lines offer fly/cruise arrangements which allow you to fly to Bali and other ports where you can play in the sun, then catch your ship on the way home or vice versa. For those seeking elegance, a few ships offer cruises to Indonesia. The *Pearl of Scandinavia*, run by Mansfield Travel in Singapore (tel: 737-9688) offers a 14-day Indonesian Islands cruise, which begins and ends in Singapore, to Penang, Belawan, Sibolga, Nias, Jakarta, Padang Bay, Bali and Surabaya. There are many discount fares when three or more people travel together and you won't just be spending long hours staring at your friends; interesting scholars, artists, writers, historians and diplomats sail as guest lecturers and travel companions.

If you like to travel with the footloose budget travellers, you can hop on a motor launch leaving Finger Pier, Prince Edward Road in Singapore for Tanjung Pinang where you can catch the *KM Tampomas*, flagship for the Indonesian National Pelni Lines which sails weekly to Medan or Jakarta. Ticketing agents will confirm the current schedules and fares.

There are many motor launches to Tanjung Pinang, one at 8.30 a.m. but none later than noon. Check at Finger Pier in the morning for schedules and tickets. The 5 to 6 hour ride costs S$65 (US$30) but a S$75 (US$35) fast boat is available. Intra Express Pte. Ltd. next to the Garuda ticketing office in United Square (tel: 254-0914) also sells tickets. It is advisable to leave Singapore two days before the *Tampomas* departs and spend time on Tanjung Pinang. *Tampomas* leaves for Jakarta every Saturday at 5 p.m. local time. It costs the same amount of money to buy a package fare from German Asian Travels in the Straits Trading Building, 14th floor, 9 Battery Road, Singapore (tel: 221-5539) which includes the boat ride from Singapore, the transfer from Tanjung Pinang to the *KM Tampomas* by sampan and accommodation on board according to the class booked. Food and drinks can be purchased on board but may be costly. You might want to pack your own food. It is an 'unforgettable' two-day trip across the Java Sea aboard a crowded ship with primitive sanitation facilities and it's recommended only for the hardy. The ship arrives in **Port Tanjung Priok, Jakarta** on Monday at 5 a.m. and turns around to head back ready for embarkation at 6 a.m. Cabins must be booked one to two weeks in advance. Deck class can be obtained at short notice. (See "Travel Essentials" for custom and health formalities).

Several other big shipping companies run ships both big and small, in and out of the hundreds of ports in Indonesia. However, most of them carry cargo with limited space for passengers and are less accommodating than the *KM Tampomas*. Check with the harbourmaster for prices. It's often cheaper to go direct to the captain himself and pay for your fare.

TRAVEL ESSENTIALS

PASSPORTS & VISAS

All travellers to Indonesia must be in possession of a passport valid for at least six months after arrival and with proof (tickets) of onward passage.

Visas have been waived for nationals of 30 countries for a visit not exceeding two months. Those countries are: Australia, Austria, Belgium, Brunei, Canada, Denmark, Finland, France, Greece, Iceland, Ireland, Italy, Japan, Liechtenstein, Luxembourg, Malaysia, Malta, Netherlands, New Zealand, Norway, Philippines, Singapore, South Korea, Spain, Sweden, Switzerland, Thailand, United Kingdom, United States of America and Germany.

Entry and exit must be through the airports or seaports of Jakarta, Bali, Medan, Manado, Biak, Pontianak Ambon, Batam, Surabaya; the Pekanbaru and Balikpapan airports; and the seaports at Semarang and Riau. For other ports of arrival and departure, visas are required.

Visas are free also for registered delegates attending a conference which has received official approval. Taiwan passport holders are also given visa free entry but only at the airports in Jakarta, Medan and Bali.

For citizens of countries other than the 30 listed above, tourist visas can be obtained from any Indonesian Embassy or consulate. Two photographs are required and a small fee is charged.

Each visitor is required to pay an airport tax of Rp 11,000 for international departures; and between Rp 800 and Rp 3,000 for domestic flights depending on the airport of departure.

Surat Jalan: A *surat jalan* is a letter from the police permitting the bearer to go to certain places. It is advisable to carry one when travelling in some of the outer islands, but in Java only in such out-of-the-way places as the Ijen plateau. If in doubt check with a good travel agent. In Jakarta a *surat jalan* may be obtained in an hour or two at Police Headquarters (Markas Besar Kepolisian Republik Indonesia) in Jalan Trunojoyo (Kebayoran Baru).

MONEY MATTERS

The exchange rate for a US$1 was about Rp 1,835 at time of press. It is advisable not to exchange large sums of money if you plan to be in Indonesia for more than a month.

Changing Money: Foreign currency, in banknotes and traveller's checks, is best exchanged at major banks or leading hotels (though hotel rates are slightly less favourable than bank rates). There are also limited numbers of registered money changers, but avoid unauthorised changers who operate illegally. Banks in many smaller towns are not necessarily conversant with all foreign banknotes, so it is advisable to change most currencies in the cities. Your *rupiah* may be freely converted to foreign currencies when you are leaving the country.

Traveller's cheques: Traveller's cheques are a mixed blessing. Major hotels, banks and some shops will accept them, but even in the cities it can take a long time to collect your money (in small towns, it is impossible). The US dollar is recommended for traveller's cheques. Credit cards are usable if you stay in the big hotels. International airline offices, a few big city restaurants and art shops will accept them, but they are useless elsewhere.

HEALTH

Yellow fever vaccinations is required if you arrive within six days of leaving or passing through an infected area. It is also advisable to be vaccinated against cholera, typhoid and paratyphoid.

If you intend staying in Indonesia for sometime, particularly outside of the big cities, gammaglobulin injections are recommended; they won't stop hepatitis, but many physicians believe that the risk of infection is greatly reduced. Diarrhoea may be a problem: it can be prevented by a daily dose of Doxycycline, an antibiotic used to prevent "traveller diarrhoea". Obtain this from your doctor at home. At the first signs of stomach discomfort, try a diet of hot tea and a little patience. Stomach upsets are often a reaction to a change in food and environment. Proprietary brands of tablets such as Lomotil and Imodium are invaluable cures. A supply of malaria suppressant tablets is also highly recommended. Make sure the suppressants are effective against all the strains of malaria. It was discovered that a malaria strain was resistant to the usual kind of malarial prophylactic (chloroquine). Consult your physician.

All water, including well water, municipal water and water used for making ice, MUST be made safe before consumption. Bringing water to a rolling boil for 10 minutes is an effective method. Iodine (Globoline) and chlorine (Halazone) may also be used to make water potable. All fruit should be carefully peeled before eaten and no raw vegetables should be eaten.

Last but not least, protect yourself against the sun. Tanning oils and creams are expensive in Indonesia, so bring your own.

GETTING AROUND

DOMESTIC AIR TRAVEL

Indonesia, for those who can afford it, is aviation country. The national carrier, Garuda, serves both international and domestic routes. It is the only carrier using jet airplanes on domestic routes. Garuda has several flights daily from Jakarta to Bali, Medan, Ujung Pandang, Manado, Balikpapan and other destinations. Shuttle flights run to Surabaya, Semarang, Bandung and Bandar Lampung.

Merpati also offers regular services to 100 destinations within Indonesia. Of special interest are the "pioneer flights" to remote destinations not served by other airlines. Merpati is particularly active in eastern Indonesia, serving the smaller islands and interiors of Sulawesi, Kalimantan and Irian Jaya. Besides Garuda and Merpati, there are also several privately owned airlines with both scheduled and charter services. Those with scheduled services include Bouraq, Mandala and Sempati.

You will most probably be using the domestic air network in Indonesia. **Garuda**, **Merpati** and **Bouraq** are the principal carriers, between them covering a veritable labyrinth of destinations. If you intend travelling by plane frequently you should obtain a copy of the timetables and latest fares from the respective company. With the information, a seemingly endless variety of itineraries can be planned. The addresses of the three domestic carriers are:

Garuda Indonesian Airways, Head Office, Jl. Merdeka Selatan 13, Jakarta 10110, Tel: 370709.

Merpati Nusantara Airlines, Jl. Angkasa 2, Jakarta, Tel: 413608.

Bouraq Indonesia Airlines, Jl. Angkasa 1-3, Jakarta, Tel: 655170.

WATER TRANSPORT

There are a few inter-Indonesian shipping routes that offer an adventurous and surprisingly inexpensive way to travel. There are two ships worth considering, though neither of them are "cruise" ships. Economy-class fares offer a viable way of spanning some of the great distances involved if you are making some extensive journeys through the archipelago.

The first one is the *KM Kerinci* (pronounced ke-rin-chee) which plies the route between Padang (west coast of Sumatra) to Jakarta and on to Ujung Pandang (Sulawesi). It sails from Padang every Monday evening at 10 p.m. and arrives in Tanjung Priok on Tuesday at 4 p.m. The second leg of the journey takes you to Ujung Pandang arriving on Thursday at 1 p.m. The boat "turns round" in Ujung Pandang leaving at 6 p.m. making the return trip to Jakarta on the same day. The 42-hour return journey gets you to Jakarta at 10 o'clock on Saturday morning. Sailing time for the final segment of the trip back to Padang is at 9 p.m. the same day and arriving in Padang on Monday at 6 a.m.

The second and truly mammoth route is the *KM Kambuna* which starts from Belawan (Medan) and then follows this itinerary: Tanjung Priok (Jakarta), Surabaya, Ujung Pandang, Balikpapan, Bitung (Manado, North Sulawesi), Balikpapan, Ujung Pandang, Surabaya, Tanjung Priok and Belawan. The whole round journey takes two weeks, though you should certainly consider building one or two segments into your own itinerary.

Another boat, *KM Rinjani*, goes all the way to Bau Bau (an island off south-east Sulawesi) and on to Ambon and Sorong. Both the *Kerinci* and the *Kambuna* are operated by the national shipping company **PELNI**. They can be contacted in Jakarta **PT PELNI**, Jl. Angkasa 18, Jakarta, tel: 417569 or 415428.

WHERE TO STAY

ACCOMMODATIONS

JAVA

Jakarta: Jakarta has come a long way since the 1960s, when the only international-class hotel in town was the Hotel Indonesia, built by the Japanese as a war reparation. Of the older, pre-war establishments, only the Transaera and the Royal remain, but neither is truly "colonial" in ambience.

Jakarta now has five five-star hotels. Two have extensive grounds and sports facilities: the **Borobudur Intercontinental** and the **Jakarta Hilton**. In addition to Olympic-size swimming pools, tennis courts, squash courts, health clubs, jogging tracks and spacious gardens, they also boast discos and a full complement of European and Asian restaurants. The other hotels in the same category, the **Mandarin,** the **Hyatt** and the **Sahid Jaya** are newer "city" hotels—providing central locations and emphasising superior service and excellent food.

There are another half-dozen or so first-class hotels in town. The **Sari Pacific** is centrally located and has a popular coffee shop and deli. The **Horison** has a restaurant specialising in seafood. The **President Hotel** is Japanese operated with several Japanese restaurants. And the venerable **Hotel Indonesia** has a supper club with nightly floor shows and a swimming pool garden open to the public (US$4 admission).

At the upper end of the moderate price range, the most centrally located hotels are the **Transaera** and the **Sabang Metropolitan** (about US$35 and up for a double). The **Transaera** is quiet, with older, spacious rooms and the **Monas** is known for its good service. The **Sabang Metropolitan** is convenient for business and shopping.

In Kebayoran Baru, you may opt to stay at the **Kemang** or the **Kebayoran Inn**, both popular with frequent visitors for their reasonable prices and quiet, residential surroundings. Or try the **Interhouse**, centrally located by Kebayoran's shopping district, Blok M. There are several hotels in town providing small but clean, air-conditioned rooms for around US$25. These include the **Menteng Hotel** and the **Marco Polo**, both in Menteng.

True budget travellers almost invariably stay at Jl. Jaksa No. 5 (**Wisma Delima**) or one of the other homestays on that street. In the US$3 to US$5 a night class, the **Borneo Hostel** around the corner and the **Pondok Soedibyo** are perhaps cleaner. The nearby **Bali International** costs more at US$10. At the **Royal** a double room with fan and breakfast costs US$18, and for a little bit more you can get an air-conditioned room at the **Srivijaya**.

WEST JAVA

Serang: The **Krakatau Guest-house** is open to the general public – air-conditioned motel-style bungalows renting for about US$20, a bit seedy now but inexpensive. The **Merak Beach Motel**, located right on the water just next to the Merak Bakauhuni ferry terminal at the far northwestern tip of the island, is clean and charges reasonably at US$25–US$30 a night for an air-conditioned room (tax and service included).

The most comfortable place on the west coast beaches to the south is the **Anyer Beach Motel**. Tidy little concrete bungalows set in a grove by a broad, secluded beach for US$30 on up to US$80 a night for a suite (plus 21 percent tax and service).

Farther south around the village of Carita are two somewhat more rustic seaside establishments: the **Selat Sunda Wisata Cottages**, a small resort with several air-conditioned bungalows by the shore (US$35 per night), and the larger **Carita Krakatau Beach Hotel** – US$36 per night, plus 21 percent tax and service.

Thousand Islands: To stay at Pulau Puteri or Pulau Melinjo, make prior bookings through any travel agent, or directly through: **P.T. Pulau Seribu Paradise**, Jakarta Theatre Building, Jl. M.H. Thamrin, Jakarta, Tel: 359333, 359334.

Pelabuhan Ratu: One hotel, called **Bayu Amrta,** rents small rooms perched on a cliff overlooking the sea for US$10 a night and has bungalows sleeping up to six people for only US$20. The place is in need of repairs, but the beach is beautiful and they serve very good seafood. Another place called **Karangsari** across the road is about the same price and slightly cleaner but without the view. Even the first-class **Samudra Beach Hotel** is not exorbitant considering the facilities, at US$35 per night for a standard double room (plus 21 percent tax and service). It is said they keep a room here for Loro Kidal, the goddess of the South Sea. Closer to town, the **Pondok Dewata** offers air-conditioned bungalows by the sea for US$23 and larger ones (with two bedrooms and four beds) for US$45, all inclusive.

Bandung: The only international class hotel in Bandung is the **Panghegar** (US$45 on up for a double, plus 21 percent tax and service). The **Savoy Homann** (US$35 plus 21 percent) is of a similar standard and very charming. Many of Bandung's hotels fall into the intermediate category. Several small guest-houses are in old Dutch mansions, including the **Soeti** and **Kwik's** (both about US$20 for a double).

For a bit more money, Cisitu's newer **Sangkuriang Guest-house** (US$30 a night plus 21 percent tax and service) is very pleasant—located in a residential neighbourhood just above the ITB university campus. The **Hotel Istana** is clean and reasonably priced at US$25 a night for a double (plus 21 percent), with an excellent restaurant. The **Hotel Trio** has spotlessly clean rooms, excellent service, a sumptuous breakfast, free transportation to and from the airport or train station, and free coffee and tea all for US$30 but is usually fully booked.

There are few good *losmen* for the budget traveller. The cheap hotels around the train station are rather dingy. The **Wisma Gelanggang** youth hostel is not too bad at US$1.50 a night for a dorm bed. For about US$12 you can get a decent room at the **Hotel Dago** or **Hotel Lugina.**

NORTH COAST

Cirebon: First choice in hotels is the venerable old **Grand Hotel**, with a variety of rooms, ranging in price from the huge air-conditioned President's Suite for US$50 all the way down to a small room in the back with a fan for only US$7.50. The **Patra Jasa Motel** has more modern (but tacky) rooms for between US$30 and US$70, as do the nearby **Omega** and **Cirebon Plaza**.

Budget travellers will find a variety of accommodation in the US$3 to US$7 range right next to each other all located along Jl. Siliwangi around the train station and farther down pass the Grand Hotel (**Hotel Baru**, **Hotel Familie**, **Hotel Semarang**, **Hotel Damai**, etc.). Also around the corner along the canal on Jl. Kalibaru (**Hotel Asia**).

The Pertamina Cirebon Country Club (Ciperna) located up on the hill beyond the airport, 11 km from Cirebon on the road to Kuningan, has an Olympic-size swimming pool and an 18-hole golf course. Bungalows may also be rented here for as little as US$7 a night with a nice view of the coast.

Pekalongan: Top-of-the-line is the **Hotel Nirwana**, near the bus terminal, with air-conditioned rooms for US$25 to US$35 a night, and a large new swimming pool. More conveniently located is the **Hayam Wuruk** right on the main street, with air-conditioned double rooms for only US$15 to US$20 including breakfast. The Hayam Wuruk also has pleasant double rooms with a fan for US$10.50, singles for US$8 – breakfast, tax and service included. Cheaper accommodation across from the train station at the western edge of town include the **Istana**, the **Gajah Mada** and the **Ramayana.**

Semarang: Semarang's best hotel is the **Patra Jasa**, located up on a hill in Candi Baru overlooking the city (US$35 to US$60 for a double, plus 21 percent tax and service). The **Candi Baru Hotel**, closer to the city but still in the hills, has spacious rooms with a view for less money (US$17 to US$28 for a double; US$40 for a suite), and in the same area, the **Green Guest-house** is a bargain at US$14 to US$18 for an air-conditioned double with breakfast (tax and service included).

Down in the centre of town, the best hotel is the **Metro Grand Park** (doubles are US$35 to US$45 plus 21 percent). The old Dutch hostelry, the **Dibya Puri**, is just across a busy intersection from here – only US$25 for an air-conditioned double, with breakfast (tax and service included), but the place is looking (and smelling) a bit wilted these days, the **Queen Hotel** around the corner on Jl. Gajah Mada is a newer place and about the same price.

Budget travellers can check out some of the hotels on Jl. Imam Bonjol around the train station, like the **Dewa Asia**, the **Tanjung** and the **Singapore**, all with rooms in the US$5 to US$10 range. The **Nam Yon Hotel**, right in the middle of the Chinatown district, also has clean rooms for as little as US$4.50 a night including breakfast – air-conditioning for only US$14 a night (tax and service included).

YOGYAKARTA

Yogya has a room for everyone, from the US$350-a-night presidential suite at the **Ambarrukmo** to the dollar-a-night closets at "Home Sweet Homestay" on Gang Sosrowijayan I.

The 4-star **Ambarrukmo Hotel**, built by the Japanese in the early 1960s, is still the only international-class luxury hotel in Yogya, with rooms going for US$65 on up (plus 21 percent tax and service). It is symbolically situated some miles to the east of town near the airport upon the grounds of the old royal *pesanggrahan* or rest house once used to entertain visiting dignitaries to the court. Some of the old buildings are still standing, including the elegant *pendapa* and the *dalem agung* ceremonial chambers.

The old **Hotel Garuda** (US$35 to US$50 a night) right on Malioboro has just added a modern seven-storey wing at the back, and has upgraded their spacious colonial suites (huge rooms and bathrooms, with high ceilings and an outer sitting-room/balcony looking out onto a central courtyard). The hotel has quite a history, as it housed several government ministries during the Indonesian revolution (1946-49).

The **Arjuna Plaza** and the **New Batik Palace hotels** are centrally located on Jl. Mangkubumi (US$25 to US$30 for a double, plus 21 percent). The **Gajah Mada Guesthouse**, with air-conditioned doubles for US$24 is a quiet place located on campus in the north of town. Mrs. Sardjito, the widow of Gajah Mada University's first rector also rents rooms at her elegant home on Jl. Cik Ditiro, opposite the Indraloka office.

Many other small hotels and guesthouses cluster along **Jl. Prawirotaman** in the south of Yogya. A few of these have air-conditioned rooms in the US$15 to US$25 range (plus 21 percent), including breakfast. Try the **Airlangga** or the **Duta**.

The guest-houses along **Jl. Prawirotaman** are all converted homes – generally quiet, clean and comfortable. Cheapest rate available here for a double is US$7.50, including tax, service and breakfast; but most are in the US$10-to-US$12 a night range. Some also have air-conditioned doubles for only US$15 a night.

The many small hotels around **Jl. Pasar Kembang** (also on Jl. Sosrowijayan and down the small lanes in between), are substantially cheaper and more central, but this is not a pleasant area. Many places here have rooms for US$3 to US$5 and even less. Try the **Kota** down at the end of Jl. Pasar Kembang – very clean.

There is even one agency, Indraloka, that will place you in the home of an English or Dutch-speaking family where you share home-cooked meals and enjoy the warm hospitality of the Javanese. **Indraloka Homestay Service**, founded and run by Mrs. B. Moerdiyono, currently costs US$21 a night for a double (plus 21 percent tax and service) including breakfast. Home-cooked lunch or dinner is an additional US$6. The families are mostly headed by Dutch educated professionals (doctors and university lecturers), and the rooms have all the western amenities and a fan. Mrs. Moerdiyono also arranges tours through Java to Bali, using her network of homestays in other cities. Write to her at Jl. Cik Ditiro 14, Yogyakarta, Tel: 0274 or 3614.

EAST JAVA

Surabaya: Hyatt Bumi (US$75 a night on up, plus 21 percent tax and service) is the only four-star luxury hotel. The **Simpang**, at the corner of Jl. Tunjungan and Jl. Pemuda, costs US$64 a night (plus 21 percent). The **Mirama** and the **Ramayana** just to the south are in US$50 to US$60 range (plus 21 percent), and the **Elmi** and the **Garden**, also in the same area, have rooms for a bit less.

The older **Majapahit Hotel** on Jl. Tunjungan (formerly the "Oranje" built in 1910) is something of a historical monument. It is the site of the famous "flag incident" that sparked off the revolutionary battle for Surabaya. Air-conditioned rooms for US$40, non-air-conditioned ones for US$24, plus 21 percent).

The **Sarkies** across the street and down Jl. Embong Malang is another older hotel owned by the Majapahit, with air-conditioned rooms for US$25 to US$30. Or try the **Royal** and Olympic for around US$20.

Budget travellers always stay at the **Bamboe Denn/Transito Inn**, with dorm beds for US$1, singles for US$2 and doubles for US$3. They have lots of travel information here to help you get around and they serve cheap breakfasts and snacks. For a bit more (US$6 to US$8) try **Wisma Ganeca** near Gubeng Station.

BALI

Sanur: There are so many excellent first-class hotels in Sanur, that you can scarcely go wrong. The main choice is between the convenience and luxury of a big four-star hotel (there are three: the **Bali Beach**, the **Bali Hyatt** and the **Sanur Beach**) or the quiet and personality of a private bungalow by the sea (at two-thirds to half the price). Reservations are advisable during the peak seasons: July to September and December to January.

For such a luxury hotel, the new **Bali Hyatt** offers a remarkably breezy, spacious Royal Hawaiian feeling, with striking public areas, clay tennis courts and hanging Babylonian gardens. The venerable **Hotel Bali Beach** (constructed by the Japanese in the early 1960s) looks more like a traditional Miami Beach luxury hotel– a 10-storey concrete block by the sea, set amidst a golf course, bowling alleys and two swimming pools. Last but not least of the three four-star establishments, the smaller **Sanur Beach Hotel**, owned by Garuda, claims to be the friendliest large hotel in Sanur.

Of the smaller cottage resorts, the **Tanjung Sari Hotel** is the hands down choice of frequent visitors. This was one of the island's first beach-bungalow establishments and is still its most charming and efficient. The nightclub, Rumours, features backgammon and a well-stocked video loft.

The **Segara Village** deserves mention for its snazzy and congenial Indonesian atmosphere. **La Taverna** gets kudos for its Italian Balinesia and attractive beach-restaurant pizzeria. **Wisma Baruna**, the smallest and oldest first-class hotel in Sanur is also very cozy, with a superb breakfast pavilion overlooking the lagoon.

In the intermediate range, **Bali Sanur Bungalows**, at the upper end of the scale, are recommended. All of the other beach bungalow establishments in this category are excellent value and generally quite pleasant, the major consideration being whether you require air-conditioning or not (rooms at the lower end of the scale have only a fan).

Abian Irama Inn, Jl. Brig. Ngurah Rai, Sanur, Denpasar, Tel: 8415. Ten minutes from the beach. Some air-conditioned rooms. US$15 to US$36 a night.

Bali Sanur Irama Bungalows (23 rooms), Jl. Tanjungsari, Sanur, P.O. Box 306, Denpasar, Tel: 8421. The least expensive of the "Bali Sanur Bungalows" group. All with air-condition and hot water. US$30 to US$35 a night.

Cheapest room in Sanur is US$7 a night. Your best bet is the **Tourist Beach Inn**, just 100 metres from the beach. Three bungalow establishments opposite the post office – **Sanur Indah**, **Taman Sari** and **Hotel Rani** – give you a bit more space, but are farther from the beach. The **Taman Agung** is the nicest budget place, with well-kept gardens and very quiet. True budget travellers will get better value for money in Kuta.

NUSA TENGGARA

Komodo: At the PPA site in Loho Liang, there are several large and comfortable native-style cabins with a total capacity of 80 beds. Each cabin has two toilet/bath rooms and an overnight stay costs about US$4. Cheap meals of rice and fish are available.

You'll have already reported to the PPA office in Labuhanbajo or Sape. On arriving in Komodo you will need to register and pay a US$1 at the PPA office in Loho Liang. Any time you leave the PPA compound you must be accompanied by a guide (one for every three visitors) whose fee is US$2.50 per day.

Flores: There are three *losmen* in Labuhanbajo that you should try first: the **Mutiara** (7 rooms), the **Makmur** (7 rooms) and the **Komodo Jaya** (4 rooms). In Ruteng, seek out the **Wisma Sindha** (20 rooms), located in the centre of town, the **Wisma Agung** (15 rooms) at between US$7 and US$12, or the **Losmen Karya** (5 rooms) at US$2 per person per night.

SUMATRA

Medan:

The list of accommodations include:

Angkasa, Jl. Sutomo 1, Tel. 321244. Rooms up to US$15.

Danau Toba International, Jl. Imam Bonjol 7, Tel: 327000. US$45 to US$55 for single. US$50 to US$60 for double.

Garuda Plaza Hotel, Jl. Sisingamangaraja 18, Tel: 326255. Rooms between US$35 and US$40.

Polonia, Jl. Jend. Sudirman 14-18, Tel: 325300 (10 lines). Rooms between US$30 and US$45.

Dirga Surya, Jl. Imam Bonjol 6, Tel: 321244 or 325660. Rooms up to US$30.

Garuda Hotel, Jl. Sisingamangaraja 27, Tel: 22775. Rooms up to US$20.

Garuda Motel, Jl. Sisingamangaraja 7, Tel: 22760 or 51203. Rooms up to US$20.

Natour's Hotel Granada, Jl. A. Yani VI/I, Tel. 326211, 344699. Rooms up to US$18.

Pardede International, Jl. Ir. Juanda 14, Tel 32866. Rooms up to US$28.

Sumatra, Jl. Sisingamangaraja 21, Tel: 24973. Rooms up to US$12.

Waiyat, Jl. Asia, Tel: 27575 or 321683. Rooms up to US$9

FOOD DIGEST

WHAT TO EAT

Each province or area in Indonesia has its very own cuisine or specialty. As Indonesia is surrounded by sea, there is naturally an abundance of seafood. Many varieties of fish, lobsters, oysters, prawns, squids, shrimps and crabs all figure in a typical diet. Steamed, grilled or fried, they are unfailingly fresh and excellent. If you intend to eat it only once during your stay, have it in the city of Cirebon, famed for its seafood. In fact, its name means "shrimp river".

Beginning with the northern tip of Sumatra, dip into some Acehnese food. Acehnese food is displayed and served cold on many small plates, in the same way as Padang food, but some say it is more delicately spiced, with a wider range of flavours. Aceh food is usually served with steamed rice and common dishes include fish (*ikan panggang*), papaya flower salad (*sambal bunga kates*) and egg cooked in spinach (*sayur bayam*). Squares of twice-cooked black rice (*pulot hitam dua masak*) serves as sweets. Tea with honey, ginger and condensed milk (*serbat*) is an alternative.

West Sumatra is noted for its Padang food. Hot and spicy, Padang specialties are renowned. Don't leave without trying mutton brain *opor*, beef *rendang* or *gulai ayam*, guaranteed to tantalise your palate.

Javanese cuisine may be divided into four categories: Sundanese (West Javanese), Central Javanese, East Javanese and Madurese cooking. For an excellent Sundanese meal of grilled carp (*ikan mas bakar*), grilled chicken (*ayam bakar*), prawns (*udang pancet*), barbequed squid (*cumicumi bakar*) and a raw vegetable salad with shrimp paste chili sauce (*lalap/sambal cobek*) or sample *ikan mas* (gold fish), one of the popular Sunda specialties. Ask for it to be served fried (*ikan mas goreng*) or wrapped in banana leaves and steamed with spices (*ikan mas pepes*).

The Central Javanese delicacies are fried chicken and *gudeg*. Javanese chickens are farmyard chickens, allowed to run free in the village. As a result they are full of flavour but very tough in comparison with factory feed chickens in the West. The Javanese boil their chickens first in a concoction of rich spices and coconut cream for several hours, before deep frying them for about a minute at very high temperatures to crisp the outer coating.

The pilgrimage point for fried chicken lover from all over Java (and all over the world) is **Nyonya Suharti's** (also known as Ayam Goreng "Mbok Berek", after the woman who invented this famous fried chikcen recipe) located 7 km (4 miles) to the east of Yogya on the road to the airport (a short distance beyond the Ambarrukmo on the same side). Nyona Suharti's chicken is first boiled and coated in spices and coconut, then fried crisp and served with a sweet chili sauce and rice. Excellent when accompanied by pungent *petai* beans and raw cabbage. Indonesians patronise the place in droves, and you can see Jakartans in the airport lounge clutching their take-away boxes of the special chicken for friends and family back home.

Gudeg is the specialty of Yogyakarta, consisting of rice with boiled young jackfruit (*nangka muda*), a piece of chicken, egg, cocount cream gray and spicy sauce with boiled buffalo hide (*sambal kulit*).

East Java and Madura are known for their soups and their *sate*. Try the *soto madura* (spicy chicken broth with noodles or rice), chicken or mutton *sate* (barbequed meat skewers). *Sate* is usually served with *longtong* (boiled rice, stuffed in banana leaves) and fragrant peanut sauce.

PARKS & RESERVES

MOUNT LEUSER NATIONAL PARK

Location: 100 km west of Medan, Sumatra, Indonesia. Open all year.

Size: 900 sq km

Access: By taxi or public bus to village of Bukit Lawang (95 km). There are several direct buses to the village where the visitor can register, or change buses at Binjai to Medan at 5.30 a.m.; a bus leaves for Brastagi via Medan.

Taxi fare: Rp 60,000 (1-4 persons).

Public transportation: Rp 4,000 (one way).

Registration: Obtain a permit from

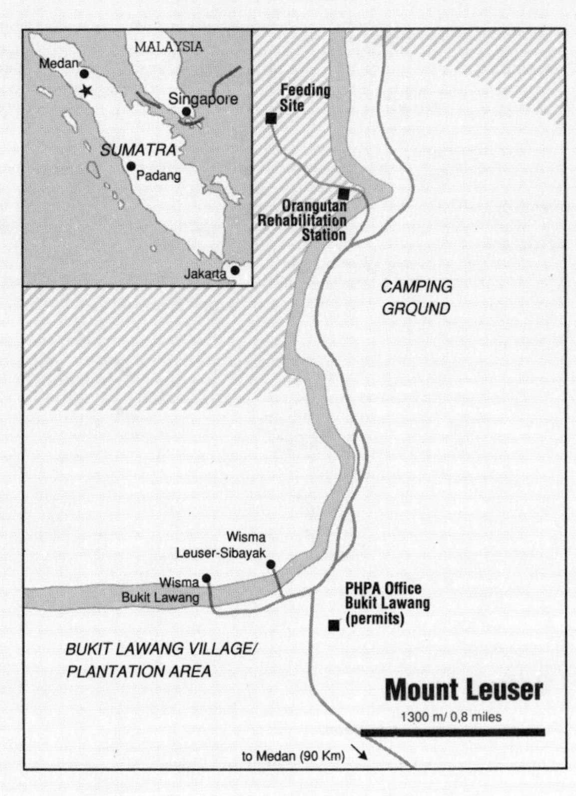

PHPA headquarters Jl Juandang Boga or PHPA Bukit Lawang Rp 3,000.

Accommodation: In Medan or Danau, try the Toba International, Jl. Imam Bonjol 7. Tel. 327 000. Less expensive accommodations are also available at Wisma Leuser-Sibayak and Wisma Bukit-Lawang).

Camping: Available in the park which has a shelter and two public toilets.

Food: Simple meals at either the guesthouses or the *warungs* in the village.

Transport in park: Jungle walks on foot.

Trails: Many.

Further advice: Visitors *must* be accompanied by a PHPA park ranger for jungle walks. Park guide services can be arranged with the officer at park headquarters in Bukit Lawang.

BERBAK WILDLIFE RESERVE

Location: 100 km east of Jambi, Sumatra, Indonesia. Open all year but access restricted during December and January.

Size: 1,900 sq km

Access: From Jambi by speedboat to Nipah Panjang at the River Berbak estuary. From this village, visitors either continue with a speedboat trip to village Air Hitam Laut a board the vessel of the PHPA. If by boat it would take about 25 hours. During the months of December and January, access to the reserve by sea may be dangerous because of the prevailing monsoons.

Fares: From Jambi to Nipah Panjang by public speedboat at Rp 10,000 per person, if speed boat needs to be chartered, at Rp 100,000.

Public transportation: From Nipah Panjang to Air Hitam Laut by chartered speedboat; prices normally range from Rp 150,000 (two days) to Rp 400,000 (one week) depending on your negotiation skills. A basic understanding of Bahasa Indonesia is essential!

Registration: Get permit from PHPA headquarters, Boga or at Kantar Sub Balai Konsenasi Sumbur Dya Alam Tenaipura, Jambi.

Accommodation: Primitive accommodation can be arranged by the Reserve superintendent at his office.

Bring sleeping mat, mosquito net and

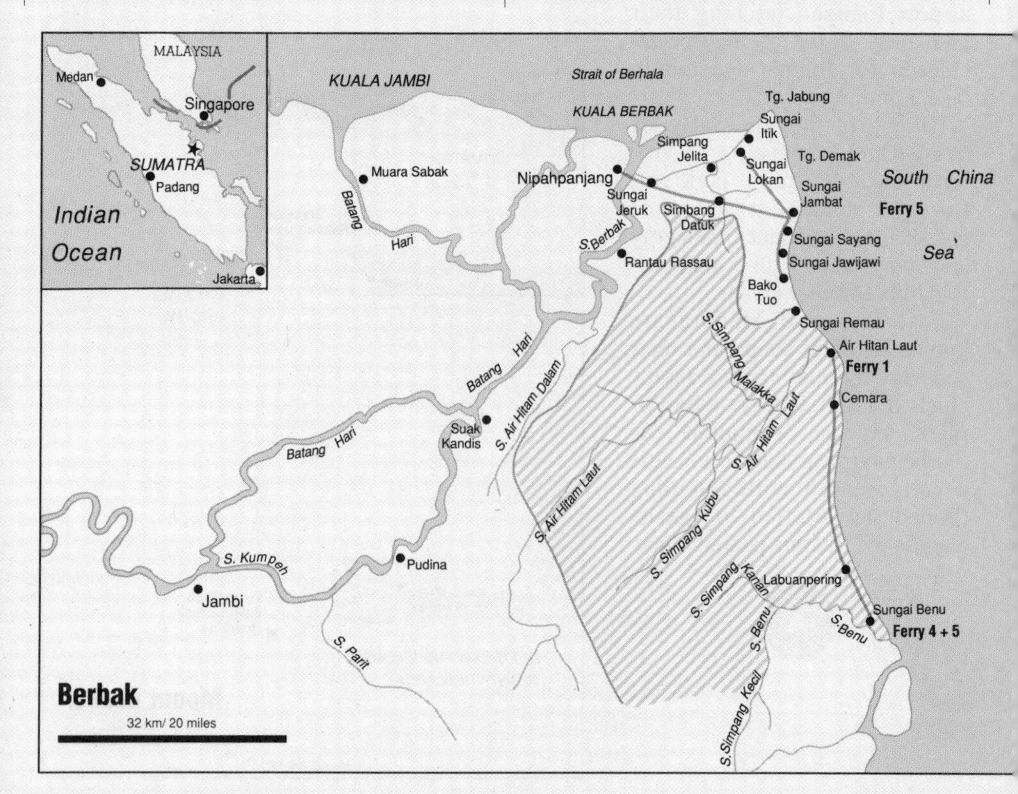

insect repellent. The same kit applies for accommodation at Nipah Panjang.

Camping: Jambi has many hotels in various categories.

Food: A number of simple *warungs* (restaurant) are available at Nipah Panjang and Air Hitam.

Transport in park: By boat. During the dry season, when the forest is not flooded, short walks are possible in the jungle. PHPA guide is obligatory and you need to pay him a daily fee Rp 5,000 for his services.

Trails: None.

SEMBILANG

Location: 100 km North of Palembang. Open all year but access restricted during December and January.

Size: 3,800 sq km

Access: By public speedboat to Sungsang 85 km down river on the Musi. From this village, rent a speedboat to explore the great many tidal mangroves of the Sembilang area.

Fare: Speedboat at Rp 6,000 one-way Palembang-Sungsang.

Public transportation: If speedboat has to be chartered from Palembang, at Rp 150,000 for a 2-day visit. Charges rise to Rp 900,000 for five days.

Registration: Obtain permit at PHPA headquarters, Boga or Sut Balai Kansenasi Sumbu Daya Alam (KSDA), Jl. H. Balian Kmb, Palembang 25977.

Accommodation: Palembang has a wide range of hotels to offer (Swarna Dwipa) Rp 50,000 for doubles.

Camping: At Sungsang stay with Bpk Nanang Zen who provides basic accommodation for Rp 2,000 and has a restaurant which offers great seafood. Try his shrimps and cockles specialty!

Transport in parks: By boat only.

Trails: None.

Further advice: A basic understanding of Bahasa Indonesia is essential. Bring plenty of sun lotion, insect repellent.

Visitors must be accompanied by a PHPA guide during their visit. Services of these guides can be arranged through the KSDA office, Palembang.

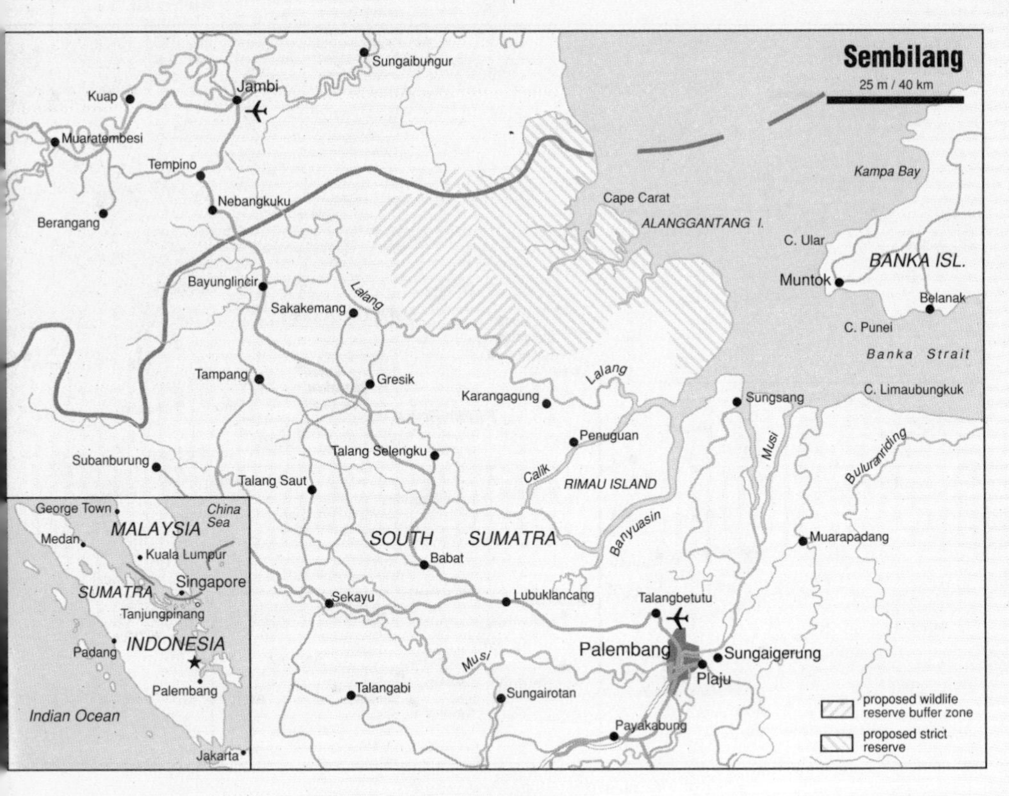

UJUNG KULON

Location: 200 km west of Jakarta; western most tip of Java. Open all year but the sea can be rough from October March.

Size: 786 sq km

Access: By road on a minibus from Jakarta to Labuan, 3-4 hours. Then charter a boat from Labuan to main guesthouse in Peucang, 5-6 hours. Fares vary.

By boat: Labuan-Krakatau, 4 hours; Peucang-Krakarau, 6 hrs.

Public transportation: Take ferry from Labuan-Sumur, charter motorbike to Taman Jaya 5, then walk into park.

Registration: Guesthouses/bungalows should be booked in advance to avoid disappointment. Contact PHPA, Taman Nasional Ujung Kulon, Jl. Ir. H. Juanda 43, Labuan. Tel. Labuan 42.

Camping: Possible but no facilities.

Food: None; bring your own food in.

Transport in park: None.

Trails: There is a well defined trail right round the peninsula and running across it are a few others. A ranger will accompany you for a fee.

Further advice: The best period to visit the park is during April-September as rough seas make it inaccessible in the wet season.

MOUNT GEDE-PANGRANGO (INCLUDES CIBODAS BOTANIC GARDENS)

Location: 100 km south of Jakarta. Closed from December to March depending on weather conditions.

Size: 150 sq km

Access: By bus/car towards Bandung on the toll-road from Jakarta. Through Cipanas, then take a turn to the right in Cimacen village. The sign is on the left.

Taxi fare: Cars can be hired in Jakarta for US$40 per day.

Registration: At park entrance.

Accommodation: The Cibodas guesthouse has 5 rooms sleeping 2-3 people each and costs $17 per room. Tel. 0255 2233.

Camping: Possible in many parts of the park. Need to pay a nominal permit fee.

Food: Food-stalls at the entrance to the park; otherwise bring your own.

Transport in park: None.

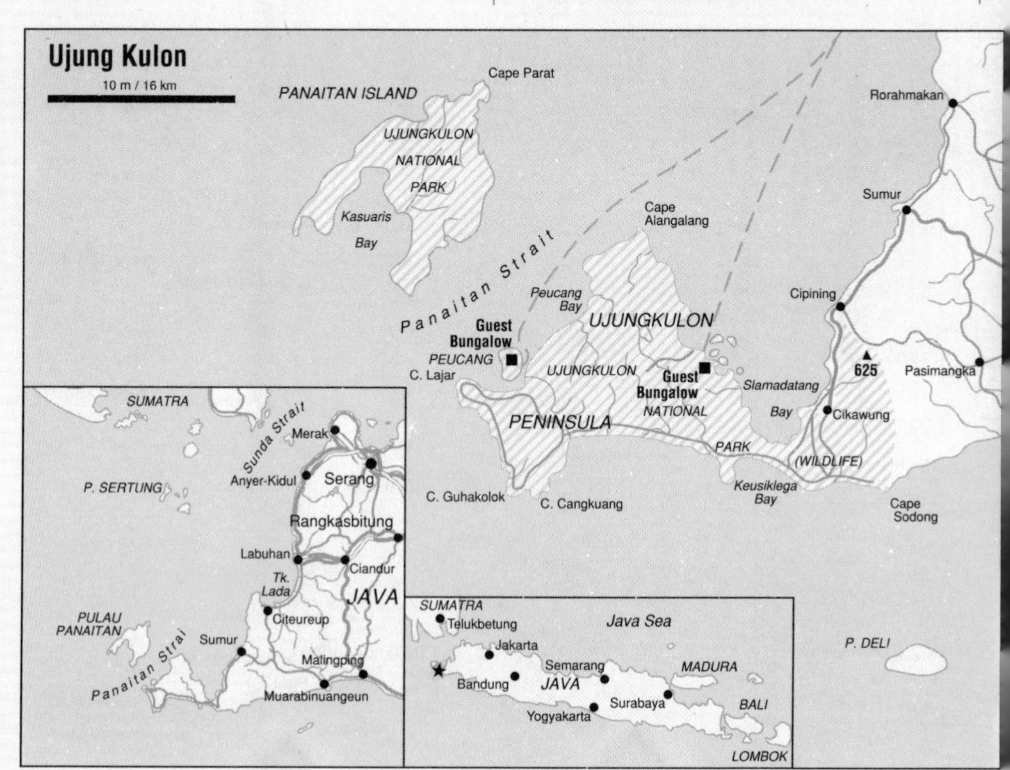

Trails: Good, leading through the forest and on the plateau via the volcanic peaks. In fact, this park is one of the few in Indonesia where you don't actually need the services of a guide!

Further advice: Avoid weekends if possible – the Park is very popular with day-trappers from Jakarta and gets very crowded.

KERINCI-SEBALT NATURE RESERVE (PROPOSED NATIONAL PARK)

Size: 7,500 ha. part of the Bukit Bavisan Mountain chain, of the provinces West Sumatra, Jambi, Bangkulu and southern Sumatra. Open all year.

Access: Bus operates from Padang to Sungai Penuh via Rainan, 8 hours; Jambi to Sungai Penuh, 18 hours. Air services company Merpati arranges regular flights Jambi-Sungai Penuh, at least twice a week. However, unfavourable weather conditions such as turbulence and fog may disrupt the flight schedule.

Public transportation: Travelling by road B. Kerinci-Seblat is an advantage, as most of the other roads are in poor condition. There are several ways to get to Sungai Penuh. By bus, a one-way trip costs Rp 12,000. The taxi fare, one-way from Padang, is Rp 100,000. By air, one-way from Jambi, the flight costs Rp. 85,000.

Registration: Through PHPA headquarters Boga or Jub Balai Konsenasi Jumbu Drya Nam at either Jambi or Padang.

Accommodation: Good hotels located in Padang and Jambi. When in Padang, try staying at Mariani Int., Jln Bundo Kandung, US$15. Tel. 22020/22634.

Camping: At Sungai Penuh.

Food: Many restaurants with simple meals in Sungai Penuh.

Transport in park: On foot or by chartered local minibuses.

Trails: Running from Sungai Penuh to Mount Kerini summit, and to Danau Mount Tujab; from Tambang Samah to Seblot and other mountains.

Further advice: Several tracks follow rivers, such as Air Rupit, Air Seblat and Air Tebat Pelapo. Visitors *must* be accompanied by a park officer.

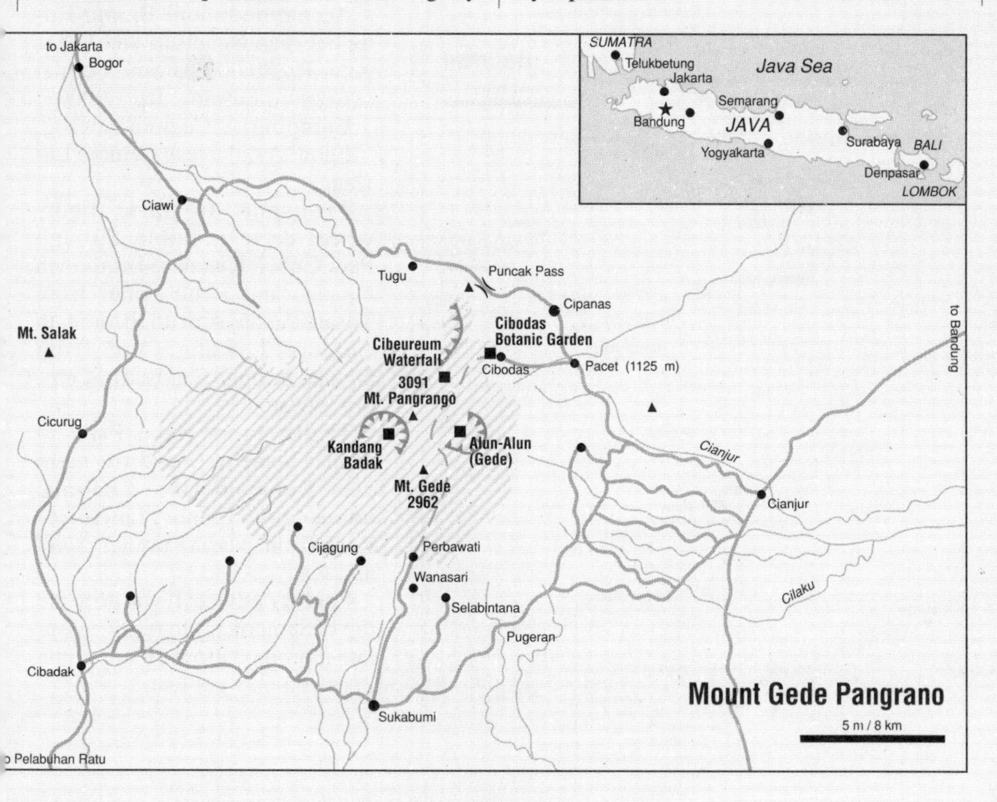

BALURAN NATIONAL PARK

Location: 250 km east of Surabaya, 22 km of ferry station Bali-Java, 166 km west of Denpasar Bali, 32 km north of Banyuwangi/ railway station. Open all year.

Size: 260 sq km

Access (car, taxi): Car rentals are available, with or without driver Rp 70,000 a day. Or hire a motorcycle for Rp 50,000. From Surabaya to Baluran, about 6 hours; from Denpasar, 5 hours. .

Taxi fare: From Denpasar Rp 75.000.

Public Transportation: Bus or train plying Surabaya-Baluran, 8 hours. Bus from Denpasar; ferry operates along the route Gilimanuk–Banyuwangi–Baluran (towards Surabaya), 5 hours. It is advisable to change buses at ferry station.

Registration: At headquarters, south end of park at provincial road Wonorejo. Contact Head of Baluran Park, Jl. Sudirman 108, Banyuwangi Tel. 0333- 41119.

Accommodation: Rp 65,000 in Banyuwangi at 30 km; Rp 15,000 in the park – very basic amenities.

Camping: In park at Wonorejo.

Food: In Banyuwangi; also a few small restaurants in Wonorejo; no food in the park.

Trails: 4 km to Talpat and another 5 km into volcano crater along the Kacip river about 18 km along East coast from Bamah to Karang teko.

Further advice: Combine visit to Baluran with brief tours to nearby parks Meru Betiri, Alas Purwo and Ijen crater lake, taking up half a day only. Also, do not miss Bali Barat Park nearby. Banyuwangi is the most central town for all these parks.

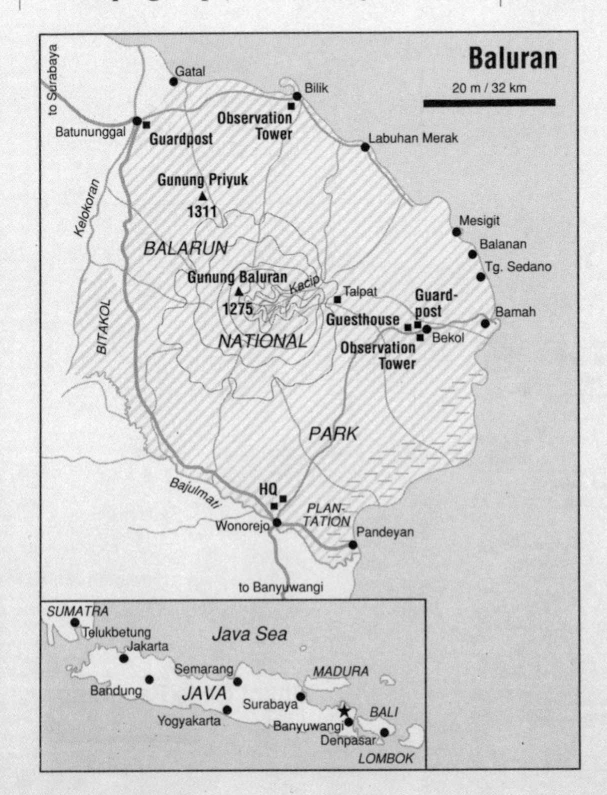

BROMO-TENGGER-SEMERU

Location: South of Surabaya, East of Malang. Open all year.

Access: Taxi from Surabaya at Rp 60,000 for a 3-hour ride; from Ngadisari by four-wheel drive, 3 km to Cemora Lawang and to Bromo 8 km, or to Pananjakan. By horse from Ngadisari to Bromo Rp 7,000

Public Transportation: Buses go to Sukaura at about 20 km from Ngadisari.

Registration: At park headquarters, about 3 km before Sukapura, on the road from Probolinggo.

Accommodation: Bromo Hotel between Ngadisari and Sukapura, Rp 80,000; Bromo Permai at Cemoro Lawang, Rp 12,000.

Camping: At Cemora Lawang.

Food: Available at Cemora Lawang.

Transport in park: Four-wheel drive possible through Park Gubuk Klatak to Bromo via Ngadas and Ranu Pane; from Tosari-Penanjakan to Bromo to Cemora Lawang; and from Ngadas through Sandsea to Cemora Lawang.

Trails: From Ranu Panu to Mount Semeru. Spend the night near Kumbolo lake. Continue journey from Cemora Lawang to Ranu Panu, passing through sand and sea.

Further advice: If you go to the top of Semeru, try to reach before late morning before clouds come in as you may get lost.

BALI BARAT NATIONAL PARK

Location: Northwestern Tip of Bali, 130 km for Denpasar. Open all year.

Size: 760 sq km

Access: Take taxi or bus to Gilimanuk, the ferry station to Java. From Gilimanuk take minibus or horse carriage to park headquarters at Cekik. A day's taxi trip for the tourist areas in south Bali costs Rp 50,000-100,000; any other offer is unreasonable.

Registration: At park headquarters, Cekik.

Accommodation: In Labuan Lalang there are several cottages for rent, managed by the park (Rp 5,000 per bed in the guesthouse or Rp 25,000 per cottage). Catering on request or at the foodstall in Teluk Terima, 1 km to the west. The coconut plantation in the enclave runs a simple hotel (Margarana Homestay) for Rp 7,500 per person per night.

Gilimanuk has several small hotels at Rp 5,000-7,500 per person per night. Catering on request or at the foodstall in Teluk Terima, 1 km to the east. A quiet place to stay is lovely Hotel Nusantara II, situated at the Bay of Gilimanuk just opposite the ferry-port: Rp 6,000. Direct views on the park at the other side of the bay.

Camping: The park's only official camping site is at Cekik. Facilities are modest and water is not always available. On special request, it is sometimes possible to camp in Labuan Lalang.

Food: Many restaurants and foodstalls in Gilimanuk. Otherwise, bring your own food.

Transport within park: Mostly on foot, but Sumber Klampok, Teluk Terima, Labuan Lalang and Goris/Banyuwedang can be reached by minibus/from Gilimanuk: take any minibus to Grogak, Seririt, Lovina Beach or Singaraja. Klatakan and Palasari can be reached by any minibus in the direction of Negara, Tabanan or Denpasar.

Outboard motorboats to Menjangan Island can be chartered at Labuan Lalang: Rp 30,000 for the first four hours (including the boat driver and the trip to and fro the island) and Rp 5,000 for each additional hour or part thereof. Charges are per boat and the number of passengers is limited to 8-10 persons maximum. A one-way trip to the island takes 30-40 minutes.

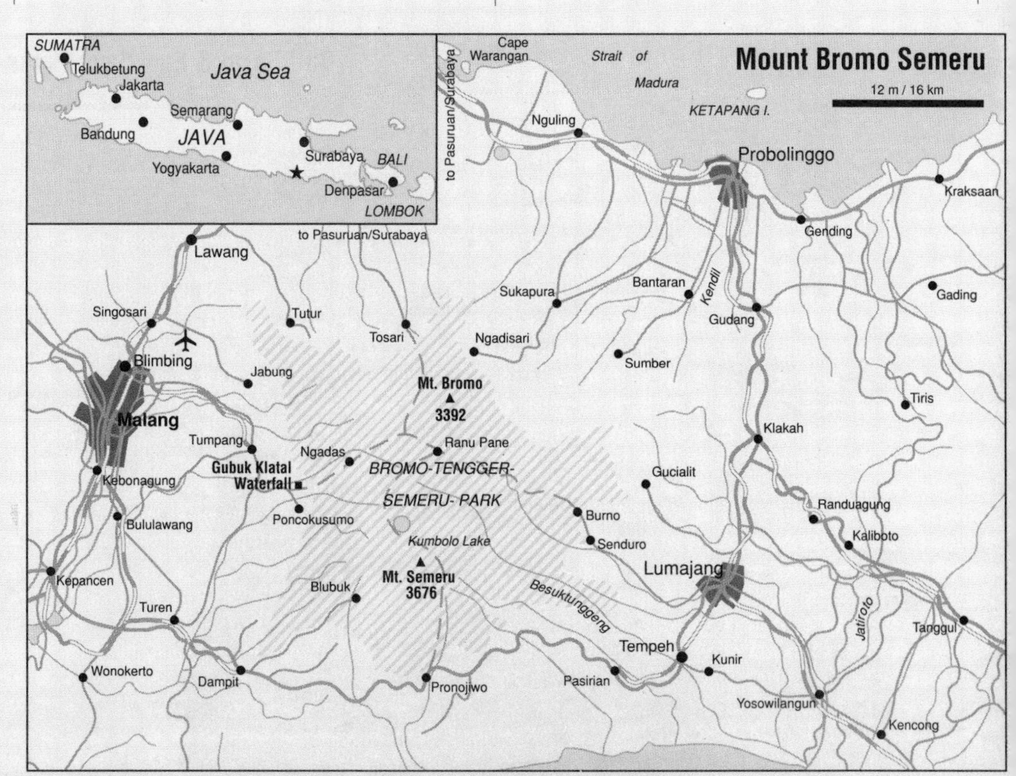

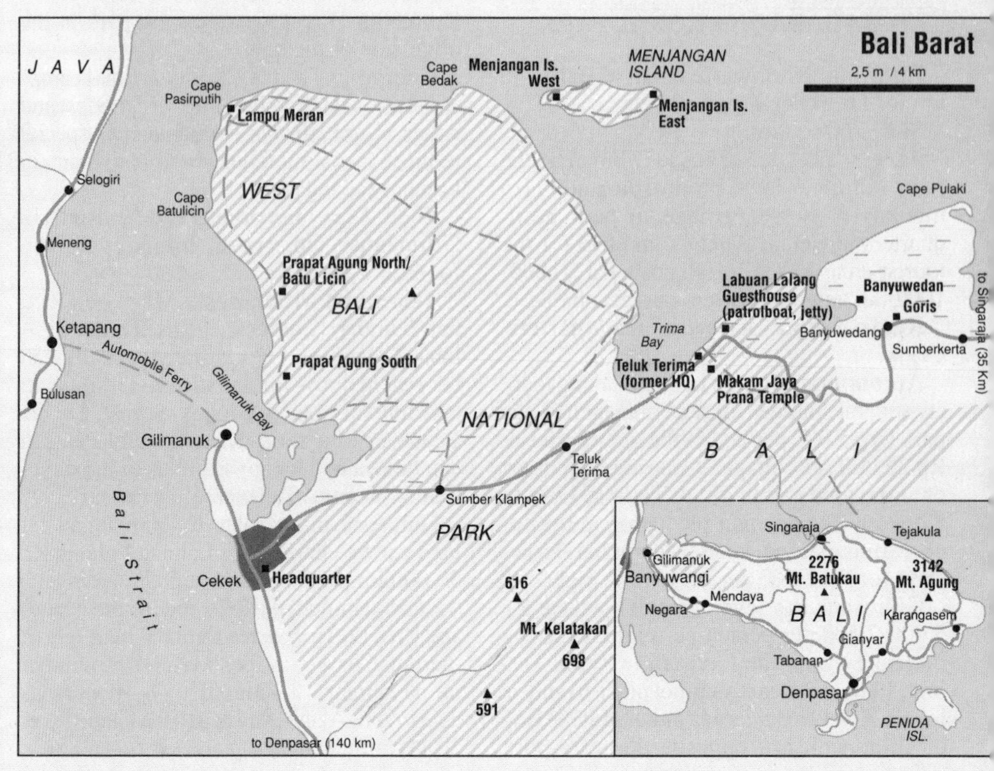
Bali Barat
2,5 m / 4 km
JAVA
Cape Pasirputih
Lampu Meran
Cape Bedak
Menjangan Is. West
MENJANGAN ISLAND
Menjangan Is. East
Selogiri
Cape Batulicin
WEST
Meneng
Prapat Agung North/ Batu Licin
BALI
Ketapang
Automobile Ferry
Gilimanuk Bay
Prapat Agung South
Bulusan
Gilimanuk
NATIONAL
Cape Pulaki
Labuan Lalang Guesthouse (patrolboat, jetty)
Banyuwedan
Goris
Trima Bay
Banyuwedang
Sumberkerta
to Singaraja (35 Km)
Teluk Terima (former HQ)
Makam Jaya Prana Temple
Teluk Terima
BALI
Bali Strait
Sumber Klampek
PARK
Cekek
Headquarter
616
Mt. Kelatakan
698
591
to Denpasar (140 km)
Singaraja
Tejakula
Gilimanuk
Banyuwangi
2276
Mt. Batukau
3142
Mt. Agung
Mendaya
Negara
BALI
Karangasem
Gianyar
Tabanan
Denpasar
PENIDA ISL.

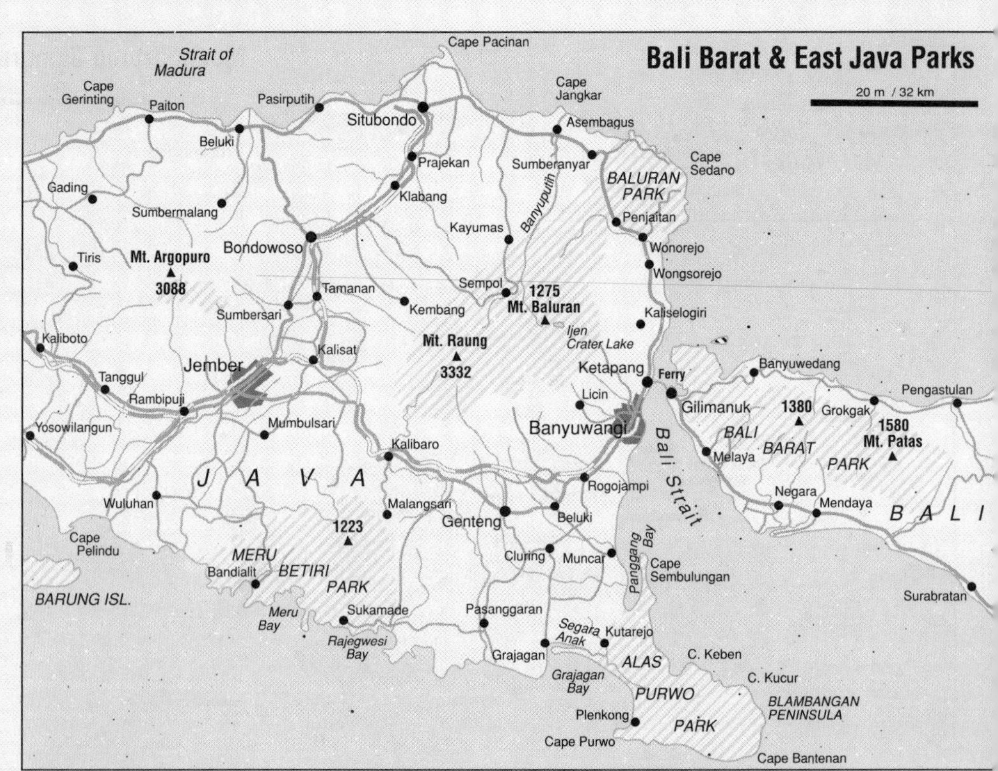
Bali Barat & East Java Parks
20 m / 32 km
Strait of Madura
Cape Pacinan
Cape Gerinting
Paiton
Pasirputih
Situbondo
Cape Jangkar
Asembagus
Beluki
Prajekan
Sumberanyar
Cape Sedano
BALURAN PARK
Gading
Klabang
Sumbermalang
Banyuputih
Penjaitan
Kayumas
Tiris
Mt. Argopuro
3088
Bondowoso
Wonorejo
Wongsorejo
Sempol
1275
Mt. Baluran
Tamanan
Kembang
Sumbersari
Kaliselogiri
Ijen Crater Lake
Kaliboto
Kalisat
Mt. Raung
3332
Ketapang
Ferry
Banyuwedang
Jember
Tanggul
Rambipuji
Licin
Gilimanuk
1380
Grokgak
Pengastulan
1580
Mt. Patas
Yosowilangun
Mumbulsari
Banyuwangi
BALI
BARAT
PARK
Kalibaro
Melaya
JAVA
Bali Strait
Rogojampi
Negara
Mendaya
Wuluhan
Malangsari
BALI
Genteng
Beluki
1223
Cape Pelindu
Cluring
Muncar
Panggang Bay
Cape Sembulungan
MERU BETIRI PARK
Bandialit
BARUNG ISL.
Surabratan
Meru Bay
Sukamade
Pasanggaran
Rajegwesi Bay
Segara Anak
Kutarejo
Grajagan
ALAS
C. Keben
Grajagan Bay
C. Kucur
PURWO
BLAMBANGAN PENINSULA
Plenkong
PARK
Cape Purwo
Cape Bantenan

DUMOGA-BONE NATIONAL PARK

Location: 200 km west of Manado, 20 km east of Gorontalo. Open all year.

Size: 3,000 sq km within park boundaries another 2,000 sq km adjoining forest.

Access: From Manado to headquarters 250 km or 4 hours by car to village Doloduo at west end Dumoga valley. To west end of park Gorontalo by plane (or by car from Manado 10-15 hours).

Taxi fare: Manado-Doloduo/Park headquarters. Rp 75,000.

Public Transportation: Manado-Kotamobagu Rp 7,5000, Kotamobagu-Doloduo (park headquarters) Rp 3,000.

Registration: Park headquarters at Doloduo (no telephone). Radio contact through Sub Balai PHPA North Sulawesi. Jl. Supratman 68, Manado. Tel. 0431-2688.

Accommodation: Kotamobagu/Gorontalo, Rp 40,000.

Camping: Allowed.

Food: Bring food from Kotamobagu or Gorontalo. Small restaurant at Doloduo at one km from park entrance.

Transport in Park: By foot at Dumoga side from Doloduo or Toraut, by car at Gorontalo side to Pinogu enclave at some 25 km inside park or 50 km from Gorontalo.

Trails: Doloduo Park headquarters to Motayangan, 10 km, and to Toraut, 15 km. Toraut Surroundings, 3-km track or longer tracks. The Tublabolo-Hunggayono-Pinogu trail is 18 km.

Further advice: Combine visit to Dumoga with visit to Gn. Ambang Reserve and sulphur-crater and Mooat lake at only 20 km east of Kotamobagu.

KOMODO

Location: Between the islands of Sumbawa (to the west) and Flores (to the east). ca 3 hours flight east of Bali. Open all year.

Size: 375 sq km

Access: By plane from Bali to Bima, overland to Sape, then by ferry (6 hrs). Or by plane to Labuan Bap and by boat from there (3-4 hours).

Fare: By air, Bali-Bima, US$43; Bali-Labuan Bajo, US$65. Chartered boat from Labuan Bajo, US$50.

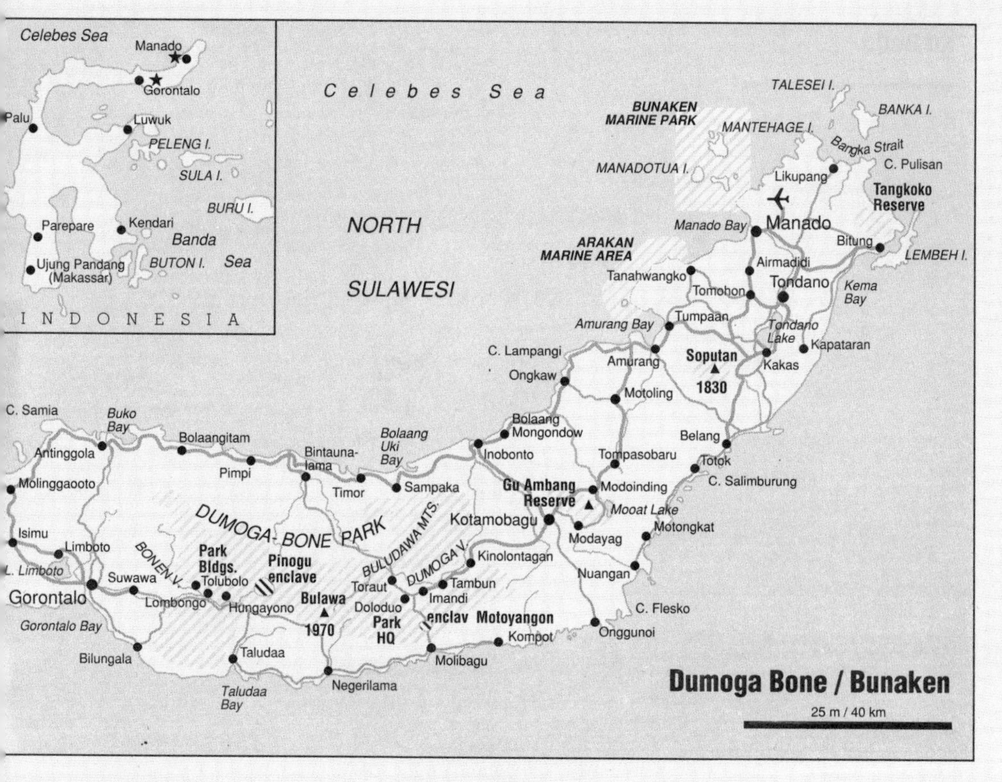

Public transportation: Ferry Sape-Komodo (Saturdays only). Ferry Labuan-Bajo-Komodo (Tuesdays only).

Registration: PHPA office on Komodo

Accommodation: Guesthouse, with 32 rooms available.

Camping: No facilities available.

Food: Limited food available – better to bring your own.

Transport in park: None.

Trails: Recommended is the one leading to the dragons' feeding site; the others are poorly marked and should only be tackled with a ranger.

Further advice: Komodo is steep and rugged with few trees, so trekking is only for the fit! The dragons are fed only on Sundays.

LANGUAGE

Indonesia's motto, *Bhinneka Tunggal Ika* (unity in diversity) is seen in its most driving, potent form in the work of language. Although there are over 350 languages and dialects spoken in the archipaelago, the one national tongue, *Bahasa Indonesia*, will take you from the northernmost tip of Sumatra through Java and across the string of islands to Irian Jaya. *Bahasa Indonesia* is both an old and new language. It is based on Malay, which has been the lingua franca throughout much of Southeast Asia for centuries, but it has changed rapidly in the past few decades to meet the needs of a modern nation.

Although formal Indonesian is a complex language demanding serious study, the con-

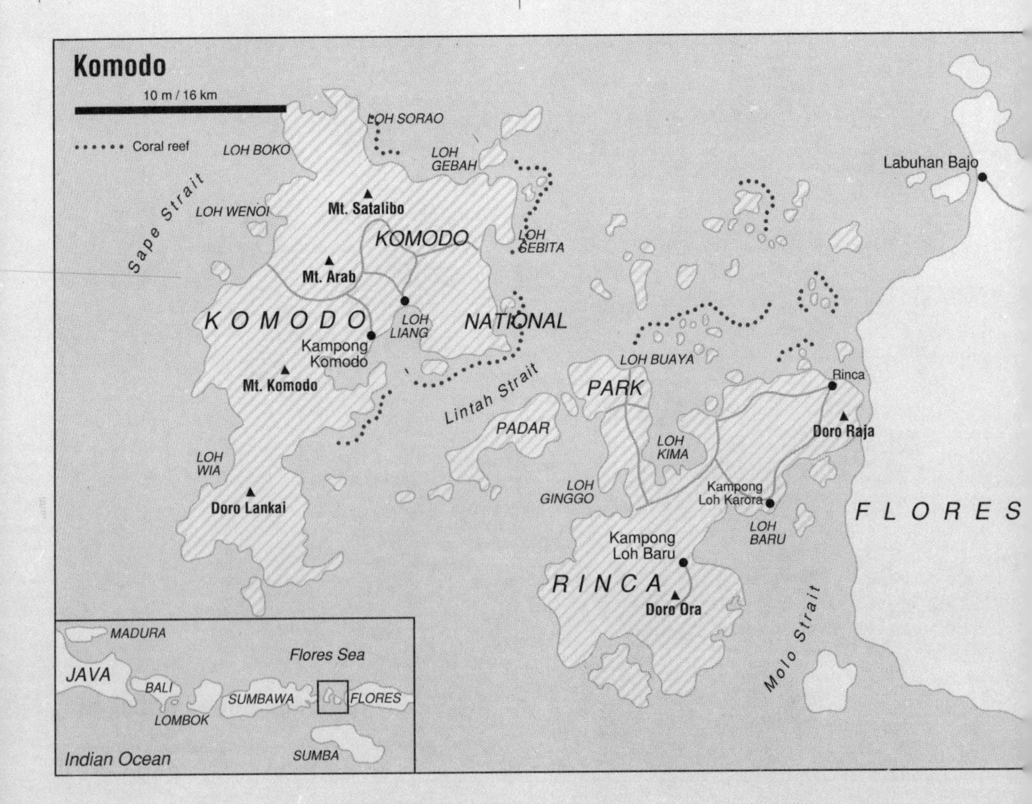

struction of basic Indonesian sentences is relatively easy. A compact and cheap book, *How to Master the Indonesian Language* by Almatseier, is widely available in Indonesia and should prove invaluable in helping you say what you want to say. Indonesian is written in the Roman alphabet and, unlike some Asian languages, is not tonal.

Indonesians always use their language to show respect when addressing others, especially when a younger person speaks to his elders. The custom is to address an elder man as *bapak* or *pak* (father) and an elder woman as *ibu* (mother), and even in the case of slightly younger people who are obviously VIPs, this form of address is suitable and correct. *Bung* (in West Java) and *mas* (in Central and East Java) roughly translate as "brother" and are used with equals, people your own age whom you don't know all that well, and with hotel clerks, taxi drivers, tour guides and waiters (it's friendly, and a few notches above "buddy" or "mate").

USEFUL ADDRESSES

DOMESTIC AIRLINES

Bouraq, Head Office, Jl. Angkasa 1-3, Kemayoran, Tel: 655279, 655170.

Garuda, Head Office, Jl. Merdeka Selatan 13, Jakarta 10110, Tel: 3801901.

Mandala, Jl. Veteran I/34, Tel: 368107.

Merpati, Head Office, Jl. Angkasa 2, Tel: 413608, 417404.

Pelita Air Service, Jl. Abdul Muis 53, Tel: 27508.

Sempati, Jl. Merdeka Timur 7, Tel: 348760, 367743.

Seulawah, Jl. Patrice Lumumba 18BD, Tel: 354207.

HOSPITALS

British Medical Scheme, Setia Building, Kuningan, Tel: 515481, 515367, 359101.

Cipto Mangunkusumo Hospital, Jl. Diponegoro, Tel: 343021, 882829.

Fatmawati Hospital, Jl. R.S. Fatmawati, Tel: 760124, 764147.

Gatot Subroto Hospital, Jl. Abdul Rachman Saleh, Tel: 371008.

Husada Hospital, Jl. Mangga Besar 137-139, Tel: 620108, 622555.

Islamic Hospital Jakarta, Jl. Letjend. Suprapto, Tel: 414208, 414989.

Medikaloka (24-Hours Emergency System), Kuningan Plaza South Tower, Gr. Fl., Jl. H.R. Rasuna Said Kav. C 11-14, Tel: 520212, 511160.

Metropolitan Medical Centre, Wisata International Hotel, Jl. M.H. Thamrin, Tel: 320408.

Mintaharja Hospital, Jl. Bendungan Hilir, Tel: 581031.

Persahabatan Hospital, Jl. Raya Persahabatan, Tel: 481708.

Pertamina Hospital, Jl. Kyai Maja 43, Tel: 775890, 775891.

St. Carolus Hospital, Jl. Salemba Raya 41, Tel: 883091, 882401.

Sumber Waras Hospital, Jl. Kyai Tapa Grogol. Tel: 596011, 591646.

Tjikini Hospital, Jl. Raden Saleh 40, Tel: 374909, 365297.

Yayasan Jakarta Hospital, Jl. Jend. Sudirman, Tel: 582241, 584576.

GETTING THERE

BY AIR

The majority of visitors arrive in and depart from Manila by air. Over 200 international flights arrive in Manila weekly. Manila International Airport and Domestic Terminal is centrally located. Several international flights also land weekly in Cebu.

BY SEA

Freighters and cruise ships take advantage of the excellent harbour of Manila Bay, although most travellers prefer to arrive by air.

The government is adamant about protecting tourists from endangering themselves by travelling on small craft between East Malaysia (Borneo) and the country's southernmost islands. The Moro rebels and continuing piracy make this exotic route unadvisable.

TRAVEL ESSENTIALS

PASSPORTS & VISAS

All tourists must have valid passports. All, except tourists from countries with which the Philippines has no diplomatic relations, stateless persons and nationals from restricted countries, may enter without visas and stay for 21 days, provided they hold onward or return tickets.

MONEY MATTERS

The monetary unit of the Philippines is the peso (P) which equals about US4¢. There are 100 centavos to a peso and the exchange rate fluctuates around P22 to US$1. The U.S. dollar, pound sterling, Swiss franc, French franc, Deutsche mark, Canadian dollar, Italian lira, Australian dollar and the Japanese yen are all easily convertible. Outside Manila, generally, the U.S. dollar is widely acceptable after the peso.

Traveller's cheques can be easily cashed, and major credit cards are accepted in Manila only. Avoid at all costs street money changers.

HEALTH

Yellow fever vaccination is necessary for those arriving from an infected area, except for children under 1 year who may be subject to isolation when necessary. When traveling in remote areas of the country it is advisable to take anti-malarial drugs.

WHAT TO WEAR

Light and loose clothes are most practical. Pack a sweater especially if you plan to go to the mountains. At formal gatherings Filipino men mostly wear the *barong tagalog* (Tagalog shirt). This is a long-sleeved shirt with side slits worn outside the pants. Traditionally it is made in white or pastels out of a very fine silk called *jusi*. The shirt is so transparent that a T-shirt is always worn underneath. The front and cuffs and sometimes the sleeves are ornately embroidered. The style is from the 19th century when only Spaniards were allowed to tuck their shirts inside their pants. Wearing the *barong* therefore distinguished a Filipino from a Spaniard and became a declaration of patriotism.

Filipino women often wear the *terno* for formal occasions. This is a long gown with huge "butterfly" sleeves which has elaborate embroidery on the skirt and bodice.

When visiting churches and mosques it is well to remember that shorts and scanty or provocative dress will be inappropriate.

WHAT TO BRING

Apart from your own special personal requirements, there is no need to bring with you any equipment other than possibly a travel plug adaptor and photographic supplies. (The standard voltage throughout the Philippines is 110 volts or 220 volts AC or a combination of both.) Medications are available at drug stores in major cities. Avoid film processing in the Philippines. Any Tourist Information Centre should be able to advise you as to where you can go to buy your special requirements.

CUSTOMS

Tourists may bring in duty-free cigarettes, alcohol and vehicles within the following stipulations. Each passenger is allowed 400 cigarettes or 100 cigars or 500 grams of pipe tobacco or an assortment of these. Two regular-sized bottles of alcoholic beverages are allowed per person, and cars and other vehicles are allowed duty free entry provided they have "Carnets de Passages in Douanes" and a letter from the Philippine Motor Association guaranteeing the exportation of the vehicle within 1 year from the date of arrival or the payment of duties and tax thereon.

PORTER SERVICES

You will never want for a porter anywhere in the Philippines. The problem is usually quite the reverse as several porters will compete to carry your bags. About P5 per bag is appropriate, more for heavy packages.

EXTENSION OF STAY

Visitors who wish to extend their stay from 21 days to 59 days should contact the Commission on Immigration and Deportation. Such extensions costs about P500.

GETTING ACQUAINTED

THE NATIONAL FLAG

The Philippine Flag is composed of three parts: a white equilateral triangle on the left and two horizontal stripes, blue and red. It is unusual insofar as it indicates whether the country is at peace or at war: in times of peace, the blue is over the red; and in war time, the red over the blue.

The eight rays of the Philippine sun, in the middle of the white triangle, represent the first eight provinces which revolted against Spanish domination. The three stars at the triangle's corners indicate the three major groups of the Philippine islands: Luzon, Visayas and Mindanao.

GOVERNMENT

After the February 1986 people power uprising, the country reverted to a democratic form of government. In February 1986, Corazon C. Aquino assumed the presidency of the Republic of the Philippines after the peaceful overthrow of the administration of Ferdinand Marcos.

Among the events that are believed to have led to this historic 4-day "people's revolution" of 1986 are the assassination in August 1983 of Benigno "Ninoy" Aquino, husband of President Aquino and the leading oppositionist at the time of his death; the controversial results of a "snap" presidential and vice-presidential election in February 1986 called by former President Marcos to seek renewal of his mandate; and the withdrawal of allegiance to the Marcos government of the then Vice Chief of Staff of the Armed Forces of the Philippines and the Defense Minister.

With the recognition by the Filipino people and the international community of the legitimacy of the Aquino government, the head of state has turned to the difficult tasks of reconstructing the economy and removing the inefficiency and corruption inherited from the previous regime.

In February 1987 the Filipino ratified a new Constitution providing for a democratic republican state and a presidential form of government. This paved the way for re-establishment and election of members of the Congress – the bicameral legislative body. Elections to the Senate and the House of Representatives were held in May 1987. Congress convened in July 1987. The democratic restructure was completed in January 1988 when the officials of the country's local government units were elected.

For administrative purposes the republic is divided into 12 regions (plus Metro Manila) comprising 73 provinces. Provinces consist of a several municipalities centered on a provincial capital. Municipalities are sub-divided into *barangays*, the smallest socio-political unit, headed by a *barangay* captain.

ECONOMY

While hardly on the verge of an economic miracle, the Philippines, despite several coup attempts, is showing signs of a sustained recovery that would have been thought improbable a few years ago. Slowly but surely the government is tackling the pressures of economic survival.

The nation does not have to look far for inspiration. The city and province of Cebu has demonstrated how hard work and enterprise can propel economic growth towards new heights, reviving confidence in a country that was hit by an economic crisis in the early 1980s.

Inspired by Cebu, the archipelago's economic planners aim to encourage the growth of small industries to cement links between industries and sectors. About one third of the country's output is industry (minerals, semiconductors, garments, footwear, wood-forest products and handicrafts) while agriculture (coconut, tobacco, sugar, abaca, bananas, rice and corn) accounts for about a quarter. Developing agro-based industries such as aquaculture and food processing will marry the two sectors.

In an effort to improve income distribution in a country where half the population lives below the poverty line, the government is presently focusing its attention on small industries rather than promoting conglomerates. Land reform, however, has so far been unsuccessful.

Pioneer industries such as agribusiness, high-tech industries and energy-related enterprises offer foreign investors the best chance to avoid weighty government regulations on foreign equity holdings while availing of tax holidays and other incentives. Lack of infrastructure and frequent power cuts still discourage foreign investment but planned industrial estates will have infrastructure and linkages to Manila.

GEOGRAPHY & POPULATION

The archipelago comprises 7,107 islands, of which 11 main islands account for more than 95 percent of the total land area of 300,439 sq km. Only about 2,000 islands are inhabited and 2,500 of them are not named. The islands are dotted with white sand beaches, lush tropical vegetation, lakes and rivers.

From north to south the islands stretch for 1,840 km and east to west for 1,104 km. The highest peak is Mount Apo in Davao province in Mindanao, at 2,953 metres. The second highest is Mount Pulog near Baguio in northern Luzon at 2,930 metres. Some 17 active volcanoes can be found, best known of which is Mayon Volcano in southern Luzon near Legaspi.

Three distinct geographical regions are found: Luzon, the largest and northernmost island is where the capital Manila is located. In the centre is the tightly packed Visayan island group comprising Negros, Cebu, Bohol, Panay, Masbate, Samar and Leyte. To the south is Mindanao, the second largest island where Davao, Zamboanga, Marawi City and Cagayan de Oro can be found. From the southwestern tip of Mindanao the islands of Sulu including Basilan, Jolo and Tawi Tawi dot the seascape down to Borneo. To the west of the Visayas lies the Palawan archipelago with more than 1,700 islands.

The country has a population of 60 million about 10 million of whom are concentrated in Metro Manila. Filipinos are basically of Malay-Polynesian origin, though there is evidence of Indian, Chinese, Spanish, Arab and North American. The population increases at about 1,500,000 annually – too rapidly for comfort – but any attempts to implement family planning programs are met with strong opposition from the Catholic Church.

TIME ZONES

The Philippines is 8 hours ahead of Greenwich Mean Time. All year round sunrise is about 6 a.m. and sunset about 6 p.m. give or take 30 minutes.

CLIMATE

The best months to travel to the Philippines are from December to May during the dry season. The climate generally is subject to monsoons, but tempered by trade winds. From June to November is the season of the southwest monsoon, which brings sultry wet weather. December to May is the season of the northeast monsoon. The temperature hovers around 21°C (80°F). January is coolest month with 25.5°C (78°F); the hottest is May with an average of 28°C (83°F). Humidity is high, ranging from 71 percent in March to 85 percent in September.

Typhoons annually take their toll in the country. They are tropical revolving storms that are called hurricanes in the Atlantic and South Pacific, cyclones in the Indian Ocean and Bay of Bengal and "willy-willies" north and west of Australia.

CULTURE & CUSTOMS

Filipinos have a justifiable reputation as one of the most hospitable people in the world, especially in rural areas where folkways survive to more traditional forms. As in most Asian societies, the guest is much honoured – the one who gets the best bed, the choicest cuts of meat, the airiest room. Many backpackers claim they never had to sleep out in the open unless it was by choice wherever there was at least a farmer's or fisherman's hut nearby.

In the cities, despite a large number of new hotels and pensions, a foreign visitor lucky enough to have the name of a local resident is usually fed and shown around, if not given a place to stay for free.

Given this situation, the foreign visitor is obliged to avoid taking advantage of this goodwill which may seem contrary to the normally more impersonal conduct of life in Western societies. As a guest, it would be best for him/her to observe the rules of local sociology because nothing is more disruptive of daily life than someone who does not try to fit into the basically community-oriented consciousness of the Filipino.

The following would be some highlights of this consciousness, useful to remember for smooth and rewarding navigation within Filipino culture.

CLANNISHNESS

This is the rule of survival, the main strength and also the source of corruption in Filipino society. Kinship ties of both blood and marriage, often up to the third degree removed, are kept well-defined and operative in all levels and facets of life. Clans operate as custodians of common experiences (many old families religiously keep family trees) and the memory of geographical and racial origins. They are also disciplinary mechanisms, placement agencies and informal social security systems. When marriages are drawn between two clans, it is hardly ever a matter of individual choice alone as much as an alliance.

Within the sometimes tyrannical embrace of the clan, members of all ages find their place in an orderly world where children are fussed over (cared for by an assortment of aunts, uncles, cousins and grandparents) and the elderly are given care and reverence all the way to the last rites.

If a foreign guest finds himself in the middle of clan hospitality, it would be considered good form to give special acknowledgement to family elders. It does not hurt to use the honorifics of *lolo* and *lola* for the grandfather, grandmother, grandaunts and granduncles of the clan. To go a step further and greet them by putting their right hand to your forehead in a time-honored Filipino gesture of respect goes a long way in establishing friendly relations.

One notable first question to a foreign guest, meaningless to many city-bred Westerners, is "Where do you come from?" The name of your country does not say much. The name of your hometown is more satisfying because even if your host does not have the vaguest notion of what or where that might be, there is something psychologically comforting for him to know that you, too, wandering so far from home, have family like himself.

SIR

Local sociologists have coined the phrase "smooth interpersonal relationships" (SIR for short) to indicate the key premise of human contact among Filipinos. What it means is that the edges of face-to-face communication must be kept smooth at all times by courtesy and gentle speech, no matter that the content must sometimes be unpleasant.

Direct confrontations are generally avoided. When forced to deliver negative messages, Filipinos are fond of emissaries and subtle indirection relying for its effective communication of the sensitivity of the other party. Many Westerners accustomed to direct statements and blunt approach find this hypocritical and cowardly. In the native context, however, smooth interpersonal relationship has its value in giving everyone room for manoeuvre and amendment of hasty statements.

Part of the ritual of SIR are polite forms of address when conversing with strangers, older people and people of high social rank. The basic forms are the words *po* and *ho* to end every sentence. Use *opo* and *oho* to say "yes" and *hindi po* and *hindi ho* to say "no". Even in face-to-face conversation, one refers to a new acquaintance, an older person or a dignitary in the third person-plural – *kayo*, *sila*. The intention is to maintain a respectful distance, beginning on the verbal level from which to slowly establish a pleasant relationship.

Another practice falling under the general heading of smooth interpersonal relationships is the custom of offering one's house and goods to anyone within the immediate vicinity. Guests arriving at or around mealtime are always asked to stay and share food so people are usually careful not to come at inconvenient times unless they are sure of their welcome. Even the merest strangers sitting together on a park bench will offer each other the contents of their lunch boxes. Friends and acquaintances sharing a ride on public transportation invariably struggle, often comically, to pay the other's fare.

There are several other sub-concepts to smooth interpersonal relationships it would be useful to remember.

One is *hiya*, literally translated as "shame" but better defined as delicacy of feeling that sensitizes one to the feeling of others. With this unspoken premise, individuals are prevented from taking each other for granted. Related to *hiya* is *pakiramdaman* or "feeling each other out". Beyond words, Filipinos usually intuit or divine what the other means to communicate.

A second strand in this old organic fabric of social relationships is *utang na loob* or debt of gratitude. Favours long past are never forgotten and always returned in an invisible bond of reciprocity that keeps the whole society functioning. Of course, *utang na loob* (both individual and collective) has also been responsible for malfunctioning bureaucracies resistant to impersonal, rational management procedures but that is the price to pay for a transition from a traditional system to a more modern one. To this day, it is a coarse individual indeed who forgets his *utang na loob*.

A third sub-concept is *pakikisama*, for which there is no exact English equivalent but can be defined as "getting along" or submitting to group will. As with all cultural traits, this one has negative applications such as when adolescents fall in with bad company, adults find themselves in a circle habitually living beyond their means, or government officials fall into webs of compromise. Positively applied, however, *pakikisama* has tremendous power to mobilize individual energies for common goals.

BAHALA NA

Perhaps the crowning glory of local sociology is this Filipino expression which one anthropologist has traced to a linguistic root of *Bathala na*, meaning "leave it to God." This is a typical Filipino reaction to crises and insoluble problems. Development experts have often enough decried "*Bahala na*" as passive and fatalist, the sole factor in the delayed maturity of the Filipino. Some other students of the Filipino character, however, praise its deeply philosophical origins. As proof, they point to how *Bahala na* at its best has successfully supported Filipino morale through the trials of his particular destiny.

TIPPING

Most major restaurants and hotels add 10 percent service charge automatically. Smaller establishments leave the tipping to your discretion, but even if service charge is levied, it is better to leave some loose change, at least. Taxi drivers, beauticians, barbers, bellboys, port attendants – in fact anyone who performs a service for you – expect a small gratuity.

ELECTRICITY

The standard voltage throughout the Philippines is 220 volts AC, 60 cycles. Many areas also have 110 volts capability.

BUSINESS HOURS

Shops open Mon-Sat, 9 a.m. or 10 a.m. to 7 p.m. In Manila many shops catering to tourists are open on Sun. The Philippine attitude of *Bahala na* (whatever happens) prevails outside Manila so shops don't usually stick to any schedule.

Government and business hours are Mon-Fri, 8 a.m. to 5 p.m. and workers break for lunch from noon to 1 p.m. Banks are open Mon-Fri, 9 a.m. to 3 p.m.

HOLIDAYS

The Gregorian calendar is used in the Philippines. The principal public and religious holidays are listed below. When a holiday falls on a Sunday, the following Monday is declared a public holiday.

January 1: New Year's Day.

Maundy Thursday: Flagellants in the streets, processions and "*Genaculos*" (passion plays).

Good Friday.

Easter Sunday.

May 1: Labour Day. A day of tribute to the Filipino worker.

May 6: *Araw ng Kagitingan* (Day of Valor). Celebrations at Fort Santiago to commemorate the bravery of the Filipinos who fought for the country's independence during World War II.

June 12: Independence Day. Military and civic parades at Rizal Park.

June 24: Manila Day (Manila only). Parade, film festivals and cultural presentations.

July 4: Fil-American Friendship Day. The country celebrates the days when it was granted independence by the United States government.

August 26: National Heroes Day. This day in 1896 marked the start of the Philippine Revolution.

September 11: Barangay Day. Re-establishment of the oldest political unit that begun during the pre-conquest years of the archipelago.

November 1: All Saints Day. A day to honor the dead.

November 30: Bonifacio Day. Celebration of the birth of the country's great Plebeian, Andres Bonifacio.

December 25: Christmas Day. Nine days of pre-dawn masses called "*Misa de Gallo*", culminating in a midnight mass on Christmas eve, which is immediately followed by a "*Noche Buena*" or midnight repast.

December 30: Rizal Day. Wreath laying ceremony at the National Hero's Monument in Rizal Park.

RELIGIOUS SERVICES

Historically, the Filipinos have embraced two of the great religions of the world – Islam and Christianity. Islam was introduced during the 14th century, shortly after the expansion of Arab commercial ventures in Southeast Asia. By the 16th century, it was extending its influence northward when the Spaniards came to curb its spread. Today, it is limited mainly to the southern region of the country.

Locally, two Filipino independent churches were organised at the turn of the 20th century and are prominent today. These are the Aglipay (Philippine Independent Church) and the Iglesia Ni Kristo (Church of Christ), founded in 1902 and 1914, respectively. The Iglesia Ni Kristo has expanded its membership considerably. In fact, the Iglesia Ni Kristo church, with its unique towering architecture, is a prodigious sight in almost all important towns, provincial capitals and major cities.

Most Christian church services are held on Sunday morning and evening while Friday is the Muslim holy day. Details of these and other services are usually available at hotel reception desks and Tourist Information Centres. (Manila, tel: 50-17-03).

COMMUNICATIONS

MEDIA

Newspapers & Magazines: Several morning and evening English-language daily newspapers are published in Manila. These include the *Philippine Star*, *Manila Bulletin*, *Manila Times*, *Malaya* and *Manila Chronicle*. The *Fookien Times* is published in English and Chinese. There are also a number of local magazines published in English: *Mr. & Ms.*, *Lifestyle Asia*, *Taipan*, *Metro*, *Computer Times*, *What's on in Manila*, *Expat*, *Mod* and *Panorama*.

A fairly large selection of foreign magazines and newspaper are available, for example: *Newsweek*, *Time*, *Asiaweek*, *Far Eastern Economic Review*, *Asian Business*, *Business Traveller*, *The Economist*, *Reader's Digest*, *Vogue*, *Yazhou Zhoukan*, plus the *Asian Wall Street Journal* and *The International Herald Tribune*. All the above are sold either in major hotels or at bookstores, supermarkets and newsstands throughout the city.

TELEVISION & RADIO

People's Television (channel 4) broadcasts a variety of programmes in color which includes the latest national and foreign news, sports events, live coverages, variety shows, foreign serials and shows, and soap operas.

There are also a number of independent television stations serving Manila. They are Radio Philippine Network (RPN), channel 9; Greater Manila Area T.V. (GMA), channel 7; Intercontinental Broadcasting Network (IBC), channel 13; Alto Broadcasting System/Chronicle Broadcasting Network (ABS/CBN), channel 2.

Most radio stations are privately owned. They broadcast music ranging from classical to hard rock, interspersed with news bulletins, commercials and entertainment talk. The most politically conscious station is Radio Veritas (AM 846 kHz).

POSTAL SERVICES

Post offices are open from 8 a.m. to 5 p.m. on weekdays and 8 a.m. to 1 p.m. on Sat and holiday. The Bureau of Post is at Plaza Lawton (Liwasang Bonifacio), Manila.

There are private agencies in Manila which deliver express mail within city boundaries. For general letter posting, there are plenty of mail boxes at the post offices, street posts and in some commercial establishments. At Manila International Airport, the post office is located at the arrival area. Stamps are sold at post offices and in some government offices, hotels, universities and commercial establishments.

TELEPHONE & TELEX

Most telephone services are operated by Philippine Long Distance Telephone Company (PLDT). There is a minimum charge of 75 centavos (three 25-centavo coins) from public pay phones. As few public telephones exist, it is a common practice to use the phones of small businesses and shops for P1.50 charge. For long distance domestic calls dial 109 to get the operator. The number of the operator for international calls is 108. Be warned that using the telephone will prove frustrating to anyone accustomed to Western style efficiency.

Most of the larger Philippine hotels have telex facilities, available to guests at a small charge. The number of hotels with a fax facility is growing rapidly in Manila.

EMERGENCIES

SECURITY & CRIME

The **Tourist Assistance Unit** (TAU) is open 24 hours to assist visitors in trouble. Telephone 50-17-28 and 59-90-31 and ask for TAU. The **Department of Tourism** is at T.M. Kalaw Street, Rizal Park, Manila.

The **Integrated National Police Headquarters** is located at Camp Datuin, Quezon City, Metro Manila, tel: 921-59-21. For urgent police assistance in Manila dial 166.

LOSS

Lost credit cards should be reported immediately to:

American Express
Tel: Manila 815-93-66

Diners Club
Tel: Manila 810-45-21

Valuables: Most larger hotels have a safe or a safe box where your valuables may be deposited. When travelling it is always advisable to conceal such valuables as cameras. Always lock your car.

MEDICAL SERVICES

It is advisable to have medical insurance when visiting the Philippines as payment must usually be guaranteed before treatment. Major Manila hospitals are:

Cardinal Santos Memorial Hospital
Wilson Street
San Juan, Metro Manila
Tel: 721-33-61

Makati Medical Centre
2 Amorsolo corner de la Rosa Street
Makati, Metro Manila
Tel: 815-99-11

Manila Doctor's Hospital
667 United Nations Avenue
Ermita, Manila
Tel: 50-30-11

St. Luke's Medical Center
279 Rodriguez Boulevard
Quezon City, Metro Manila
Tel: 78-09-71

Note: As far as possible avoid medical treatment in the Philippines as many travellers and expatriates have reported unpleasant experiences.

GETTING AROUND

DOMESTIC TRAVEL

Transportation around the archipelago emanates from the country's hub, Manila. Flying is quick and cheap and Philippine Airlines, Aerolift and Pacific Airways Corporation, the country's domestic carriers cover the country with their routes. For those with time to spare transportation possibilities include bus, train, car and boat travel.

Philippine Airlines is the major domestic (and international) carrier operating to over 40 domestic points. Up-and-coming airlines competing with the flag carrier include Aerolift, (Daet, Cebu, Boracay, Bohol, Dipolog, Lubang and Busuanga) and Pacific Airways Corporation (Lubang, Boracay and Busuanga). Listed below are the telephone numbers of the three airlines.

Aerolift
Tel: 817-23-61

Pacific Airways Corporation
Tel: 832-27-31

Philippine Airlines
Tel: 832-09-91

WATER TRANSPORT

Manila and Cebu are the two centers of shipping. Be advised that inter-island boat travel will only suit those prepared to "rough it". The effort is rewarding however, as some of the ports served by the steamers have hardly changed in decades – and the bonus is that seemingly half the local populace greets arriving boats down at the wharf. Tickets on major lines (e.g. Manila-Cebu) can be booked through travel agencies. Again, the low fares are a pleasant surprise.

PUBLIC TRANSPORT

In Metro Manila, the bus and jeepney rates are P1 for the first 4 km plus 25 centavos for every kilometre thereafter.

Tricycles (motorcycle with a side-car attached) are sometimes available for short trips on the side streets.

There are also air-conditioned Love Buses with terminals in Escolta in Binondo, Manila, The Centre Makati and Ali Mall in Cubao, Quezon City. Rates are P8.

Taxi fare is P2 at flagdown for the first 500 metres and P1 for every 250 metres thereafter. Taxis can be found almost everywhere, especially near hotels, shopping centers and cinemas. Always have small change available and pay in pesos.

Metrorail, the overhead railway system, charges a flat rate of P3.50 at any point along Taft and Rizal Avenue, from Baclaran to Monumento in Caloocan City.

The Pasig River ferry operates from Lawton near the Central Post Office to Guadalupe near Makati. The fare is P9.50 one way.

TRAINS

Train travel is only for the very brave with lots of time to spare. Only one line operates out of Manila. The train from Manila's Tutuban Station in Tondo runs south to Legaspi City from where you can visit Mayon Volcano in the Bicol region.

BUSES

The central Luzon region near Manila and areas surrounding provincial capitals have a reasonable road system. Dozens of bus companies operate services to the main tourist centers of Luzon and fares are low by Western standards, e.g. The 5-hour journey from Manila to Baguio costs less than P100.

PRIVATE TRANSPORT

You can rent station wagons, bantams, coasters, buses, jeepneys and air-conditioned limousines. Cars may be rented with or without a driver. Charges vary according to type of vehicle.

Hourly rates are available, charges at one sixth the daily rate. In excess of 6 hours, daily rates apply. A valid foreign or international driver's licence is acceptable.

WHERE TO STAY

The Philippines offers a wide range of accommodation to suit every budget from beach resorts and pension houses to apartments and luxury hotels.

Manila boasts 10 de luxe hotels. The city has been well organised for tourism and conventions. The Department of Tourism and the Philippine Convention and Visitor's Bureau have an excellent network of promotion offices all over the world to provide information and in some cases even make hotel reservations. For detailed information and rates contact the Tourist Information Centre, Department of Tourism, 2nd floor, T.M. Kalaw Street, Rizal Park, Manila, tel: 50-17-03.

FOOD DIGEST

WHAT TO EAT

Nowhere else is the Philippines' long history of outside influences more evident than in its food...the experience of a lifetime. Philippine cuisine, an intriguing blend of Spanish, Malay and Chinese influences, is noted for the use of fruits, local spices and seafoods.

Filipinos eat rice three times a day, morning, noon and night. It is a "must" to sample a Filipino breakfast of fried rice, *longaniza* (native sausage) and fish, which is normally salted and dried accompanied by tomatoes and *patis* (fish sauce) on the side.

When ordering it's best to watch the Filipinos. Even before the food arrives, sauce dishes are brought in and people automatically reach for the vinegar bottle with hot chili, or the soy sauce which they mix with *kalamansi* (small lemons). Grilled items are good with crushed garlic, vinegar and chili. It's a good idea to start a meal with *sinigang*, a clear broth slightly soured with small nature fruit and prepared with *bangus* (milkfish) or shrimp.

Some typical Philippine dishes worth trying are *tinola* which is made with chicken and *pancit molo* which is dumpling of pork, chicken and mushrooms cooked in chicken or meat broth. *Adobo* is pork in small pieces, cooked for a long time in vinegar with other ingredients such as chicken, garlic and spices and then served with rice.

A typical feast dish, *lechon* is suckling pig stuffed with tamarind leaves and roasted on lighted coals until the skin is crackling and the meat tender. *Sinanglay*, another festive dish, is fish or piquant crabs with hot pepper wrapped in leaves of Chinese cabbage, and then cooked in coconut milk.

PARKS & RESERVES

MOUNT ISAROG NATIONAL PARK

Location: 12 km west of Naga City in the Province of Camarines Sur. Open all year.

Size: 101 sq km

Access: The park can be reached by numerous bus routes, the most direct being a ride of 5 km from Carolina to Panicuason and a 7-km ride via Pili to Curry.

Accommodation: Duplex resthouse and picnic facilities.

Local park administration: The Regional executive Director, DENR Regional office, Legaspi City.

QUEZON NATIONAL PARK

Location: 164 km south of Manila and 17 km east of Lucena in Quezon.

Access: At the end of the South expressway out of Manila turn left at the St. Tomas intersection towards Lucena. Go through Alaminos, bypassing San Pablo, and follow the road to Pagbilao. About 14½ km beyond Pagbilao and 5 km past the turn off to Padre Burgos, the highway diverges. Take the old road which winds into the park.

Accommodation: Camping is allowed at Cam Trining. Rustic overnight facilities are available at the Golden Shower Hotel or you could return to Pagbilao and stay at resort facilities by the beach.

Trails: From the parkway proceed up the road some 350 metres and take a path to the right through a collection of "spider lilies". From here it is a 1½- to 2-hour hike to Mount Pinagbanderahan through tropical forest.

Local park administration: Regional Executive Director, DENR Regional Office, Quezon City, Manila.

MOUNT APO NATIONAL PARK

Location: Mindanao, 25 km west of Davao City, in Regions 11 and 12. Open all year.

Size: 128 sq km

Access: The foothills of the park are accessible via well maintained roads from the cities of Davao and Cotabato.

Registration: Advisable to register at the Municipal Building, Kidapawen.

Accommodation: Camp site at Lake Agco, including a log cabin.

Food: Bring your own.

Trails: Four routes via: Kidapawen, North Cotabato; Makilala, North Cotabato; Kapatagan, Davao del Sur; and Kabarisan, Davao City.

Local park administration: Department of Environment & Natural Resources (DENR), in Davao City and Cotabato City.

MOUNT CANLAON NATIONAL PARK

Location: 25 km southeast of Bacolod City in the provinces of Negros Oriental and Occidental. Open all year.

Size: 246 sq km

Access:

1) From Bacolod City to Mambucal (34 km) and a further 4 km on a rough road into the Park;

2) From Bacolod City to Guintubdan via La Carlota City on asphalted road (71 km) and 15 km of rough road; and

3) From Bacolod City to Masulog traversing the national highway (113 km).

Registration: At the entrance at Wasay and Guintubdan.

Accommodation: In Bacolod City.

Trails and facilities: Scenic drive of 6.5 km from the park boundary to the proposed Nature Centre Site; access road of about 4.2 km; various nature trails.

Local park administration: Provincial DENR office, Bacolod City.

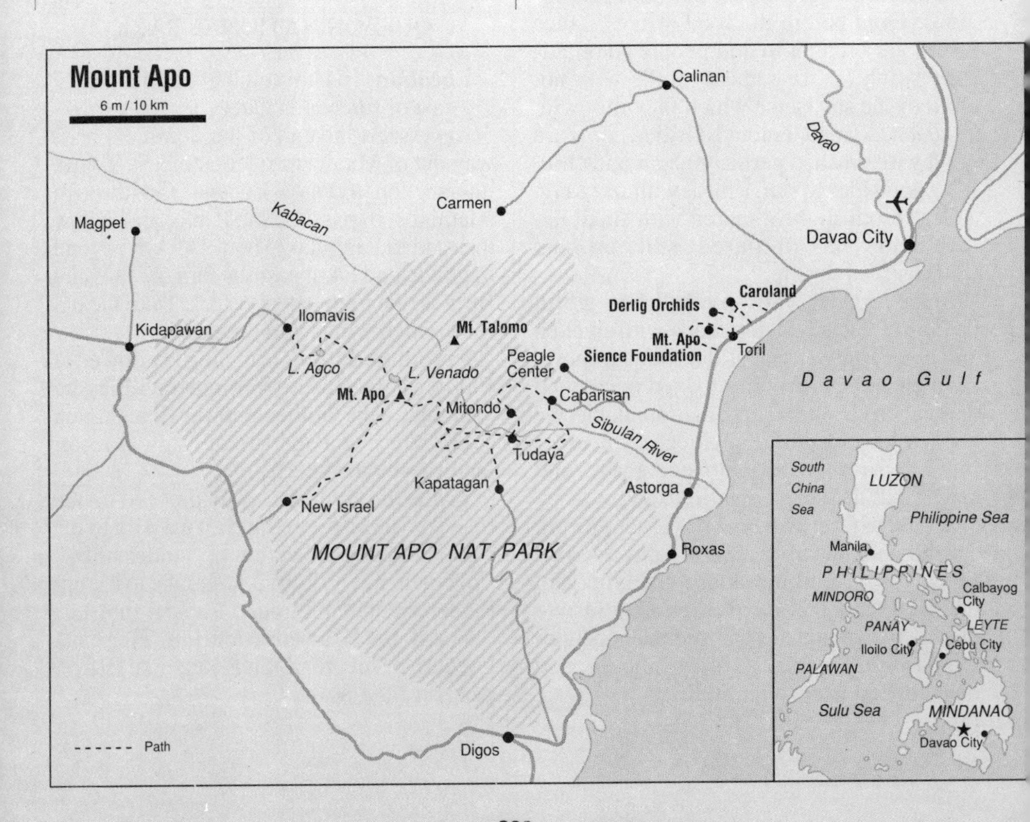

IGLIT-BACO NATIONAL PARK

Location: 35 km northeast of San Jose City in the Municipality of Bongabong, Oriental Mindoro Province. Open all year.

Size: 754 sq km

Access: The park is accessible by aircraft (40 minutes) from Manila via San Jose, Occidental Mindoro or by boat (2-3 hours) from Batangas City to Calapan, Oriental Mindoro. The trip from Calapan to Bongabong takes 1-2 hours via a long, bumpy road. From the nearest town, the interior of the park can be reached only through wild trails.

Trails: See page 290.

Accommodation: None; advisable to use facilities in San Jose or Bongabong.

Local park administration: The Regional Executive Director, DENR Regional Office, Quezon City.

ST PAUL'S SUBTERRANEAN RIVER NATIONAL PARK

Location: Located on the west coasting Palawan, about 81 km north or Pureto Princesa. Open all year. (Map on page 400)

Size: 39 sq km

Access: See page 291.

Accommodation: Limited accommodation is available, along with an administrative building, view decks, park stations, picnic and camping areas and nature trails.

Local park administration: The Park Superintendent, Department of Environment and Natural Resources, Puerto Princesa, Palawan.

MOUNT PULOG NATIONAL PARK

Location: 140 km north of Manila; 26 km northeast of Baguio City. (Map on page 400)

Size: 116 sq km

Access: From Baguio City, proceed to Ambuklao towards Kabayan.

Accommodation/facilities: There are no facilities or amenities in the Park; nature trails are maintained by the local inhabitants.

Local park administration: Regional Executive Director, DENR, Cordillera Autonomous Region, Baguio City.

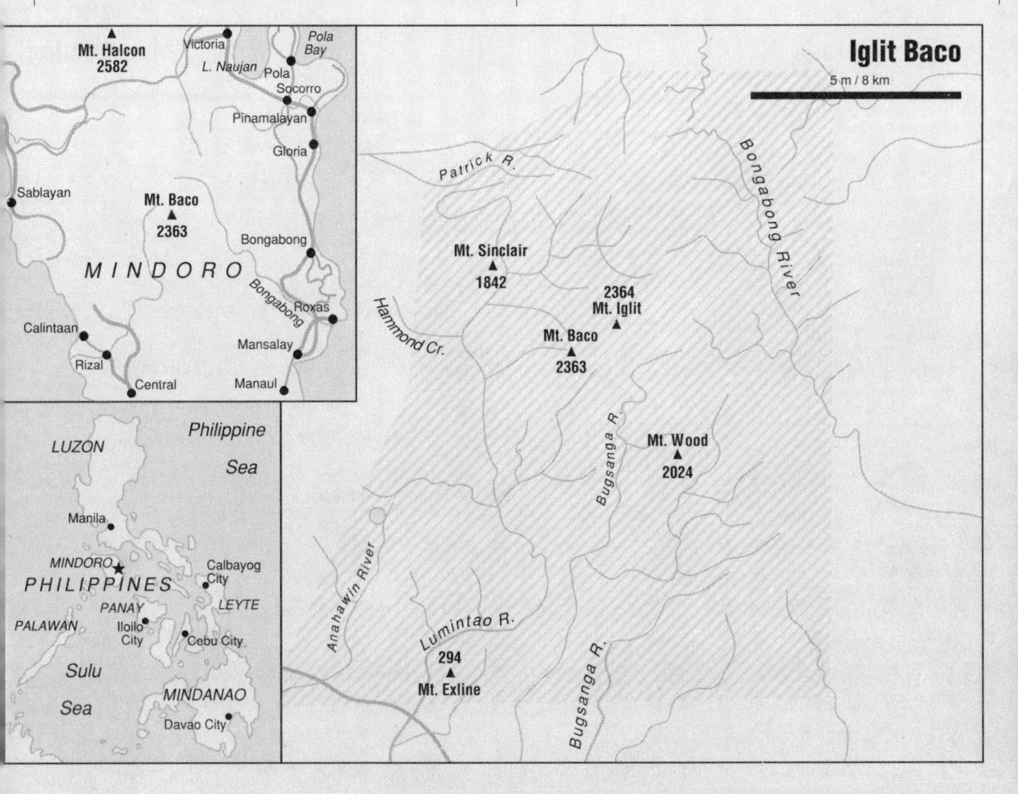

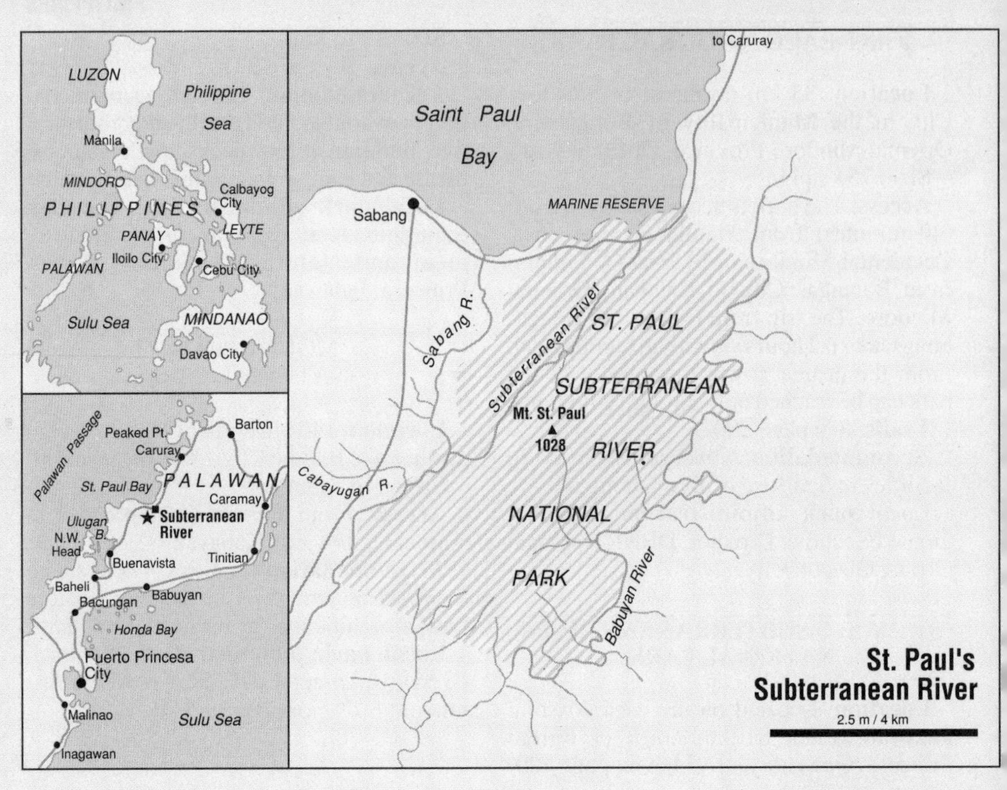
LUZON
Philippine
Sea
Manila
MINDORO
PHILIPPINES
Calbayog City
LEYTE
PANAY
Iloilo City
Cebu City
PALAWAN
Sulu Sea
MINDANAO
Davao City
Palawan Passage
Peaked Pt.
Barton
Caruray
St. Paul Bay
PALAWAN
Caramay
Ulugan B.
N.W. Head
Subterranean River
Buenavista
Tinitian
Baheli
Babuyan
Bacungan
Honda Bay
Puerto Princesa City
Malinao
Sulu Sea
Inagawan
to Caruray
Saint Paul Bay
Sabang
MARINE RESERVE
Sabang R.
Subterranean River
ST. PAUL
SUBTERRANEAN
RIVER
NATIONAL
PARK
Mt. St. Paul
1028
Cabayugan R.
Babuyan River
St. Paul's
Subterranean River
2.5 m / 4 km

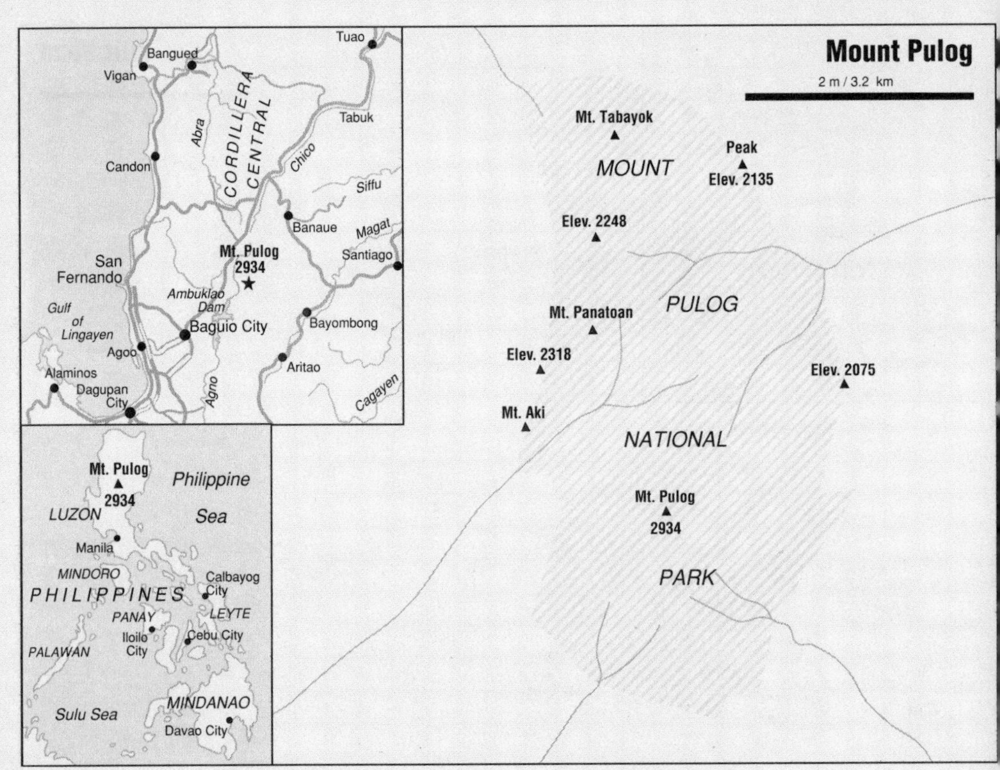
Tuao
Bangued
Vigan
Tabuk
Abra
CORDILLERA CENTRAL
Chico
Candon
Siffu
Banaue
Magat
Santiago
Mt. Pulog
2934
San Fernando
Ambuklao Dam
Gulf of Lingayen
Baguio City
Bayombong
Agoo
Aritao
Alaminos
Dagupan City
Agno
Cagayen
Mt. Pulog
2934
Philippine
Sea
LUZON
Manila
MINDORO
PHILIPPINES
Calbayog City
LEYTE
PANAY
Iloilo City
Cebu City
PALAWAN
Sulu Sea
MINDANAO
Davao City
Mount Pulog
2 m / 3.2 km
Mt. Tabayok
MOUNT
Peak
Elev. 2135
Elev. 2248
Mt. Panatoan
PULOG
Elev. 2318
Elev. 2075
Mt. Aki
NATIONAL
Mt. Pulog
2934
PARK

A list of the scientific names of Southeast Asian wildlife mentioned in this book:

MAMMALS

Anoa
(Bubalus depressicornis and B. quarlesis)

Babirusa
(Babyrousa babyrousa)

Badger, Ferret
(Melogale orientalis)

Badger, Hog
(Arctonyx collaris)

Banteng
(Bos javanicus)

Bat, Black-capped Fruit
(Chironax melanocephalus)

Bat, Bornean Horseshoe
(Rhinolophus borneansis)

Bat, Cantor's Roundleaf
(Hipposideros galeritas)

Bat, Cave Nectar
(Eunycteris spelea)

Bat, Common Roundleaf Horseshoe
(Hipposideros galnitas)

Bat, Diadem Roundleaf Horseshoe
(Hipposideros diadema)

Bat, Greater Roundleaf Horseshoe
(Hipposideros armiger)

Bat, Grey Fruit
(Aethalops alecto)

Bat, Hairy-winged
(Harpiocephalus harpia)

Bat, Malayan Fruit
(Cyanopterus brachyotis)

Bat, Naked
(Cheiromeles torquatos)

Bat, Narrow-winged Brown
(Philetor brachypterus)

Bat, Orange Tube-nosed
(Murina cyclotis)

Bat, Wrinkled-lipped
(Tadarida plicata)

Bear, Asiatic Black
(Selenarctos thibetanus)

Beji
(Lipotes vexillifer)

Binturong
(Arctictis binturong)

Buffalo, Water
(Bubalus bubalis)

Cat, Fishing
(Felis viverrina)

Cat, Leopard
(Felis planiceps)

Civet, Large Indian
(Viverra zibetta)

Civet, Palm
(Paradoxunes hermaphroditus)

Civet, Small-toothed Palm
(Arctogalidia trivirgata)

Cuscus, Sulawesi
(Phalange celebensis)

Deer, Barking (Muntjac)
(Muntiacus muntjak)

Deer, Bawean
(Cervus kubli)

Deer, Rusa
(Cervus timorensis)

Deer, Sambar
(Cervus unicolor)

Dog, Wild (Dog, Red)
(Cuon alpinus)

Dolphin, Indo-Pacific Hump-backed
(Sousa chiniesis)

Dolphin, Irrawaddy
(Orcaella brevinostris)

Dolphin, Spotted
(Stenella spp.)

Dugong
(Dugong dugon)

Elephant, Indian
(Elephas maximus)

Fox, Flying
(Pteropus vampyrus)

Gibbon, Bornean
(Hylobates muelleri)

Gibbon, Dark-handed
(Hylobates agilis)

Gibbon, Javan
(Hylobates moloch)

Gibbon, Pileated
(Hylobates pileatus)

Gibbon, White-handed
(Hylobates lar)

Gumben
(Dasyurus albopunctatus)

Lemur, Flying
(Cynocephalus variegatus)

Leopard
(Phantera pardus)

Leopard, Clouded
(Neofelis nebulosis)

Linsang
(Prinodor linsang)

Loris, Slow
(Nycticebus concang)

Macaque, Assamese
(Macaque assamensis)

Macaque, Black
(Macaca hecki; M. nigrescens; M. nigra)

Macaque, Long-tailed
(Macaque fascicularis)

Macaque, Pig-tailed
(Macaque nemistrina)

Macaque, Rhesus
(Macaque mullata)

Macaque, Stump-tailed
(Macaque arctoides)

Marten, Yellow-throated
(Martes flavigula)

Mole, Short-tailed
(Talpa micrura)

Mongoose, Javan
(Herpestes javanicus)

Mongoose, Short-tailed
(Herpestes brachyrus)

Monkey, Banded Leaf
(Presbytis femoralis)

Monkey, Black Leaf
(Presbytis obscura)

Monkey, Grey Leaf
(Presbytis hosei)

Monkey, Javan Leaf
(Presbytis comata)

Monkey, Phayre's Leaf
(Presbytis phayrei)

Monkey, Proboscis
(Nasalis lavoatus)

Monkey, Red Leaf
(Presbytis rubicunda)

Monkey, Silvered Leaf
(Presbytis cristata)

Monkey, Thomas Leaf
(Presbytis thomasi)

Moonrat
(Echinosorex gymnurus)

Mousedeer, Large
(Tragulus napu)

Mousedeer, Small
(Tragulus javanicus)

Orang-utan
(Pongo pygmaeus)

Otter, Common
(Lutra lutra)

Otter, Hairy-nosed
(Lutra sumatrana)

Otter, Oriental Small-clawed
(Aanyx cinerea)

Otter, Smooth
(Lutra perspicillata)

Pangolin, Chinese
(Manis pentadactyla)

Pangolin (Scaly Anteater)
(Manis javanica)

Pig, Bearded
(Sus barbatus)

Pig, Javan
(Sus verrucosus)

Pig, Wild
(Sus scrofa)

Porcupine, Brush-tailed
(Athenus macrouma)

Porcupine, Common
(Hystrix brachyra)

Porcupine, Malayan
(Hystrix bracyura)

Rabbit, Short-eared
(Nesolagis netscheri)

Rat, Cloud
(Creteromys schadenbergi)

Rat, Kerinci
Rat, Kinabalu (Rat, Mountain Giant)
(Sundamys infraluteus)

Rat, Large Bamboo
(Rhizomys sumatrensis)

Rhinoceros, Javan
(Rhinoceros sundaicus)

Rhinoceros, Sumatran
(Dicerorhinus sumatrensis)

Serow
(Capricornis sumatrensis)

Shrew, Szechwan Burrowing Siamang
(Hylobates syndactylus)

Squirrel, Black-banded
(Calloscurius nigrovittatus)

Squirrel, Black Giant
(Ratufa bicolor)

Squirrel, Bornean Mountain Ground
(Dremomys everetti)

Squirrel, Common Giant
(Squirrel Cream-coloured Giant)
(Ratufa affinis)

Squirrel, Giant Flying
(Squirrel, Red Giant Flying)
(Petaurista petaurista)

Squirrel, Grey-bellied
(Callosciurus caniceps)

Squirrel, Himalayan Striped
(Tamiops maclellandei)

Squirrel, Kinabalu
(Callosciurus baluensis)

Squirrel, Mountain Red-bellied
(Callosciurus flavimanus)

Squirrel, Plain Pigmy
(Exilisciurus exilis)

Squirrel, Plantain
(Callosciurus notatus)

Squirrel, Prevost's
(Callosciurus prevosti)

Squirrel, Red-checked Ground
(Dremomys rufigenis)

Squirrel, Slender
(Sundasciurus tenius)

Squirrel, Spotted Giant
(Petaurista elegans)

Squirrel, Tentineck's
Squirrel, Whitehead's Pigmy
(Exilisciurus whiteheadi)

Sunbear, Malayan
(Helarctos malayanus)

Susu, Ganges
(Platanista gangetica)

Susu, Indus
(Platanista minor)

Tapir, Malayan
(Tapirus indicus)

Tarsier, Philippine
(Tarsius syrichta)

Tarsier, Sulawesi
(Tarsius spectrum)

Tarsier, Western
(Tarsius bancanus)

Tiger
(Phantera tigris)

Treeshrew, Common
(Tupaia glis)

Treeshrew, Large
(Tupaia tana)

Treeshrew, Mountain
(Tupaia montana)

Treeshrew, Szechuan
(Tupaia chinensis)

Vole, Pere David's
(Eothenomys melanogaster)

Weasel, Malay
(Mustela nudipes)

Whale, Sperm
(Physeter macrocephalus)

BIRDS OF SOUTHEAST ASIA

Adjutant, Lesser
(Leptoptilus javanicus)

Argus, Crested
(Rheinardia ocellata)

Argus, Great
(Argusianus argus)

Babbler, Abbott's
(Trichastoma abbtti)

Babbler, Golden
(Stachyris poliocephala)

Babbler, Grey-throated
(Stachyris nigriceps)

Babbler, Horsefield's
(Trichastoma sepiarium)

Babbler, Moustached
(Malacopteron magnirostre)

Babbler, Short-tailed
(Trichastoma malaccense)

Babbler, Sooty-capped
(Malacopteron affine)

Babbler, White-hooded
(Gampsorchynchus rufulus)

Barbet, Black-browed
(Megalaima oorti)

Barbet, Blue-crowned & Golden-naped)
(Megalaima pulcherrima)

Barbet, Blue-eared
(Megalaima australis)

Barbet, Blue-throated
(Megalaima asiatica)

Barbet, Brown-throated
(Megalaima carvina)

Barbet, Coppersmith
(Megalaima haemacephala)

Barbet, Fire-tufted
(Psilopogon pyrolophus)

Barbet, Gold-whiskered
(Megalaima chrysopogon)

Barbet, Golden-throated
(Megalaima franklinii)

Barbet, Great
(Megalaima virens)

Barbet, Javan (Barbet, Black-banded)
(Megalaima javensis)

Barbet, Lineated
(Megalaima lineata)

Barbet, Red-crowned
(Megalaima rafflesii)

Bee-eater, Blue-bearded
(Nyctyornis athertoni)

Bee-eater, Blue-tailed
(Merops philippinus)

Bee-eater, Chestnut-headed
(Merops leschenaulti)

Bee-eater, Purple-bearded
(Meropogon forsteni)

Bee-eater, Red-bearded
(Nyctyornis amictus)

Bittern, Cinnamon
(Ixobrychus cinnamomeus)

Bittern, Yellow
(Ixobrychus sinensis)

Blackbird, Mountain
(Turdus poliocephalus)

Blackeye, Mountain
(Chlorocharis emilliae)

Bleeding-Heart, Luzon
(Gallicolumba luzonica)

Booby, Brown
(Sula leucogaster)

Bristlehead, Bornean
(Pityriasis gymnocephala)

Broadbill, Black-and-Red
(Cymbirhynchus macrorhynchus)

Broadbill, Green
(Calyptomena viridis)

Broadbill, Long-tailed
(Psarosomus dalhousiae)

Broadbill, Silver-breasted
(Serilophus lunatus)

Broadbill, Whitehead's
(Calyptomena whiteheadi)

Bulbul, Cream-vented
(Pycnonotus simplex)

Bulbul, Grey-bellied
(Pycnonotus cyaniventris)

Bulbul, Grey-eyed
(Hypsipetes propinquus)

Bulbul, Olive-winged
(Pycnonotus plumosus)

Bulbul, Pale-faced (Bulbul Flavescent)
(Pycnonotus flavescens)

Bulbul, Puff-throated
(Criniger pallidus)

Bulbul, Red-eyed
(Pycnonotus brunneus)

Bulbul, Red-whiskered
(Pycnonotus jocosus)

Bulbul, Sooty-headed
(Pycnonotus aurigaster)

Bulbul, Straw-headed
(Pycnonotus zeylanicus)

Bulbul, Yellow-vented
(Pycnonotus goiavier)

Bullfinch, Brown
(Pyrrhula nipalensis)

Bushlark, Singing
(Mirafra javanica)

Cisticola, Bright-capped
(Cisticola exilis)

Cisticola, Zitting
(Cisticola jundicis)

Cockatoo, Philippine
(Cockatoo, Red-vented)
(Cacatua haematuropygia)

Cockatoo, Sulphur-crested
(Cacatua galerita)

Cockatoo, Yellow-crested
(Cacatua sulphurea)

Coucal, Greater
(Centropus sinensis)

Coucal, Lesser
(Centropus bengalensis)

Coucal, Sulawesi Forest
(Coucal, Celebean)
(Centropus celebensis)

Crag-Martin, Dusky
(Hirundo concolor)

Crake, Red-legged
(Rallina fasciata)

Crake, Ruddy-breasted
(Porzana fusca)

Crake, White-browed
(Porzana cinerea)

Crow, Large-billed
(Corvus enca)

Cuckoo-Dove, Barred
(Macropygia unchall)

Cuckoo-Dove, Little
(Macropygia ruficeps)

Cuckoo-Shrike, Caerulean Cutia
(Cutia nipalensis)

Darter, Oriental
(Anhinga melanogaster)

Dollarbird
(Eurystomus orientalis)

Dove, Marche's Fruit
(Ptilinopus marchei)

Dove, Merril's Fruit
(Ptilinopus merilli)

Dove, Peaceful (Dove, Zebra)
(Geopelia striata)

Dove, Spotted
(Streptopelia chinensis)

Dove, Yellow-breasted Fruit
(Ptilinopus occipitalis)

Dowitcher, Asian
(Limnodromus semiplamatus)

Drongo, Greater Racket-tailed
(Dicrurus paradiseus)

Drongo, Lesser Racket-tailed
(Dicrurus remifer)

Drongo, Spangled
(Dicrurus hottentotus)

Eagle, Black
(Ictinaetus malayensis)

Eagle, Lesser Fish
(Ichthyophaga nana)

Eagle, Philippine
(Pithecophaga jefferyi)

Eagle, Greater Spotted
(Aquila clanga)

Egret, Cattle
(Bubulcus ibis)

Egret, Great
(Egretta alba)

Egret, Little
(Egretta garzetta)

Fairy-Bluebird, Asian
(Irena puella)

Falcon, Peregrine
(Falco peregrinus)

Falconet, Black-thighed
(Microhierax fringillarius)

Fantail, Pied
(Rhipidura javanica)

Fantail, Red-tailed
(Rhipidura phoenicura)

Fantail, White-throated
(Rhipidura albicollis)

Finfoot, Masked
(Heliopais personata)

Fireback, Crested
(Lophura ignita)

Fireback, Crestless
(Lophura erythrophthalma)

Fish-Eagle, Grey-headed
(Icthyophaga ichthyaetus)

Fish-Eagle, Lesser
(Icthyophaga nana)

Fish-Owl, Buffy
(Ketupa ketupu)

Flowerpecker, Crimson-headed
(Flowerpecker, Scarlet-headed)
(Dicaeum trochileum)

Flowerpecker, Orange-bellied
(Dicaeum trigonostigma)

Flowerpecker, Scarlet-backed
(Dicaeum cruentatum)

Flycatcher, Grey-headed
(Culicicapa ceylonensis)

Flycatcher, Little Pied
(Ficedula westermanni)

Flycatcher, Mangrove Blue
(Cyornis rufigastra)

Flycatcher, Mugimaki
(Ficedula mugimaki)

Flycatcher, Narcissus
(Ficedula narcissina)

Flycatcher, Rufous-browed
(Ficedula solitaria)

Flycatcher, Snowy-browed
(Ficedula hyperythra)

Flycatcher, Verditer
(Muscicapa thalassina)

Flycatcher, White-gorgetted
(Ficedula monileger)

Flycatcher-Shrike, Bar-winged
(Hemipus picatus)

Forktail, Lesser
(Enicurus velatus)

Forktail, Slaty-backed
(Enicurus schistaceus)

Friarbird, Noisy
(Philemon corniculatus)

Frigatebird, Christmas Island
(Fregata andrewsi)

Frigatebird, Greater
(Fregata minor)

Frigatebird, Lesser
(Fregata ariel)

Frogmouth, Kinabalu
Fulvetta, Brown
(Alcippe brunneicauda)

Fulvetta, Mountain
(Alcippe peracensis)

Fulvetta, Rufous-winged
(Alcippe castaneceps)

Goldenback, Common
(Dinopium javanense)

Goldenback, Greater
(Chrysocolaptes lucidus)

Goshawk, Crested
(Accipiter trivirgatus)

Goshawk, Spot-tailed
(Sparrowhawk, Spot-tailed)
(Accipiter trinotatus)

Grebe, Little
(Podiceps ruficollis)

Greenshank, Nordmann's
(Tringa guttifer)

Guiaiabero
(Bollopsittacus lunulatus)

Hawk, Bat
(Machaerhampus alcinus)

Hawk-Owl, Brown
(Ninox scutulata)

Hawk-Eagle, Blyth's
(Spizaetus alboniger)

Hawk-Eagle, Changeable
(Spizaetus cirrhatus)

Heron, Great-billed
(Ardea sumatrana)

Heron, Grey
(Ardea cinerea)

Heron, Little
(Butorides striatus)

Heron, Purple
(Ardea purpurea)

Hill-Partridge, Sumatra
(Arborophila orientalis)

Honey-Buzzard, Crested
(Pernis ptilorhynchus)

Hornbill, Black
(Anthracoceros malayanus)

Hornbill, Blyth's
(Rhyticeros plicatos)

Hornbill, Brown
(Ptilolaemus tickelli)

Hornbill, Bushy-crested
(Annorhinus galeritus)

Hornbill, Great
(Buceros bicornis)

Hornbill, Helmeted
(Rhinoplax vigil)

Hornbill, Indian Pied
(Anthracoceros albirostris)

Hornbill, Rhinoceros
(Buceros rhinoceros)

Hornbill, Rufous-necked
(Aceros nipalensis)

Hornbill, Southern Pied
(Anthracoceros convexus)

Hornbill, Tarictic
(Penelopides panini)

Hornbill, White-crowned
(Berenicornis comatus)

Hornbill, Wreathed
(Rhyticeros undulatus)

Hornbill, Wrinkled
(Rhyticeros corrugatus)

Ibis, Black-headed
(Threskiornis melanocephola)

Iora, Common
(Aegithina tiphia)

Jay, Eurasian
(Garrulus glandarius)

Junglefowl, Green
(Gallus varius)

Junglefowl, Red
(Gallus gallus)

Kingfisher, Bar-headed
(Actenoides princeps)

Kingfisher, Black-capped
(Halcyon pileata)

Kingfisher, Blue-banded
(Alcedo euryzonia)

Kingfisher, Blue-eared
(Alcedo meninting)

Kingfisher, Brown-winged
(Pelargopsis amauroptera)

Kingfisher, Collared
(Halycon chloris)

Kingfisher, Javan
(Halycon cyaniventris)

Kingfisher, Philippine Forest
(Ceyx melanurus)

Kingfisher, Ruddy
(Halcyon coromanda)

Kingfisher, Rufous-backed
(Ceyx rufidorsus)

Kingfisher, Sacred
(Halycon sancta)

Kingfisher, Small
(Alcedo caerulescens)

Kingfisher, Spotted Wood
(Halycon lindsayi)

Kingfisher, Stork-billed
(Pelagopsis capensis)

Kingfisher, White-throated
(Halycon smyrnensis)

Kite, Black
(Milvus migrans)

Kite, Black-shouldered
(Elanus caeruleus)

Kite, Brahminy
(Haliastur indus)

Koel, Black-billed
Koel, Common
(Eudynamis scolapacea)

Laughingthrush, Black
(Garrulax lugubris)

Laughingthrush, Chestnut-capped
(Garrulax mitratus)

Laughingthrush, Chestnut-crowned
(Garrulax erythrocephalus)

Laughingthrush, Red-fronted
(Garrulax rufifrons)

Leaf-Warbler, Ashy-throated
(Phylloscopus maculipennis)

Leaf-Warbler, Orange-browed
Leafbird, Blue-winged
(Chloropsis cochinchinensis)

Leafbird, Orange-bellied
(Chloropsis hardwickii)

Lorikeet, Ornate
Magpie, Blue
(Urocissa erythrorhyncha)

Magpie, Green
(Cissa chinensis)

Maleo (Gray's Brush Turkey)
(Macrocephalon maleo)

Malkoha, Chestnut-bellied
(Phaenicophaeus sumatranus)

Malkoha, Green-billed
(Phaenicophaeus tristis)

Megapode, (Incubator Bird)
(Megapodius freycinet)

Mesia, Silver-eared
(Leiothrix argentauris)

Minivet, Ashy
(Pericrocotus divaricatus)

Minivet, Scarlet
(Pericrocotus flammeus)

Minivet, Sunda
(Pericrocotus miniatus)

Minla, Blue-winged
(Minla cyanouroptera)

Minla, Chestnut-tailed
(Minla strigula)

Monarch, Black-naped
(Hypothymis azurea)

Moorhen, Common
(Gallinula chloropus)

Munia, Chestnut
(Lonchura malacca)

Munia, Dusky
(Lonchura fuscans)

Munia, Scaly-breasted
(Lonchura punctulata)

Munia, White-headed
(Lonchura maja)

Myna, Black-winged
(Sturnus melanopterus)

Myna, Common
(Acridotheres tristis)

Myna, Golden-crested
(Ampeliceps coronatus)

Myna, Hill
(Gracula religiosa)

Myna, Jungle
(Acridotheres fuscus)

Myna, White-vented
(Acridotheres javanicus)

Nightjar, Large-tailed
(Nightjar, Long-tailed)
(Caprimulgus macrurus)

Nightjar, Savanna
(Caprimulgus affinis)

Niltava, Large
(Niltava grandis)

Nuthatch, Blue
(Sitta azurea)

Nuthatch, Velvet-fronted
(Sitta frontalis)

Openbill, Asian
(Anastomus oscitans)

Oriole, Black-hooded
(Oriolus xanthornus)

Oriole, Black-naped
(Oriolus chinensis)

Oriole, Maroon
(Oriolus traillii)

Osprey
(Pandion haliaetus)

Paradise-Flycatcher, Asian
(Terpsiphone paradisi)

Paradise-Flycatcher, Japanese
(Terpsiphone atrocaudata)

Parakeet, Long-tailed
(Psittacula longicauda)

Parakeet, Racket-tailed
(Prioniturus platurus and P. flavicans)

Parakeet, Red-breasted
(Psittacula alexandri)

Parrot, Blue-crowned Hanging
(Loriculus galgulus)

Parrot, Vernal Hanging
(Loriculus vernalis)

Partridge, Chestnut-bellied
(Arborophila javanica)

Partridge, Rofous-throated
(Arborophila rufogularis)

Peacock-Pheasant, Malaysian
(Polyplectron malacense)

Peacock-Pheasant, Mountain
(Polyplectron inopinatum)

Peacock-Pheasant, Palawan
(Polyplectron emphanum)

Peafowl, Green
(Pavo muticus)

Pelican, Spot-billed
(Pelecanus philippensis)

Pheasant, Argus
(Argusianus argus)

Pheasant, Silver
(Lophura nycthemera)

Piculet, Rufous
(Sasia abnormis)

Piculet, Speckled
(Picumnus innominatus)

Piculet, White-browed
(Sasia ochracea)

Pigeon, Green Imperial
(Ducula aenea)

Pigeon, Green-winged
(Chalcophaps indica)

Pigeon, Mountain Imperial
(Ducula badia)

Pigeon, Nicobar
(Caleonas nicobarica)

Pigeon, Pied Imperial
(Ducula badia)

Pigeon, Pink-necked
(Treron vernans)

Pigeon, Thick-billed
(Treron curvirostra)

Pigeon, Wedge-tailed Green
(Treron sphenura)

Pigeon, Yellow-footed
(Treron phoenicoptera)

Pigeon, Yellow-vented
(Treron slimundi)

Pintail, Northern
(Anas acuta)

Pipit, Richard's
(Anthus novaeseelandiae)

Pitta, Banded
(Pitta guajana)

Pitta, Black-and-Crimson
Pitta, Blue
(Pitta cyanea)

Pitta, Blue-headed
(Pitta baudi)

Pitta, Blue-winged
(Pitta moluccensis)

Pitta, Eared
(Pitta phayrei)

Pitta, Garnet
(Pitta granatina)

Pitta, Giant
(Pitta caerulea)

Pitta, Gurney's
(Pitta gurneyi)

Pitta, Hooded
(Pitta sordida)

Pitta, Mangrove
(Pitta megaryncha)

Pitta, Red-breasted
(Pitta, Blue-breasted)
(Pitta erythrogaster)

Pitta, Rusty-naped
(Pitta oatesi)

Pitta, Schneider's
(Pitta schneideri)

Plover, Lesser Golden
(Pluvialis dominica)

Plover, Mongolian
(Charadrius mongolus)

Pond-Heron, Javan
(Ardeola speciosa)

Prinia, Bar-winged
(Prinia familiaris)

Prinia, Hill
(Prinia atrogularis)

Prinia, Yellow-bellied
(Prinia flaviventris)

Racket-tail, Green
(Prioniturus luconensis)

Racket-tail, Luzon Mountain
(Prioniturus montanus)

Racket-tail, Palawan
(Prioniturus platenae)

Rail, Barred
(Gallirallus torquatus)

Rail, Buff-banded
(Gallirallus philippensis)

Rail-Babbler, Malaysian
(Eupetes macrocerus)

Reed-Warbler, Black-browed
(Acrocephalus bistrigiceps)

Reed-Warbler, Great
(Acrocephalus arundinaceus)

Robin, Magpie
(Copsychus saularis)

Robin, Siberian Blue
(Erithacus cyane)

Robin, White-tailed
(Cinclidium leucurum)

Rock-Thrush, Blue
(Monticola solitarius)

Roller, Indian
(Coracias benghalensis)

Roller, Purple-winged
Sandpiper, Curlew
(Calidris ferruginea)

Sandpiper, Spoon-billed
(Eurynorhynchus pygmeus)

Scimitar-Babbler, Chestnut-backed
(Pomathorhinus schisticeps)

Scimitar-Babbler, Red-bellied
(Pomathorhinus ochraceiceps)

Scops-Owl, Collared
(Otus bakkamoena)

Scops-Owl, Mentawai
Scops-Owl, Sulawesi
(Otus manadensis)

Sea-Eagle, White-bellied
(Haliaeetus leucogaster)

Serpent-Eagle, Crested
(Spilornis cheela)

Serpent-Eagle, Sulawesi
(Spilornis rufipectus)

Shama, Rufous-tailed
(Copsychus pyrropygus)

Shama, White-browed
(Copsychus luzoniensis)

Shama, White-rumped
(Copsychus malabaricus)

Shortwing, Lesser
(Brachypteryx leucophrys)

Shortwing, White-browed
(Brachypteryx montana)

Shrike, Long-tailed
(Lanius schach)

Shrike-Babbler, Black-eared
(Pteruthius melanotis)

Shrike-Babbler,White-browed
(Pteruthius flaviscapis)

Sibia, Black-headed
(Heterophasia melanolevea)

Sibia, Long-tailed
(Heterophasia picaoides)

Spiderhunter, Grey-breasted
(Arachnothera affinis)

Spiderhunter, Little
(Arachnothera longirostra)

Spiderhunter, Streaked
(Arachnothera magna)

Spiderhunter, Whitehead's
(Arachnothera juliae)

Starling, Bali (Rothschild's Myna)
(Leucopsar rothschildi)

Starling, Philippine Glossy
(Aplonis panayensis)

Stork, Milky
(Ibis cinereus)

Stork, Painted
(Ibis leucocephalus)

Stork, Storm's
(Ciconia stormii)

Stork, White-necked
Stork, Wooly-necked
(Ciconia episcopus)

Sunbird, Black-throated
(Aethopyga saturata)

Sunbird, Brown-throated
(Anthreptes malacensis)

Sunbird, Copper-throated
(Nectarinia calcostheta)

Sunbird, Crimson
(Aethopyga siparaja)

Sunbird, Gould's
(Aethopyga gouldiae)

Sunbird, Green-tailed
(Aethopyga nipalensis)

Sunbird, Olive-backed
(Nectarinia jugularis)

Sunbird, Ruby-checked
(Anthreptes singalensis)

Swallow, Barn
(Hirundo rustica)

Swallow, Pacific
(Hirundo tahitica)

Swallow, Red-rumped
(Hirundo daurica)

Swamphen, Purple
(Porphyrio porphyrio)

Swiftlet, Black-nest
(Collocalia maxima)

Swiftlet, Edible-nest
(Collocalia fuciphaga)

Swiftlet, White-bellied
(Collocalia esculenta)

Tailorbird, Ashy
(Orthotomus ruficeps)

Tailorbird, Common
(Orthotomus sutorius)

Tailorbird, Dark-necked
(Orthotomus atrogularis)

Tailorbird, Rufous-tailed
(Orthotomus sericus)

Teal, Common
(Anas crecca)

Teal, Grey
(Anas gibberifrons)

Tern, Bridled
(Sterna anaethetus)

Tern Little
(Sterna albifrons)

Tern, White-winged
(Chlidonias leucopterus)

Tesia, Slaty-bellied
(Tesia olivea)

Thick-Knee, Great
(Esacus magnirostris)

Thrush, Ashy
Thrush, Blue Whistling
(Myophonus caeruleus)

Thrush, Chestnut-capped
(Zoothera interpres)

Thrush, Eye-browed
(Turdus obscurus)

Thrush, Malayan Whistling
(Myophonus robinsoni)

Thrush, Orange-headed
(Zoothera citrina)

Thrush, Siberian
(Zoothera sibirica)

Tit, Great
(Parus major)

Tit, Pigmy
(Psaltria exilis)

Tit, Yellow-cheeked
(Parus spilonotus)

Tit-Babbler, Striped
(Macronous gularis)

Tree-Partridge, Red-breasted
(Partridge, Red-breasted)
(Arborophila hyperythra)

Tree-Sparrow, Eurasian
(Passer montanus)

Treeduck, Lesser
(Dendrocygna javanica)

Treeduck, Whistling
(Treeduck, Wandering)
(Dendocygna arcuata)

Treepie, Racket-tailed
(Crypsirina temia)

Treepie, Rufous
(Dendrocitta vagabunda)

Treeswift, Crested
(Hemiprocne coronata)

Treeswift, Grey-rumped
(Hemiprocne longipennis)

Triller, Black-breasted
(Chlamydochaera jeffreyi)

Triller, Pied
(Lalage nigra)

Trogon, Cinnamon-rumped
(Harpactes orthophaeus)

Trogon, Diard's
(Harpactes diardii)

Trogon, Philippine
(Harpactes ardeus)

Trogon, Red-naped
(Harpactes kasumba)

Trogon, Scarlet-rumped
(Harpactes duvaucelii)

Trogon, Whitehead's
(Harpactes whiteheadi)

Turtle-Dove, Javan
(Streptopelia bitorquata)

Vulture, Red-headed
(Sarcogyps calvus)

Vulture, White-rumped
(Gyps bengalensis)

Wagtail, Yellow
(Motacilla flava)

Warbler, Arctic
(Phylloscopus borealis)

Warbler, Eastern Crowned
(Phylloscopus coronatus)

Warbler, Kinabalu Friendly
(Bradypterus accentor)

Warbler, Striated
(Megalurus palustris)

Waterhen, Isabelline
(Moorhen, Celebes)
(Amaurornis isabellina)

Waterhen, White-breasted
(Amaurornis phoenicurus)

Weaver, Baya
(Ploceus philippinus)

Whimbrel
(Numenius phaeopus)

Whistler, Mangrove
(Pachycephala cinerea)

White-eye, Mangrove
(Zosterops chloris)

White-eye, Mountain
(Zosterops montanus)

Wood-Duck, White-winged
(Duck, White-winged)
(Cairinia scutulata)

Wood-Owl, Spotted
(Strix seloputo)

Wood-Partridge, Crested
(Rollulus rouluol)

Wood-Partridge, Crimson-headed
(Haematortyx sanguiniceps)

Wood-shrike, Large
(Tephrodornis virgatus)

Wood-Swallow, Ashy
(Artamus fuscus)

Wood-Swallow, White-breasted
(Artamus leucorhynchus)

Woodpecker, Banded
(Picus miniaceus)

Woodpecker, Brown-capped
(Picoides moluccensis)

Woodpecker, Great Slaty
(Mulleripicus pulverulentus)

Woodpecker, Laced
(Picus vittatus)

Woodpecker, White-bellied
(Dryocopus javensis)

Wren-Babbler, Black-throated
(Napothera atrogularis)

Wren-Babbler, Bornean
(Ptilocichla leucogrammica)

REPTILES & AMPHIBIANS

Chikchak (Gecko)
(Gehydra dubia)

Cobra
(Naja naja)

Crocodile, Estuarine
(Crocodilis porosus)

Dragon, Komodo
(Varanus komodiensis)

Frog, Glandular
(Rana glandulora)

Frog, Flying
(Rhacophorus spp.)

Ghavial, False
(Tomista schlegeli)

Lizard, Changeable
(Calotes versicolor)

Lizard, Common Flying
(Draco volans)

Lizard, Crested
(Calotes cristatellus)

Lizard, Monitor
(Varanus indicus)

Phyton, Reticulated
(Phyton reticulata)

Pit-Viper
(Trimeresurus spp.)

Racer, Cave
(Elaphe taeuiura)

Skink
(Mabuya spp.)

Snake, Oriental Whip
(Ahaetulla prasina)

Toad, Malayan Giant
(Bufo Asper)

Tortoise, Brown
(Manouria emys)

Turtle, Green
(Chelonia mydas)

Turtle, Hawksbill
(Eretmochelys imbricata)

Turtle, Leatherback
(Dermochelys coriacea)

Turtle, Loggerhead
(Caretta caretta)

Turtle, Olive Ridley
(Lepidochelus olivacea)

Alcala, A.C. *Philippine Land Vertebrates*. Field Biology, New Day Publishers, Quezon City, Manila, 1976.

Barlor, H.S., D. and Abrera, B. *An Introduction to the Moths of South East Asia*. Malaysian Nature Society, Kuala Lumpur. 1982.

Cochrane, J. (ed), *Wild Places of Indonesia*, Yayasan Indonesia Hijau, Bogor. 1985.

Conservation Data Centre, (1989). *Birds of Khao Yai National Park*: Check-list. Conservation Data Centre, Department of Biology, Faculty of Science, Mahidol Univ., Bangkok. 1989.

Davison, G.W. H. *Endau Rompia, A Malaysia Heritage*, Malaysian Nature Society 1988.

Francis, C.M. and Smythies, B.E., *Pocket Guide to the Birds of Borneo*. Sabah Society and WWF. 1989.

Gouzales, P.C. Rees, *Birds of the Philippines*. Haribon Foundation, Philippines. 1988.

Holmes, D. and Nash, S. *The Birds of Java and Bali*. Oxford University Press. 1989.

Indonesia: National Parks & Nature Reserves. Directorate General of Tourism, Jakarta.

Issac-Williams, M.H. *An Introduction to the Orchids of Asia*. Angus and Robertson Publ., North Ryde, Australia. 1988.

Jacobs, M. *The Plant World on Luzon's Highest Mountains*. Rijks Herbarium. Leiden. 1972.

Jacobs, M. *The Tropical Rain Forest*. Springer. 1988.

Kind, B., Woodcock, M., Dickinson, E.C. *A Field Guide to the Birds of South-East Asia*. Collins London. 1976.

MacKinnon, J. *Field Guide to the Birds of Java and Bali*, Gaja Mada University Press, Yogyakarta, 1988.

MacKinnon, J., & MacKinnon, K., *Review of the Protected Areas System in the Indo-Malayan Realm*, IUCN, Gland, Switzerland & Cambridge, UK. 1986.

Medwocy, L. *The Wild Mammals of Malaya (Peninsular Malaysia) and Singapore*. Oxford University Press. 1983.

Payne, J., Francis, C.M., Phillipps, K. *A Field Guide to the Mammals of Borneo*. Sabah Society and WWF. 1985.

Payne, J. *Orang-utan Conservation in Sabah*. Kuala Lumpur: WWF Malaysia. 1988.

Payne, J and Andau, M. *Orange-utan, Malaysia's Mascot*. Kuala Lumpur: Berita Publishing Sdn Bhd. 1989.

Rabor, D.S. *Philippine Reptiles and Amphibians*. Foundation for the Advancement of Science Education, Inc., Abiva Publishing House, Inc., Manila. 1981.

Round, P. *Resident Forest Birds in Thailand*. ICBP Monograph, No. 2. 1989.

Smythies, B.E., *The Birds of Borneo*. The Sabah Society, Kota Kinabalu, Sabah, reprinted 1980.

Teoh Eng Soon. *Orchids of Asia*. Times Books International, Singapore. 1980.

Tweedie, MWF and Harrison, JL. *Malayan Animal Life*. Longmans, Kuala Lumpur. 1954.

White, A.T. *Philippine Coral Reefs. A Natural History Guide*. New Day Publishers, Quezon City, Manila. 1987.

Whitemore, T.C. *Tropical Rain Forests of the Far East*. Oxford University Press. 1984.

World Rainforest Movement. *The Battle for Sarawak's Rainforests*. World Rainforest Movement and Sahabat Alam Malaysia, 1989.

Yong Hoi Sen: *Malaysia Butterflies – An Introduction*. Tropical Press, 29 Jalan Riang, Kuala Lumpur 22-03. 1983.

ART/PHOTO CREDITS

Photographs

286R	**Apa Photo Agency**
134R, 144/145, 165, 181R, 251, 268	**Bernard, Hans-Ulrich**
284/285	**Blair Seitz/Apa Agency**
154	**Briffet, Clive**
128	**Brockelman, Warren**
91	**Chou Loke Ming**
37, 125, 269, 270, 275	**Cochrane, Janet**
14/15, 16/17, 18/19, 24/25, 29, 30/31, 32, 44/45, 47, 48, 49, 50, 53, 54, 55, 56, 57, 58, 59, 62, 77, 82R, 85, 86, 92/93, 96, 100/101, 102/103, 104/105, 106/107, 123, 129, 130, 136, 146/147, 176, 188, 202/203, 212, 226, 227, 238, 242/243, 244/245, 246, 249, 250, 253, 255, 257, 258, 259, 260, 261, 262, 264, 265, 273, 274, 276, 277, 279, 280, 281, 290, 291, 296	**Compost, Alain**
162, 206, 209L	**Conrad, Ralf**
178	**Davison, Geoffrey**
38	**Eu, Geoffrey**
191, 282/283, 287, 289	**Evrard, Alain**
164	**Fletcher & Baylis**
172, 190	**Gocher, Jill**
36, 204/205, 207, 286L	**Höfer, Hans**
28, 67, 74, 118, 119, 155, 156, 157, 163, 210	**Jeyarajasingam, Allen**
166, 222	**Kavanagh, M. (WWF)**
173	**Klock, Gerd**
148, 293	**Kumar, Kunda**
209R, 213, 214L&R, 215L&R, 216, 218, 219	**Lamb, Anthony**
76	**Lambert, Frank (WWF)**
177	**Lawrence, Max E.**

Art/Photo Credits

Photographs	
228, 229	**Marsh, Clive**
170, 171	**Mortimer, Jeanne (WWF)**
26, 295	**Opitz, Norbert**
221	**Payne, Junaidi**
39, 208, 211, 217	**Phillips, Anthea**
292	**Pierre Alain Petit**
41, 75, 141, 143	**Round, Phil**
168	**Scriven, Ken (WWF)**
cover, 46, 52, 160, 161, 167, 294	**Sreedhavan, Slim (WWF)**
33, 63, 64, 65, 66, 69, 72, 73, 94, 95, 99, 142, 150, 151, 152, 158, 174, 175, 179, 181L, 182/183, 184/185, 186L&R, 187, 189, 192, 193, 194, 195, 196, 197, 198, 199, 200, 201, 220, 223, 230, 231, 232, 234, 235, 236, 237, 239, 240, 241, 271, 272, 288	**Strange, Morten**
34, 112, 114, 120, 121, 126, 131	**Suwanarungsi, Duangdao**
90	**Tackett, Denise**
88	**Tackett, Larry P.**
Photographs	
22/23, 60/61, 70, 71, 97R, 110/111, 116, 117, 122, 132, 133, 134L, 135, 137, 138, 180	**Tsuji, Atsuo**
124	**Ulrich, Ward**
40, 42, 43, 68, 140, 248, 252, 254, 263, 268	**Verheught, Wim**
169, 224	**WWF Collection**
225	**Whitaker, R. (WWF)**
51, 266	**Wind, Jan**
20/21, 78/79, 80, 82L, 83, 84L&R, 87, 97L, 153, 159	**Yong Hoi Sen**
Maps	**Berndtson & Berndtson**
Illustrations	**Klaus Geisler**
Visual Consulting	**V. Barl**

A

B

C

D

E–F

G

H

I–J

K

L

M

N–O

P